AF893289

Methods in Molecular Biology

Series Editor
John M. Walker
School of Life and Medical Sciences
University of Hertfordshire
Hatfield, Hertfordshire, AL10 9AB, UK

For further volumes:
http://www.springer.com/series/7651

Cell-Based Assays Using iPSCs for Drug Development and Testing

Edited by

Carl-Fredrik Mandenius

Division of Biotechnology, Department of Physics, Chemistry and Biology (IFM), Linköping University, Linköping, Sweden

James A. Ross

Tissue Injury and Repair Group, University of Edinburgh, Edinburgh, UK

Humana Press

Editors
Carl-Fredrik Mandenius
Division of Biotechnology
Department of Physics
Chemistry and Biology (IFM)
Linköping University
Linköping, Sweden

James A. Ross
Tissue Injury and Repair Group
University of Edinburgh
Edinburgh, UK

ISSN 1064-3745 ISSN 1940-6029 (electronic)
Methods in Molecular Biology
ISBN 978-1-4939-9476-2 ISBN 978-1-4939-9477-9 (eBook)
https://doi.org/10.1007/978-1-4939-9477-9

© Springer Science+Business Media, LLC, part of Springer Nature 2019
This work is subject to copyright. All rights are reserved by the Publisher, whether the whole or part of the material is concerned, specifically the rights of translation, reprinting, reuse of illustrations, recitation, broadcasting, reproduction on microfilms or in any other physical way, and transmission or information storage and retrieval, electronic adaptation, computer software, or by similar or dissimilar methodology now known or hereafter developed.
The use of general descriptive names, registered names, trademarks, service marks, etc. in this publication does not imply, even in the absence of a specific statement, that such names are exempt from the relevant protective laws and regulations and therefore free for general use.
The publisher, the authors, and the editors are safe to assume that the advice and information in this book are believed to be true and accurate at the date of publication. Neither the publisher nor the authors or the editors give a warranty, express or implied, with respect to the material contained herein or for any errors or omissions that may have been made. The publisher remains neutral with regard to jurisdictional claims in published maps and institutional affiliations.

Cover Caption: Morphogenesis of hiPSC-derived endothelial cells. hiPSC derived endothelial cells (left image) forming micro tubes in matrigel (right image). Live cells were stained with DRAQ5™ (red, nuclear) and /or calcein AM (green). [Palakkan, Nanda & Ross]

This Humana imprint is published by the registered company Springer Science+Business Media, LLC, part of Springer Nature.
The registered company address is: 233 Spring Street, New York, NY 10013, U.S.A.

Preface

In 2012 Shinya Yamanaka and John Gurdon were awarded the Nobel Prize for Physiology or Medicine for their work demonstrating that mature cells can be reprogrammed to a stem cell phenotype. John Gurdon's key achievement in 1962 provided the first experimental evidence of reprogramming when he transplanted amphibian somatic cell nuclei into enucleated oocytes. This breakthrough work showed that each nucleus of a differentiated cell retains a complete set of genetic instructions for the whole organism. Shinya Yamanaka and his team, inspired by this discovery and subsequent experimental work, demonstrated that forced expression of four transcription factors could induce pluripotency in somatic cells. Subsequently, effective reprogramming methods in mouse cells (2006) and in human cells (2007) were developed. The reprogrammed cells were termed induced pluripotent stem cells (iPSCs). Since this seminal work there has been an explosion of research and development in the field of iPSC biology with reprogramming of somatic cells from different tissues and from different disease genotypes together with a gradual increase in the expertise required to differentiate iPSCs toward specialized cell fates. This progress has led to the increasing use of iPSCs for disease modeling, drug discovery, and safety.

The development of complex iPSC-derived cellular models and employing these models during the process of drug discovery in order to identify the most efficacious drugs requires enormous efforts that are difficult to accomplish within a single research institute or pharmaceutical company. For this reason, scientific consortia such as StemBANCC are instrumental in fostering a close collaboration between academic groups, pharmaceutical groups, and the biotech industry. These large collaborative enterprises are founded on complementary scientific endeavors, drug discovery efforts, and toxicology expertise to develop human iPSC-derived cellular models for drug discovery, efficacy, and safety purposes. A key area which required critical development was the development of differentiation protocols which could enable the production of key cell types from different organs and provide the basis for disease modeling and drug development. Another key area was the need to develop robust cell-based assays of particular utility in the drug development industry but also of great utility to researchers interested in the biology of particular tissues or cell types. The major goal of this book is to present a set of differentiation and assay protocols useful to researchers involved in disease modeling, drug discovery, and the cell biology of a variety of tissues.

The introductory chapter describes the requirements and preconditions for using human induced pluripotent cell lines in assay development within the pharmaceutical industry. The subsequent chapters provide protocols for the bulk culture of iPSCs and iPSC-derived cardiomyocytes by bioreactor technologies and detailed protocols for the differentiation of iPSC-derived endoderm, hepatocytes, vascular endothelial cells, cardiomyocytes, renal podocytes, neurons, and astrocytes. There are detailed protocols for high-throughput metabolic and proteomic assays, high content assays of mitochondrial function and autophagy, electrophysiology assays, and specialized lab-on-a-chip and fluorescent assays.

We are indebted to all the contributing authors for sharing their expertise, to Professor John M. Walker for inviting us to edit this volume of the MiMB series and his support during the preparation of the book, and to Dr. Patrick J. Marton and Professor Kursad

Turksen for their support. Finally, we are particularly indebted to the support received from the Innovative Medicines Initiative Joint Undertaking under grant no. 115439, resources of which are composed of financial contribution from the European Union's Seventh Framework Programme (FP7/2007-2013) and EFPIA companies.

Linköping, Sweden ***Carl-Fredrik Mandenius***
Edinburgh, UK ***James A. Ross***

Contents

Preface v
Contributors . . . . *xi*

1 Cell-Based Assays Using Differentiated Human Induced Pluripotent Cells 1
Zameel Cader, Martin Graf, Mark Burcin, Carl-Fredrik Mandenius, and James A. Ross

PART I DIFFERENTIATION OF IPC TO TARGET CELLS FOR ASSAYS

2 Monolayer Generation of Vascular Endothelial Cells from Human Pluripotent Stem Cells 17
Klaus Christensen, Filip Roudnicky, Mark Burcin, and Christoph Patsch

3 Generating Human iPSC-Derived Astrocytes with Chemically Defined Medium for In Vitro Disease Modeling 31
Katharina Janssen, Lamiaa Bahnassawy, Claudia Kiefer, Jürgen Korffmann, Georg C. Terstappen, Viktor Lakics, Miroslav Cik, and Peter Reinhardt

4 Human Induced Pluripotent Stem Cell-Derived Definitive Endoderm Bulk Culture and Hepatic Differentiation 41
Anwar A. Palakkan, Jyoti Nanda, and James A. Ross

5 Production of Cardiomyocytes from Human Pluripotent Stem Cells by Bioreactor Technologies 55
Caroline Halloin, Michelle Coffee, Felix Manstein, and Robert Zweigerdt

6 Improved Protocol for Cardiac Differentiation and Maturation of Pluripotent Stem Cells 71
Imen Jebeniani, Shunli Ding, and Michel Pucéat

7 Human Pluripotent Stem Cell Expansion in Stirred Tank Bioreactors 79
Felix Manstein, Caroline Halloin, and Robert Zweigerdt

8 A Protocol for One-Step Differentiation of Human Induced Pluripotent Stem Cells into Mature Podocytes 93
Cormac Murphy, Elisabeth Feifel, Paul Jennings, Gerhard Gstraunthaler, and Anja Wilmes

9 Differentiation of hiPS Cells into Definitive Endoderm for High-Throughput Screening 101
Teresa Bluhmki, Stefanie Traub, Eva Schruf, James Garnett, Florian Gantner, Daniel Bischoff, and Ralf Heilker

Part II High Throughput Metabolomics and Proteomic Cell-Based Assays

10 LC-MS/MS-Based Metabolomics for Cell Cultures 119
Anna Artati, Cornelia Prehn, and Jerzy Adamski

11 Preparation of iPSCs for Targeted Proteomic Analysis 131
Anna Baud, Wendy E. Heywood, Daniel Little, Paul Gissen, and Kevin Mills

12 Measurement of Bile Acids as a Marker of the Functionality of iPSC-Derived Hepatocytes 141
Francesca Mazzacuva, Maëlle Lorvellec, Agostino Cilibrizzi, Kevin Mills, Peter Clayton, and Paul Gissen

13 Mass Spectrometry Measurement of Albumin–Alpha Fetoprotein Ratio as an Indicator of iPSC-Derived Hepatocyte Differentiation 149
Francesca Mazzacuva, Maëlle Lorvellec, Agostino Cilibrizzi, Kevin Mills, Wendy E. Heywood, Peter Clayton, and Paul Gissen

Part III High Content Imaging

14 Seeding Induced Pluripotent Stem Cell-Derived Neurons onto 384-Well Plates 159
Daniel Little, Christin Luft, Oliver Pezzini-Picart, Olukunbi Mosaku, Robin Ketteler, Michael J. Devine, and Paul Gissen

15 High-Content Autophagy Analysis in iPSC-Derived Neurons Using Immunofluorescence 165
Daniel Little, Christin Luft, Olukunbi Mosaku, Robin Ketteler, Michael J. Devine, and Paul Gissen

16 High-Content Analysis of Mitochondrial Function in iPSC-Derived Neurons 175
Daniel Little, Christin Luft, Olukunbi Mosaku, Robin Ketteler, Michael J. Devine, and Paul Gissen

17 Evaluating the Effect of Drug Compounds on Cardiac Spheroids Using the Cardiac Cell Outgrowth Assay 185
Jonas Christoffersson, Florian Meier, Henning Kempf, Kristin Schwanke, Michelle Coffee, Mario Beilmann, Robert Zweigerdt, and Carl-Fredrik Mandenius

Part IV Electrophysiology Assays

18 Assay Procedures for Compound Testing of hiPSC-Derived Cardiomyocytes Using Multiwell Microelectrode Arrays 197
Udo Kraushaar and Elke Guenther

19 Multielectrode Array (MEA)-Based Detection of Spontaneous Network Activity in Human iPSC-Derived Cortical Neurons................ 209
Valeria Kizner, Sandra Fischer, and Maximilian Naujock

Part V Miscellaneous Cell-Based Assays

20 Assessing Neuronal Excitability on a Fluorometric Imaging Plate Reader (FLIPR) Following a Defined Electrostimulation Paradigm 219
Valeria Kizner, Sandra Fischer, Benjamin Jähnke, and Maximilian Naujock

21 Fabrication of a Microfluidic Cell Culture Device Using Photolithographic and Soft Lithographic Techniques 227
Jonas Christoffersson and Carl-Fredrik Mandenius

22 Using a Microfluidic Device for Culture and Drug Toxicity Testing of 3D Cells 235
Jonas Christoffersson and Carl-Fredrik Mandenius

23 hiPS Cell-Derived Neurons for High-Throughput Screening................. 243
Stefanie Traub and Ralf Heilker

Index *265*

Contributors

JERZY ADAMSKI • *Research Unit Molecular Endocrinology and Metabolism, Helmholtz Zentrum München, German Research Center for Environmental Health (GmbH), Neuherberg, Germany; German Center for Diabetes Research (DZD e.V.), Neuherberg, Germany; Lehrstuhl für Experimentelle Genetik, Technische Universität München, Freising-Weihenstephan, Germany; Department of Biochemistry, Yong Loo Lin School of Medicine, National University of Singapore, Singapore*

ANNA ARTATI • *Research Unit Molecular Endocrinology and Metabolism, Helmholtz Zentrum München, German Research Center for Environmental Health (GmbH), Neuherberg, Germany*

LAMIAA BAHNASSAWY • *Neuroscience Discovery, AbbVie Deutschland GmbH & Co KG, Ludwigshafen am Rhein, Germany*

ANNA BAUD • *Centre for Translational Omics, UCL Institute of Child Health, London, UK*

MARIO BEILMANN • *Boehringer Ingelheim Pharma GmbH and Co. KG, Nonclinical Drug Safety Germany, Biberach, Germany*

DANIEL BISCHOFF • *Drug Discovery Sciences, Boehringer Ingelheim Pharma GmbH & Co. KG, Biberach, Germany*

TERESA BLUHMKI • *Drug Discovery Sciences, Boehringer Ingelheim Pharma GmbH & Co. KG, Biberach, Germany*

MARK BURCIN • *Roche Pharmaceutical Research and Early Development, Roche Innovation Center Basel, Basel, Switzerland*

ZAMEEL CADER • *Translational Molecular Neuroscience Group, Nuffield Department of Clinical Neurosciences, University of Oxford, John Radcliffe Hospital, Oxford, UK*

KLAUS CHRISTENSEN • *Roche Pharmaceutical Research and Early Development, Roche Innovation Center Basel, Basel, Switzerland*

JONAS CHRISTOFFERSSON • *Division of Biotechnology, Department of Physics, Chemistry and Biology (IFM), Linköping University, Linköping, Sweden*

MIROSLAV CIK • *Neuroscience Discovery, AbbVie Deutschland GmbH & Co KG, Ludwigshafen am Rhein, Germany*

AGOSTINO CILIBRIZZI • *Institute of Pharmaceutical Science, King's College London, London, UK*

PETER CLAYTON • *Great Ormond Street Institute of Child Health, University College London, London, UK*

MICHELLE COFFEE • *Leibniz Research Laboratories for Biotechnology and Artificial Organs (LEBAO), Department of Cardiac, Thoracic, Transplantation and Vascular Surgery, Hannover Medical School, Hannover, Germany; REBIRTH-Cluster of Excellence, Hannover Medical School, Hannover, Germany*

MICHAEL J. DEVINE • *MRC Laboratory for Molecular Cell Biology, University College London, London, UK*

SHUNLI DING • *INSERM U1251 Université Aix-Marseille, MMG, Marseille, France*

ELISABETH FEIFEL • *Division of Physiology, Medical University Innsbruck, Innsbruck, Austria*

SANDRA FISCHER • *CNS Diseases Research, Boehringer Ingelheim Pharma GmbH & Co. KG, Biberach an der Riß, Germany*

FLORIAN GANTNER • *Translational Medicine and Clinical Pharmacology, Boehringer Ingelheim Pharma GmbH & Co. KG, Biberach, Germany*
JAMES GARNETT • *Immunology & Respiratory Diseases Research, Boehringer Ingelheim Pharma GmbH & Co. KG, Biberach, Germany*
PAUL GISSEN • *MRC Laboratory for Molecular Cell Biology, University College London, London, UK; Great Ormond Street Institute of Child Health, University College London, London, UK; NIHR Great Ormond Street Hospital Biomedical Research Centre, University College London, London, UK*
MARTIN GRAF • *Roche Pharmaceutical Research and Early Development, Roche Innovation Center Basel, Basel, Switzerland*
GERHARD GSTRAUNTHALER • *Division of Physiology, Medical University Innsbruck, Innsbruck, Austria*
ELKE GUENTHER • *AIT Austrian Institute of Technology GmbH, Vienna, Austria*
CAROLINE HALLOIN • *Leibniz Research Laboratories for Biotechnology and Artificial Organs (LEBAO), Department of Cardiac, Thoracic, Transplantation and Vascular Surgery, Hannover Medical School, Hannover, Germany; REBIRTH-Cluster of Excellence, Hannover Medical School, Hannover, Germany*
RALF HEILKER • *Drug Discovery Sciences, Boehringer Ingelheim Pharma GmbH & Co. KG, Biberach, Germany*
WENDY E. HEYWOOD • *Great Ormond Street Institute of Child Health, University College London, London, UK; Centre for Translational Omics, UCL Institute of Child Health, London, UK*
BENJAMIN JÄHNKE • *CNS Diseases Research, Boehringer Ingelheim Pharma GmbH & Co. KG, Biberach an der Riß, Germany*
KATHARINA JANSSEN • *Neuroscience Discovery, AbbVie Deutschland GmbH & Co KG, Ludwigshafen am Rhein, Germany; BG Neurochirurgie und Neurotraumatologie, Berufsgenossenschaftliches Universitätsklinikum Bergmannsheil gGmbH, Bochum, Germany*
IMEN JEBENIANI • *INSERM U1251 Université Aix-Marseille, MMG, Marseille, France*
PAUL JENNINGS • *Division of Molecular and Computational Toxicology, Vrije Universiteit Amsterdam, Amsterdam, The Netherlands*
HENNING KEMPF • *Leibniz Research Laboratories for Biotechnology and Artificial Organs (LEBAO), Hannover Medical School, Hannover, Germany*
ROBIN KETTELER • *MRC Laboratory for Molecular Cell Biology, University College London, London, UK*
CLAUDIA KIEFER • *Neuroscience Discovery, AbbVie Deutschland GmbH & Co KG, Ludwigshafen am Rhein, Germany*
VALERIA KIZNER • *CNS Diseases Research, Boehringer Ingelheim Pharma GmbH & Co. KG, Biberach an der Riß, Germany*
JÜRGEN KORFFMANN • *Neuroscience Discovery, AbbVie Deutschland GmbH & Co KG, Ludwigshafen am Rhein, Germany*
UDO KRAUSHAAR • *Department of Electrophysiology, NMI Natural and Medical Sciences Institute at the University of Tuebingen, Reutlingen, Germany*
VIKTOR LAKICS • *Neuroscience Discovery, AbbVie Deutschland GmbH & Co KG, Ludwigshafen am Rhein, Germany*
DANIEL LITTLE • *MRC Laboratory for Molecular Cell Biology, University College London, London, UK*

Maëlle Lorvellec • *MRC Laboratory for Molecular Cell Biology, University College London, London, UK*
Christin Luft • *MRC Laboratory for Molecular Cell Biology, University College London, London, UK*
Carl-Fredrik Mandenius • *Division of Biotechnology, Department of Physics, Chemistry and Biology (IFM), Linköping University, Linköping, Sweden*
Felix Manstein • *Leibniz Research Laboratories for Biotechnology and Artificial Organs (LEBAO), Department of Cardiac, Thoracic, Transplantation and Vascular Surgery, Hannover Medical School, Hannover, Germany; REBIRTH-Cluster of Excellence, Hannover Medical School, Hannover, Germany*
Francesca Mazzacuva • *Mass Spectrometry Facility, King's College London, London, UK; Great Ormond Street Institute of Child Health, University College London, London, UK*
Florian Meier • *Boehringer Ingelheim Pharma GmbH and Co. KG, Nonclinical Drug Safety Germany, Biberach, Germany*
Kevin Mills • *Great Ormond Street Institute of Child Health, University College London, London, UK; Centre for Translational Omics, UCL Institute of Child Health, London, UK*
Olukunbi Mosaku • *MRC Laboratory for Molecular Cell Biology, University College London, London, UK*
Cormac Murphy • *Division of Molecular and Computational Toxicology, Vrije Universiteit Amsterdam, Amsterdam, The Netherlands*
Jyoti Nanda • *Tissue Injury and Repair Group, University of Edinburgh, Edinburgh, UK*
Maximilian Naujock • *CNS Diseases Research, Boehringer Ingelheim Pharma GmbH & Co. KG, Biberach an der Riß, Germany*
Anwar A. Palakkan • *Tissue Injury and Repair Group, University of Edinburgh, Edinburgh, UK*
Christoph Patsch • *Roche Pharmaceutical Research and Early Development, Roche Innovation Center Basel, Basel, Switzerland*
Oliver Pezzini-Picart • *MRC Laboratory for Molecular Cell Biology, University College London, London, UK*
Cornelia Prehn • *Research Unit Molecular Endocrinology and Metabolism, Helmholtz Zentrum München, German Research Center for Environmental Health (GmbH), Neuherberg, Germany*
Michel Pucéat • *INSERM U1251 Université Aix-Marseille, MMG, Marseille, France*
Peter Reinhardt • *Neuroscience Discovery, AbbVie Deutschland GmbH & Co KG, Ludwigshafen am Rhein, Germany*
James A. Ross • *Tissue Injury and Repair Group, University of Edinburgh, Edinburgh, UK*
Filip Roudnicky • *Roche Pharmaceutical Research and Early Development, Roche Innovation Center Basel, Basel, Switzerland*
Eva Schruf • *Immunology & Respiratory Diseases Research, Boehringer Ingelheim Pharma GmbH & Co. KG, Biberach, Germany*
Kristin Schwanke • *Leibniz Research Laboratories for Biotechnology and Artificial Organs (LEBAO), Hannover Medical School, Hannover, Germany*
Georg C. Terstappen • *Neuroscience Discovery, AbbVie Deutschland GmbH & Co KG, Ludwigshafen am Rhein, Germany; OxStem Limited, Oxford, UK*
Stefanie Traub • *Trenzyme GmbH, Konstanz, Germany*

ANJA WILMES • *Division of Molecular and Computational Toxicology, Vrije Universiteit Amsterdam, Amsterdam, The Netherlands*
ROBERT ZWEIGERDT • *Leibniz Research Laboratories for Biotechnology and Artificial Organs (LEBAO), Department of Cardiac, Thoracic, Transplantation and Vascular Surgery, Hannover Medical School, Hannover, Germany; REBIRTH-Cluster of Excellence, Hannover Medical School, Hannover, Germany*

Chapter 1

Cell-Based Assays Using Differentiated Human Induced Pluripotent Cells

Zameel Cader, Martin Graf, Mark Burcin, Carl-Fredrik Mandenius, and James A. Ross

Abstract

This chapter describes the requirements and preconditions for using human induced pluripotent cell lines in assay development within the pharmaceutical industry. The joint collaborative effort between academic and pharma partners within the StemBANCC consortium which enabled the implementation of iPSC-derived cellular models for drug discovery is highlighted. This large collaborative scientific network has successfully derived a significant number of well-characterized patient-specific iPSC lines and established disease-relevant cellular assays, both of which are requirements for enabling pharmaceutical companies to develop more efficacious and safer medicines.

Key words Induced pluripotent cell, iPSC, Disease models, Differentiation protocols

1 Creation of a Cell Bank with Patient-Specific Induced Pluripotent Cell Lines

Biomedical research has made huge progress in deciphering disease mechanism across many therapeutic areas, not least through the advent of genetic studies that have identified either highly deleterious gene mutation causing rare disease or susceptibility loci and gene pathways implicated in common disease. However, for some of the most prevalent and disabling conditions of our time such as dementia, pain, autism, mental health disorders, and diabetes, there remain large gaps in our understanding of disease processes and a paucity of effective treatments. In particular, progress has been hampered by the lack of access to human tissues involved in these disorders. Animal research has partly overcome this bottleneck, but it is increasingly apparent that findings in animal models frequently do not translate into humans.

In 2007, a revolution in biomedical research was sparked, with the discovery that adult cells could be reprogrammed to a pluripotent state—the induced pluripotent stem cell (iPSC). The lack of

Carl-Fredrik Mandenius and James A. Ross (eds.), *Cell-Based Assays Using iPSCs for Drug Development and Testing*, Methods in Molecular Biology, vol. 1994, https://doi.org/10.1007/978-1-4939-9477-9_1,
© Springer Science+Business Media, LLC, part of Springer Nature 2019

ready access to patient and disease-relevant tissue to study disease processes had finally been overcome, through the ability to generate human cellular models directly from the patient population of interest. The key advantage of iPSC is that the genomic background of the patient is captured such that all the disease susceptibility factors or causal mutations are also present in the cellular model. Furthermore, the process of reprogramming from adult somatic cell to iPSC erases many aspects of the epigenetic memory of the adult cell and therefore iPSC-based disease models enable research on the genetic causation of disease without being confounded by epigenetic changes in the adult cell. Conversely, through manipulation of the cell culture environment to expose cells to factors considered important to disease development, such as inflammation, toxins or metabolic stressors, gene–environment interactions, and associated epigenetic effects can be recapitulated and interrogated.

Human iPSCs capture the genetic predispositions to disease and, to take full advantage of this property, it is necessary to identify clinically well-characterized patients. Patients with monogenic forms of disease, while rare, are highly valuable in understanding disease mechanisms since genotype to phenotype correlations are robust due to the highly penetrant nature of these type of mutations. This, in turn, increases the likelihood of successfully identifying disease-relevant and consistent cellular phenotypes in iPSC models. Furthermore, using genome engineering approaches, a disease mutation can be introduced onto a control iPSC line or the mutation "corrected" in a patient line to assess the causal role of the mutation in a particular phenotype. Common disorders which have a complex genetic basis are at the outset less tractable. While a monogenic disorder may be driven by a single cellular pathway, which is overtly dysfunctional, in a polygenic condition, multiple pathways are likely to be implicated, none of which may be disturbed at baseline. Nevertheless, under sufficient "stress," these disease pathways may be unmasked, and this might represent some of the simpler necessary gene–environment interactions of complex traits.

In order to facilitate these type of studies, collections of iPSC lines from patients with the same disease and/or disease-related genotype are required along with iPSC lines from subjects judged as healthy volunteers. The challenge is to reduce the technical variance arising from iPSC reprogramming and subsequent differentiation to ensure biological signals are detectable above the background noise. Reprogramming iPSC even today is a technically demanding and costly process. Hence, there have been several collaborative efforts to generate collections of iPSC lines from healthy volunteers and patients. These efforts have included the recently completed IMI-StemBANCC consortium supported by the EFPIA and the EU. StemBANCC have generated over

600 patient-specific iPSC lines for which the standard operating protocols ensured a consistent approach to sampling adult somatic cells from research volunteers followed by a standardized reprogramming workflow using the nonintegrating Sendai virus. All the lines generated by StemBANCC have been checked for successful reprogramming, clearance of Sendai virus and genomic stability. The lines have now been passed to biorepositories such as EBiSC, which are able to store and distribute the lines to researchers on an international basis. The work within StemBANCC which has utilized these cell lines for the establishment of cell-based assays for drug development has resulted in the protocols compiled in this book.

2 Use of Induced Pluripotent Stem Cells in Drug Development in the Pharmaceutical Industry

Discovering the most efficacious drugs for patients is very challenging due to the complexity of many diseases. For this reason, the pharmaceutical industry is continuously exploring novel ways of conducting drug discovery. Traditionally, drugs were developed in a linear approach while the assays employed in drug discovery have been based on artificially immortalized cell lines. These lines have been used in high-throughput screening to identify therapeutic agents that modulate the activity of the target. These recombinant cellular assay formats have the advantage of enabling large screening campaigns for testing millions of compounds in a reproducible manner. However, as the methodology is designed to be compatible with high-throughput screening and, as such, the cellular models are often not predictive enough of the disease biology. Assays using primary cell cultures offer an alternative to recombinant cell lines; however, the limited ability of these cells to be expanded make it difficult to use these cells in large drug-screening campaigns. Furthermore, donor variability often does not provide robust and reproducible results compatible with drug discovery requirements. In order to overcome these limitations, the pharmaceutical industry has started to use iPSC-derived cellular models to develop more disease-relevant assays for drug screening. The availability of such models provides the opportunity to screen drug candidates that cause a desirable disease-reversing change in phenotype. Furthermore, by using such "diseased" cell types, the effect of a drug candidate on a specific disease mechanism could be better understood in a pathway-centric view. In particular, the opportunity to use patient-specific cellular models carrying the disease-relevant genetic background provides great promise. Genome editing is another recent breakthrough technology that has further enhanced our abilities to develop customized cellular assays for drug discovery. Genome editing techniques in pluripotent stem cells allows, for the first time, the use of this approach in human disease-relevant models. By using genome editing, scientists are able to harbor specific disease-causing mutations while, at the same time, revert these mutations to obtain "wild-type" isogenic control cells. For phenotypic screening efforts such isogenic iPSC control lines provide important information to link an identified phenotype to a patient-specific mutation.

Despite being of great promise the use of pluripotent stem cell-derived disease-relevant assays is still limited within high-throughput screening campaigns. To best utilize their potential, the pharmaceutical industry uses pluripotent stem cell models primarily for focused screening campaigns or in secondary screening and, especially, during hit validation, qualification, and profiling. Here, disease-relevant assays help to identify the most promising drug candidates with the highest potential to translate into clinical efficacy. The throughput limitations can be overcome by iterative screening campaigns using artificial intelligence (AI). Such a screening approach will allow the coverage of the chemical space of millions of molecules by screening only a fraction of them. Furthermore, iPSC-derived cellular models play an important role in cases where there is a need to identify human-specific modalities such as antibodies or RNA therapeutics. In some cases, the biology of a specific target and its role in disease development is not fully understood. In such cases, disease-relevant cellular assays can be utilized to provide a better understanding of the mechanistic pathways in a specific disease. Phenotypic screens play a key role in identifying drug candidates that can revert a disease phenotype without knowing the specific target. Here, pluripotent stem cell assays are not only used to conduct the phenotypic screen, but also, in combination with genome editing, help to identify the target of a drug candidate during target identification. The safety of drug candidates is one of the most important requirements during drug discovery. For this reason, the pharmaceutical industry undertakes strong efforts to avoid unwanted side-effects by evaluating the safety of potential drug candidates. Pluripotent stem cell-derived models have been widely used in recent years to identify such safety aspects in a more reliable way and as early as possible during drug discovery.

Developing complex iPSC-derived cellular models and using them during drug discovery for identifying the most efficacious drugs requires enormous efforts that are difficult to accomplish within a single pharmaceutical company. For this reason, scientific consortia such as StemBANCC are instrumental in fostering a close precompetitive collaboration between academic groups, pharmaceutical groups, and the biotech industry. These large collaborative enterprises are founded on complementary scientific and drug discovery expertise to codevelop the different tasks necessary to use iPSC-derived cellular models for drug discovery. The use of well-defined processes to reprogram, cultivate, store, quality control, track, and distribute iPSC lines has ensured a high degree of robustness and translatability during assay development. Access to high-quality iPSC lines derived from patients that broadly cover different disease backgrounds has a strong impact on establishing cellular assays with the potential to better translate from the drug discovery phase toward the clinic. It is especially useful to have access to clinical data including information on disease progression in the patient. By following these principles, academic scientists and pharma partners have established differentiation protocols for several cell types, have implemented multiple disease-relevant assays and analytical methods which are compatible with drug discovery. In particular, the requirement for stringent quality control, defined cultivation, and differentiation methods in a defined workflow required the establishment of standardized operating procedures throughout the consortium. Such an approach has facilitated the standardization of reproducible assay conditions for drug discovery. At multiple stages during this process, quality requirements, such as short tandem repeat (STR) profiling, pluripotent marker validation, genomic integrity, and cell banking, have been implemented by the consortium within a standardized workflow. Furthermore, consortia such as StemBANCC, with a broad spectrum of scientific expertise, have the unique ability to perform multisite studies that otherwise would be too complex to be accomplished by a single group. Such large multisite efforts were utilized to conduct multidisciplinary studies that monitored variabilities in research site, iPSC line or scientific experimenter. Results from such studies specifically impact drug discovery ensuring that protocols and assay formats generate reproducible results across laboratories, a key requirement for the establishment of robust and reproducible assay conditions.

3 Differentiation of iPSCs for Assay Development

In recent years there has been significant progress in developing human induced pluripotent cell (hiPSC) differentiation protocols for numerous cell lineages. This elaboration of differentiation protocols has been driven by a demand for more physiological in vitro models which recapitulate human diseases, and which can provide human-relevant information for drug toxicology. Significant advances can be attributed to consortia such as StemBANCC in this respect. In this volume a collection of the differentiation protocols utilized in the StemBANCC collaboration are described.

A primary consideration for the use of hiPSCs is the ability to produce these cells in large quantities. The normal culture method for hiPSCs is both time-consuming and the number of cells produced is limited. However, Manstein and colleagues [1] describe a detailed protocol for hiPSCs cultivation as matrix-free, cell-only aggregates in defined and xenobiotic-free culture medium in stirred tank bioreactors. The process commences with frozen stock hiPSCs which have been pre-expanded in conventional culture dishes and results in 150 ml cultures in stirred tank bioreactors. This technology is designed to produce up to 500 million hiPSCs within 7 days which is a staggering improvement over the conventional 2D production of these cells. The strategy includes perfusion-based cell feeding which facilitates process control, automation and much higher cell yields. This is an important step for generating large quantities of hiPSCs for subsequent differentiation into different lineages and the use of such cells in cell-based assays and, in the longer term, regenerative medicine.

A key aspect of utilizing hiPSC-derived cells for drug screening is the ability to produce large quantities of the desired cell type such as definitive endoderm which can then be further differentiated to, for example, hepatocyte or pancreatic lineages. To date, several directed differentiation protocols have been described which generate definitive endoderm from hiPSCs, but protocols which produce sufficient cells for drug development and high-throughput screening have not been reported. Bluhmki et al. [2] describe a large-scale expansion of hiPSCs for high-throughput approaches together with an optimized stepwise differentiation of hiPSCs into definitive endoderm cells which express the classical definitive endoderm markers including SOX17, CXCR4, and EpCAM. In an alternative approach, Palakkan and colleagues [3] have developed a method for the bulk culture of definitive endoderm cells generated from hiPSCs which can be stored and differentiated to hepatocytes. This method describes the production of hiPSC-derived definitive endoderm cells which can be immunomagnetically sorted based on their expression of CXCR4. The CXCR4-sorted cells retain the ability to proliferate for extended periods and can be cryopreserved. The definitive endoderm cells were subsequently utilized to generate functional hepatocytes expressing albumin and α-fetoprotein in different multiwell formats. This provides a method to reliably produce more consistent hepatocytes in greater quantities. This approach has enabled the development of high-throughput screening strategies to screen small molecules which can enhance the maturity of the iPSC-derived hepatocytes. The provision of large quantities of hepatocytes will prove invaluable for in vitro disease models and drug screening.

The development of methods for generating vascular endothelial cells from human iPSCs in numbers which permit high-throughput screening is invaluable, not only for screening this particular cell type but also for the facility to include vascular endothelial cells in organoid constructs. Christensen et al. [4] describe an efficient and scalable protocol to differentiate hiPSCs into vascular endothelial cells using defined culture conditions. The combinatorial use of small molecule compounds, growth factors as well as morphogens directs hiPSCs toward an endothelial cell fate within 6 days. The hiPSCs are driven through mesodermal cell commitment in the priming step, and, in the induction step, vascular endothelial growth factor A (VEGF-A) and the small molecule forskolin promote the conversion of the mesodermal progenitors into vascular endothelial cells. Although the protocol described has the capacity to generate endothelial cells with efficiencies of up to 80%, an additional VE-cadherin immunomagnetic purification step provides a virtually pure population of endothelial cells.

One key area for the development of differentiation protocols is for cells present in the kidney. There has been the recent development of protocols which drive the differentiation of hiPSCs toward renal epithelial cells. However, within the glomerulus, podocytes are highly specialized visceral epithelial cells which form part of the glomerular filtration barrier. The culture of human podocyte cells is extremely challenging for both primary and immortalized cells, as the cells are nonproliferative and are subject to rapid dedifferentiation. Therefore, iPSC-derived podocytes are extremely desirable for in vitro studies of renal function and high-throughput screening. Although there have been several differentiation protocols described recently which drive hiPSC toward podocytes, these protocols are quite complex and take considerable time. Murphy and colleagues [5] describe a simple and rapid one-step protocol that drives hiPSCs toward podocyte-like cells in 10 days.

A fundamental requirement in cardiovascular research, for both disease-modeling and high-throughput screening of drugs, is the development of improved protocols for cardiac differentiation from hiPSCs and subsequent maturation of these cells. Here, Jebeniani et al. [6] report a cost-effective and simple protocol for the differentiation of specific ventricular cardiomyocytes. These cells are elongated, do not spontaneously beat, and do not feature any Ca^{2+} transient, an index of their stage of maturation toward adult cardiac cells. The differentiated cells represent a suitable model for screening both the efficiency and toxicology of drugs. Another key requirement for cardiovascular research is the production of cardiomyocytes in sufficient quantities for high-throughput screening. In this regard, Halloin and colleagues [7] describe a protocol for the production of cardiomyocytes from hiPSCs using bioreactor technologies. The strategy combines both hiPSC expansion in suspension culture, directly followed by cardiomyocyte differentiation using small molecule-Wnt pathway modulators. Further details provide for an expansion of the protocol to enable parallel inoculation of 4 × 150 ml parallel bioreactor differentiations, potentially producing more than 240×10^6 cardiomyocytes in 22 days. Such an approach would provide sufficient cardiomyocytes for in vitro assays for both drug discovery and safety pharmacology.

This approach may also provide a superior cell source for future cell therapies aimed at the repair of damaged heart tissue.

An area of huge progress in recent years is a greater understanding of the function of different cell types in the brain. The use of hiPSC-derived brain cell lineages promises to provide better cell models for drug discovery and contribute to our understanding of neurodegenerative and neuropsychiatric diseases. The neuronal differentiation protocol described by Traub and Heilker [8] enables the scalable production of neurons in sufficient numbers to meet the demands of high-throughput screening strategies. The protocol provides details of a combination of growth factor-driven expansion and inhibition of notch (GRINCH) to enable the scalable production of neurons. The subsequent GRINCH cells are matured in 384-well microplates and display neuronal markers and electrophysiological activity. In addition, Janssen and colleagues describe a protocol for the differentiation of expandable astrocytes under chemically defined conditions using neural precursor cell starting cultures which provide sufficient cells for in vitro disease modeling. Most of the published astrocyte differentiation protocols for hiPSCs are prolonged and require extensive manual intervention and/or growth as cell aggregates. The existing protocols are inefficient for use with liquid-handling devices and the production of sufficient cells for drug testing. The differentiation protocol described in this volume is suitable for the production of large quantities of astrocytes.

As the number of differentiation protocols expands for an increasing variety of cell lineages, hiPSCs will become increasingly important in the provision of relevant in vitro cell systems for disease modeling, drug screening, and understanding of cell function in different organs of the body.

4 Development of Protocols for iPSC-Based Assays

The access to human iPSCs differentiated to organ and tissue cells paves the way to development of new and better in vitro assays for pharmaceutical research in academia and industry. These assays all rely on the same procedural scheme (Fig. 1):

1. Cells from an iPSC lines in a cell bank are differentiated to tissue and organ cells.
2. The derived cells are purified and qualified for assay purposes.
3. The cells are conditioned, contained and maintained before and during the required assay time.
4. Cells are exposed to test compounds.
5. The cell response is observed.
6. The response is evaluated, and a readout presented.

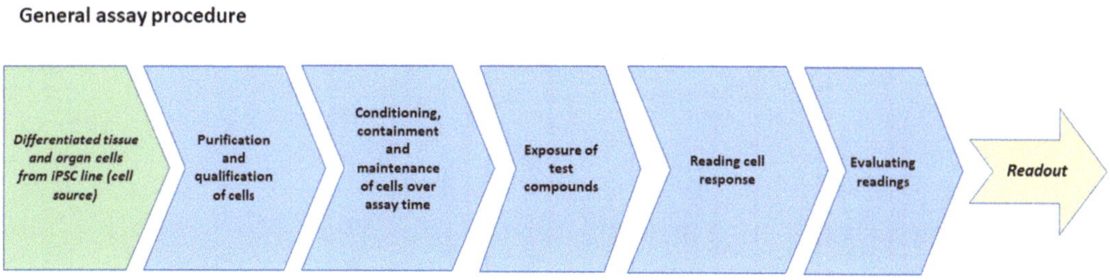

Fig. 1 A general scheme for the assay procedure workflow as described in the iPSC assay protocols

This scheme applies to all the assays described in this book. The cell sources and differentiation methods and protocols are commented on in the above text. The assays can be divided into methodologies for generating the readouts:

1. Cell-based assays related to mass spectrometric measurements and readouts.
2. Cell-based assays related to high-content imaging, where advanced imaging instrumentation is employed in combination with powerful imaging software.
3. Cell-based assays based on readouts from electrodes, represented here by multielectrode arrays.
4. Cell-based assays where the assay is carried out in a microfluidic setup where the main objective is to provide a tissue-like 3D environment for the in vitro assay.
5. Cell-based assays using other spectrometric detection methods such as fluorescence fluorimetry.

Figure 2 provides an overview of the assay protocols presented and the links to the various differentiation protocols.

4.1 Assays Related to Liquid Chromatography Combined with Tandem Mass Spectrometry (LC-MS/MS)

LC-MS/MS assays can be nontargeted to include the whole metabolome or proteome of the cell sample or targeted to specific sets of metabolites or peptides/proteins. Such assays have great potential as a human model system in regenerative medicine, disease modeling, and drug screening.

Artati et al. [10] present convenient protocols for metabolomic analysis of metabolites applicable to a variety of biological specimens (e.g., cells, body fluids, tissues, exhaled air, plants) using LC-MS/MS-based analyses for cell metabolomics. Using their protocols, over 1000 endogenous and exogenous metabolites can be detected, annotated, and quantified relatively by a nontargeted analysis approach, whereas targeted metabolomic analysis enables quantification of 188 endogenous metabolites. The protocols exemplify applications to metabolomics analysis in human iPSC systems.

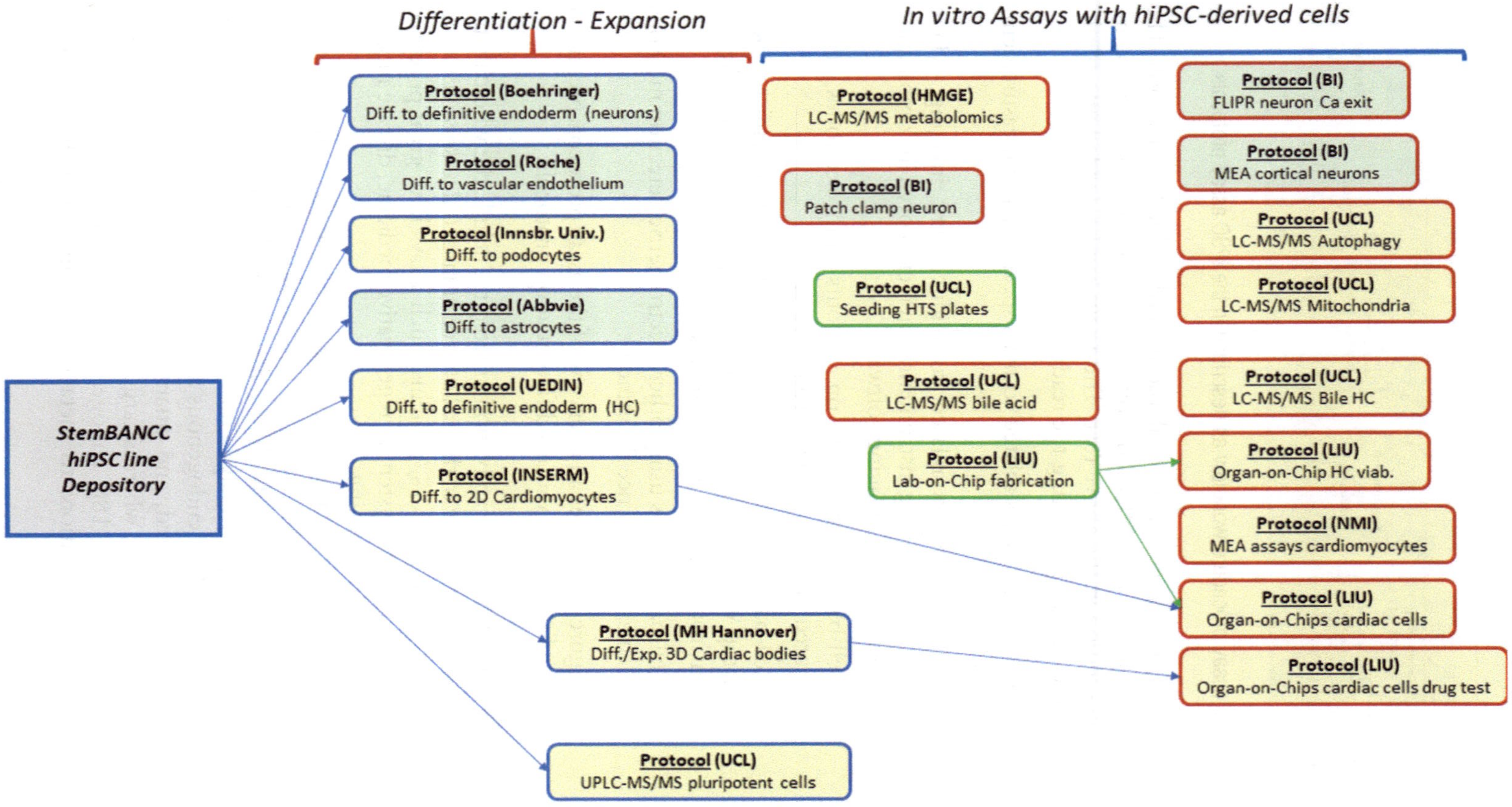

Fig. 2 Overview of the protocols for differentiation of iPSCs [1–9] and in vitro assays [10–22]

Another metabolomic assay protocol is provided by Mazzacuva and colleagues [11] for measurement of bile acids as a marker of the functionality of iPSC-derived hepatocytes. The bile acid profiling is proposed as a novel approach to assess the maturation of hepatocyte-like cells. The main advantages of the method are the simplicity and rapidity as well as the possibility to localize enzyme deficiencies by quantifying the accumulation of specific intermediates involved in various synthetic pathways.

Another example from Mazzacuva and colleagues [12] is an LC-MS/MS method for the assessment of the ratio between albumin and α-fetoprotein as an indicator of iPSC-derived hepatocyte differentiation. During the differentiation process from iPSCs to hepatocytes it is of great value to monitor the level of cellular maturation. The assay evaluates the stage of differentiation based on the monitoring of the ratio between the two plasma proteins typically secreted by immature (α-fetoprotein) and mature (albumin) hepatocytes. The LC-MS/MS analysis allows for a precise, selective, and reproducible quantitation of low abundance proteins.

A concrete example of the application of targeted proteomic analysis using LC-MS/MS for iPSCs is provided by Baud and colleagues [13]. They describe a protocol based on the preparation of iPSC for subsequent measurement of pluripotency markers which allows classification of the cells into either pluripotent or nonpluripotent cells.

4.2 Assays Related to Imaging

High-content imaging (HCI) has become a popular technique with significant relevance in drug development studies. A few of the protocols in the book apply HCI to the iPSC lines.

One such example is the high-content autophagy analysis of iPSC-derived neurons using immunofluorescence. Autophagy, the process by which cellular proteins and organelles are degraded and recycled, is essential to the survival of cells. Defective autophagic degradation can be linked to many neurodegenerative diseases including the lysosomal storage diseases. Little and colleagues [14] describe a high-content imaging assay to detect defects in the autophagy pathway in iPSC-derived neurons, where immunofluorescence is used to stain autophagosomes and the changes in autophagosome levels in response to modulators are measured in an automated image analysis procedure.

A similar example of HCI is the analysis of mitochondrial function in iPSC-derived neurons. Mitochondrial dysfunction is linked to many neurological diseases; therefore, sensitive assays for measuring the mitochondrial functions are of great value in research and drug testing. In the assay from Little and colleagues [15] HCI simultaneously measures mitochondrial membrane potential, morphology, and cell viability in live iPSC-derived neurons through the use of fluorescent dyes.

HCI can also be combined with Organ-on-Chip technology for studying cells in 3D format under fluidic conditions. In a protocol by Christoffersson et al. [16] this is done with spheroids of hiPSC-derived cardiomyocytes for evaluating the effects of drug compounds. The outgrowth of cardiomyocytes is recorded by the HCI as a readout of the effect of a drug. The cell culture model mimics the physiology and both the mechanical and chemical cues that are present in the cardiac tissues in a parallel throughput chip format.

4.3 Assays Related to Microfluidics

The microfluidic technique provides ample possibilities for development of cell-based assays where cells are maintained in 3D scaffolds or other constructs under fluidic conditions. Nowadays, plastic microfluidic devices are commercially available but can also easily be made in a laboratory facility with basic equipment for photolithographic techniques. Christoffersson and Mandenius [17] describe a convenient fabrication protocol based on photolithography and soft lithography for microfluidic devices suited for 3D cultures. Well-defined microstructures are created by exposing a photoresist to UV light under a photolithographic mask in which the desired flow pattern is engraved. The subsequent UV cross-linking generates a reusable master that serves as a template for an elastomer that reciprocally recaptures the structures of the master in optically clear and oxygen-permeable rubber-like materials. The application of a device from the fabrication procedure is exemplified in an assay protocol with liver cells [18].

4.4 Assays Related to Electrical and Fluorometric Procedures

Assessing neuronal excitability on a fluorometric imaging plate reader (FLIPR) following a defined electrostimulation paradigm is described by Kizner and colleagues [19]. A FLIPR-based calcium assay enables the detection and characterization of neuronal excitability by using electrical field stimulation to evoke and record action potential-driven calcium transients in iPSC-derived cortical forebrain neurons. By carrying out the procedure in a 384-well plate format a high-throughput measurement is accomplished.

Assay procedures for compound testing of hiPSC-derived cardiomyocytes using multiwell microelectrode arrays are described by Kraushaar and Guenther [20]. The cardiac action potential requires a precise timing of activation and inactivation of ion channel subtypes. Deviations, for example due to blockage of specific voltage-gated potassium channels, can result in life-threatening arrhythmias. The microelectrode arrays (MEA) record, noninvasively, cardiac field action potential using two parallelized MEA systems for compound tests. With iPSC-derived cardiomyocytes (hiPS-CM) these assays can support the design of patient-derived disease models. MEA-based detection can also be used for characterization of a neuronal culture. Kizner and colleagues describe this for assaying spontaneous network activity in cortical neurons [21]. The protocol is adapted to iPSC-derived cortical neurons cultured on a 24-well multiwell-MEA system.

4.5 Screening Assays

The procedure suggested by Traub and Heilker [8] for high-throughput screening of human iPSC-derived neurons encompasses a cellular amplification phase for hiPS-derived neural progenitor (NP) cells followed by measurement of the phosphorylation and downstream signaling of the endogenously expressed TrkB. By using the amplified luminescent proximity homogeneous assay (Alpha) and the high-throughput reverse transcription polymerase chain reaction, the procedure can be adapted for pharmaceutical drug targets. Together with the availability of iPSC-derived neurons, these detection technologies open new experimental routes with great potential for early drug discovery with models for neurodegenerative and neuropsychiatric diseases.

Together, the cell-based assay protocols described in this volume provide a collection of cell-based assays of particular utility in the drug development industry. The assays demonstrate how iPSC cells can be directly applied to disease models and drug screening procedures. The protocols are also viable platforms for further assay development. Critical for the success of this approach is the access to iPS cells from a range of normal individuals and individuals with disease. The established cell banks for storage and distribution of iPSC lines are instrumental in this regard.

Acknowledgments

The research leading to these results has received support from the Innovative Medicines Initiative Joint Undertaking under (grant no. 115439), resources of which are composed of financial contribution from the European Union's Seventh Framework Programme (FP7/2007-2013) and EFPIA companies. This publication reflects only the author's views, and neither the IMI JU nor EFPIA nor the European Commission is liable for any use that may be made of the information contained therein.

References

1. Manstein F, Halloin C, Zweigerdt R (2019) Human pluripotent stem cell expansion in stirred tank bioreactors. In: Mandenius C-F, Ross JA (eds) Cell-based assays using iPSCs for drug development and testing. Springer, New York
2. Bluhmki T, Traub S, Schruf E, Garnett J, Gantner F, Daniel Bischoff D, Heilker R (2019) Differentiation of hiPS cells into definitive endoderm for high throughput screening. In: Mandenius C-F, Ross JA (eds) Cell-based assays using iPSCs for drug development and testing. Springer, New York
3. Palakkan AA, Nanda J, Ross JA (2019) Human induced pluripotent stem cell-derived definitive endoderm bulk culture and hepatic differentiation. In: Mandenius C-F, Ross JA (eds) Cell-based assays using iPSCs for drug development and testing. Springer, New York
4. Christensen K, Roudnicky F, Bucin M, Patsch C (2019) Monolayer generation of vascular endothelial cells from human iPSCs. In: Mandenius C-F, Ross JA (eds) Cell-based assays using iPSCs for drug development and testing. Springer, New York
5. Murphy C, Feifel E, Jennings P, Gstraunthaler G, Wilmes A (2019) A protocol for the one-step differentiation of human induced pluripotent stem cells into mature podocytes. In: Mandenius C-F, Ross JA (eds)

Cell-based assays using iPSCs for drug development and testing. Springer, New York

6. Jebeniani I, Ding S, Puceat M (2019) Improved protocol of cardiac differentiation and maturation of pluripotent stem cells. In: Mandenius C-F, Ross JA (eds) Cell-based assays using iPSCs for drug development and testing. Springer, New York
7. Halloin C, Coffee M, Manstein F, Zweigerdt R (2019) Production of cardiomyocytes from human pluripotent stem cells by bioreactor technologies. In: Mandenius C-F, Ross JA (eds) Cell-based assays using iPSCs for drug development and testing. Springer, New York
8. Traub S, Heilker R (2019) hiPS Cell–derived neurons for high-throughput screening. In: Mandenius C-F, Ross JA (eds) Cell-based assays using iPSCs for drug development and testing. Springer, New York
9. Janssen K, Bahnassawy L, Kiefer C, Terstappen GC KJ, Lakics V, Cik M, Reinhardt P (2019) (Abbvie, Germany) Human iPSC-derived astrocytes for in vitro disease modelling. In: Mandenius C-F, Ross JA (eds) Cell-based assays using iPSCs for drug development and testing. Springer, New York
10. Artati A, Prehn C, Adamski J (2019) LC-MS/MS-based metabolomics for cell cultures. In: Mandenius C-F, Ross JA (eds) Cell-based assays using iPSCs for drug development and testing. Springer, New York
11. Mazzacuva F, Lorvellec M, Cilibrizzi A, Mills K, Clayton P, Gissen P (2019) Measurement of bile acids as a marker of the functionality of iPSC-derived hepatocytes. In: Mandenius C-F, Ross JA (eds) Cell-based assays using iPSCs for drug development and testing. Springer, New York
12. Mazzacuva F, Lorvellec M, Cilibrizzi A, Mills K, Heywood W, Clayton P, Gissen P (2019) Mass spectrometry measurement of albumin/alpha fetoprotein ratio as an indicator of iPSC-derived hepatocyte differentiation. In: Mandenius C-F, Ross JA (eds) Cell-based assays using iPSCs for drug development and testing. Springer, New York
13. Baud A, Heywood WE, Little D, Gissen P, Mills K (2019) Preparation of iPSCs for targeted proteomic analysis. In: Mandenius C-F, Ross JA (eds) Cell-based assays using iPSCs for drug development and testing. Springer, New York
14. Little D, Luft C, Mosaku O, Ketteler R, Devine MJ, Gissen P (2019) High content autophagy analysis in iPSC-derived neurons using immuno-fluorescence. In: Mandenius C-F, Ross JA (eds) Cell-based assays using iPSCs for drug development and testing. Springer, New York
15. Little D, Luft C, Mosaku O, Ketteler R, Devine MJ, Gissen P (2019) High content analysis of mitochondrial function in iPSC-derived neurons. In: Mandenius C-F, Ross JA (eds) Cell-based assays using iPSCs for drug development and testing. Springer, New York
16. Christoffersson J, Meier F, Kempf H, Schwanke K, Coffee M, Beilmann M, Zweigerdt R, Mandenius CF (2019) Evaluating the effect of drug compounds on cardiac spheroids using the cardiac cell outgrowth assay. In: Mandenius C-F, Ross JA (eds) Cell-based assays using iPSCs for drug development and testing. Springer, New York
17. Christoffersson J, Mandenius CF (2019) Fabrication of a microfluidic cell culture device using photolithographic and soft lithographic techniques. In: Mandenius C-F, Ross JA (eds) Cell-based assays using iPSCs for drug development and testing. Springer, New York
18. Christoffersson J, Mandenius CF (2019) Using a microfluidic device for culture and drug toxicity testing of 3D cells. In: Mandenius C-F, Ross JA (eds) Cell-based assays using iPSCs for drug development and testing. Springer, New York
19. Kizner V, Fischer S, Naujock M (2019) Multi-electrode array (MEA) based detection of spontaneous network activity in human iPSC-derived cortical neurons. In: Mandenius C-F, Ross JA (eds) Cell-based assays using iPSCs for drug development and testing. Springer, New York
20. Kraushaar U, Guenther E (2019) Assay procedures for compound testing of hiPSC-derived cardiomyocytes using multi-well microelectrode arrays. In: Mandenius C-F, Ross JA (eds) Cell-based assays using iPSCs for drug development and testing. Springer, New York
21. Kizner V, Fischer S, Benjamin Jähnke B, Naujock M (2019) Assessing neuronal excitability on a fluorometric imaging plate reader (FLIPR) following a defined electrostimulation paradigm. In: Mandenius C-F, Ross JA (eds) Cell-based assays using iPSCs for drug development and testing. Springer, New York
22. Little D, Luft C, Pezzini-Picart O, Mosaku O, Ketteler R, Devine MJ, Gissen P (2019) Seeding iPSC-derived neurons onto 384-well plates. In: Mandenius C-F, Ross JA (eds) Cell-based assays using iPSCs for drug development and testing. Springer, New York

Part I

Differentiation of iPC to Target Cells for Assays

Chapter 2

Monolayer Generation of Vascular Endothelial Cells from Human Pluripotent Stem Cells

Klaus Christensen, Filip Roudnicky, Mark Burcin, and Christoph Patsch

Abstract

The use of human pluripotent stem cells (hPSCs) for modeling human diseases and therapeutic applications requires differentiation methods that generate physiologically relevant cell types in a robust and standardized way. Herein, we describe an efficient and scalable monolayer protocol to convert pluripotent stem cells into vascular endothelial cells using defined culture conditions.

The combinatorial use of small molecule compounds, growth factors as well as morphogens directs human pluripotent stem cells toward endothelial cells within 6 days. The protocol has the capacity to generate endothelial cells with high efficiencies of up to 80%. An additional immunomagnetic cell purification step that is based on the surface marker VE-cadherin results in a virtually pure population of endothelial cells. In a subsequent expansion step human PSC-derived endothelial cells can be further propagated, while maintaining their endothelial identity. Thus, our differentiation protocol enables the generation of hPSC-derived endothelial cells at a scale that is relevant for drug discovery campaigns or clinical applications.

Key words Pluripotent stem cells, Induced pluripotent stem cells, Stem cell differentiation, Vascular endothelial cells

1 Introduction

The use of human pluripotent stem cells (hPSCs) [1, 2] for disease modeling, drug discovery as well as cellular therapies requires rapid, robust, and efficient differentiation strategies. Here, we describe a protocol meeting these requirements for the differentiation of hPSCs into endothelial cells, the main component of blood vessels. Based on our previously described method to generate vascular endothelial cells from hPSC [3]we extended the current protocol by an additional expansion step. The combined use of a TGFβ1-inhibitor, VEGF, and serum in the expansion medium supports the maintenance of endothelial cell identity while increasing their expansion rate. By this method we are able to generate large

Carl-Fredrik Mandenius and James A. Ross (eds.), *Cell-Based Assays Using iPSCs for Drug Development and Testing*, Methods in Molecular Biology, vol. 1994, https://doi.org/10.1007/978-1-4939-9477-9_2,
© Springer Science+Business Media, LLC, part of Springer Nature 2019

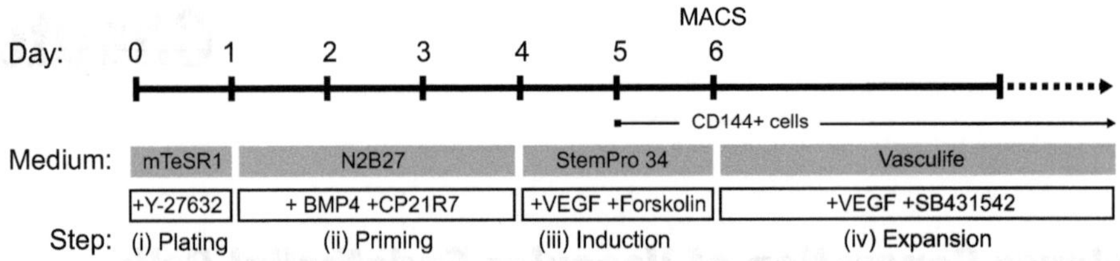

Fig. 1 Schematic illustration of the endothelial cell differentiation from hPSCs. The protocol comprises four distinct steps: (1) plating of human PSCs as single cells; (2) priming cells by inducing mesodermal lineage, (3) induction of endothelial cell conversion from progenitor cells, and (4) expansion of CD144^{+} endothelial cells. The combinatorial use of small compounds, growth factors and morphogens readily yields in endothelial cells already on day 5. However, highest efficiencies are reached on day 6 (up to 80%)

numbers of functional endothelial cells applicable for cell therapies or drug screening campaigns.

The in vitro conversion of hPSCs into endothelial cells is subdivided into four consecutive steps recapitulating the early developmental processes occurring during embryogenesis (Fig. 1). (1) First hPSCs are plated (plating step) and (2) cultured for 4 days as a monolayer in a chemically defined Priming Medium supplemented with a glycogen synthase kinase 3β (GSK3β) inhibitor and bone morphogenetic protein4 (BMP4) inducing mesodermal cell commitment (priming step). Thus, mimicking mammalian development, where vascular progenitors emerge from the lateral and posterior mesoderm [4, 5]. The importance of canonical Wnt signaling in mesoderm commitment during gastrulation is described by several studies [6–8]. (3) In the next step the Priming Medium is replaced by the Induction Medium supplemented with vascular endothelial growth factor A (VEGF-A) and the small molecule forskolin promoting the conversion of mesodermal progenitors into vascular endothelial cells (induction step). The VEGF mediated de novo formation of vascular endothelial cells in vitro reflects its key regulator role of mammalian vasculogenesis [9, 10] (4) After incubating the culture for 2 days with Induction Medium, vascular endothelial cells can be enriched by immunomagnetic separation of CD144^{+} cells (Fig. 2a, b). CD144^{+} vascular endothelial cells can be further expanded in Expansion Medium supplemented with serum, VEGF, and the transforming growth factor beta 1 (TGF-β1) SB-431542 inhibitor (expansion step). This is in agreement with several studies describing TGF-β inhibition to enhance stem cell-derived endothelial cell growth and integrity [11, 12].

Here, we describe the rapid and highly efficient generation of vascular endothelial cells from of hPSCs using defined culture conditions. Our protocol utilizes GSK3 beta inhibition and BMP4 treatment to convert hPSCs into mesodermal cells that when exposed to VEGF convert into functional endothelial cells.

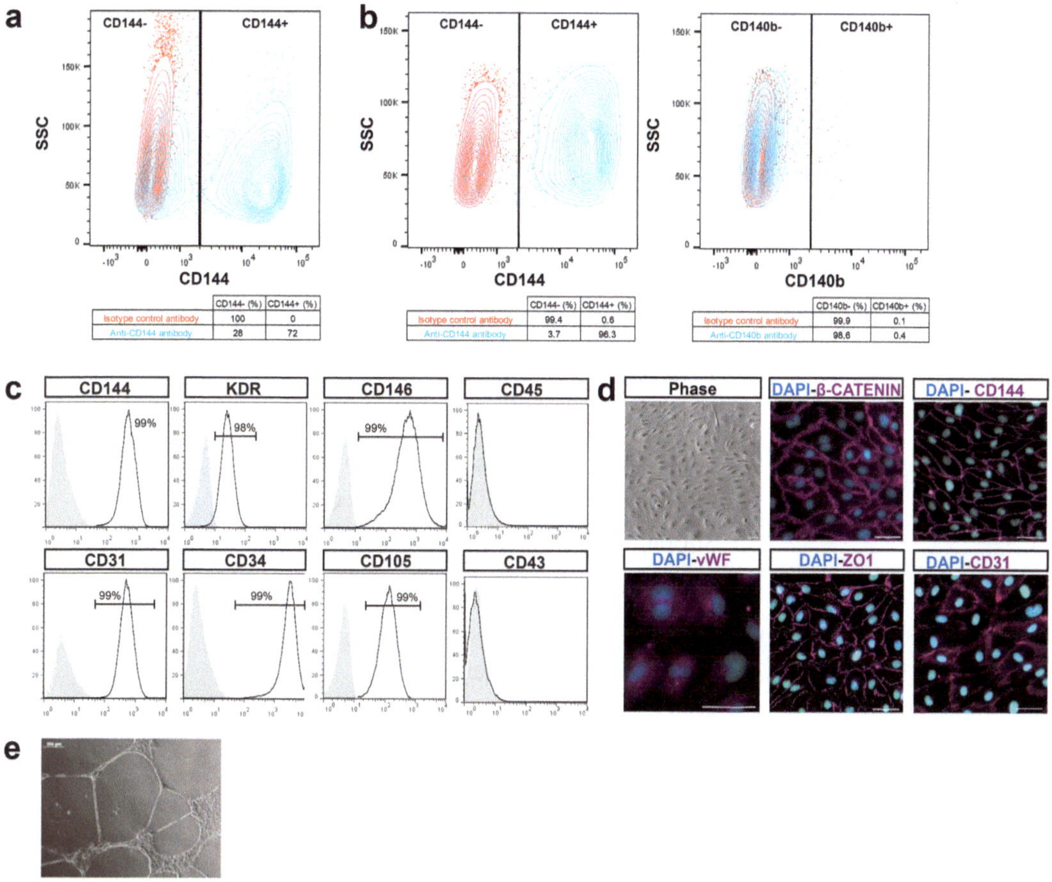

Fig. 2 In vitro characterization of hPSC-derived vascular endothelial cells. Representative fluorescence-activated cell sorting plots stained for CD144 (VE-Cadherin) before and after magnetic-activated cell sorting. (**a**) $CD144^+$ and $CD144^-$ cell population before sorting. (**b**) $CD144^+$ enriched cell fraction after sorting stained for the vascular endothelial cell specific marker CD144 and the vascular smooth muscle cell specific marker $CD140b^+$ (PDGFRB). (**c** and **d**) Vascular endothelial cell marker expression confirmed by immunostainings for CD144, CD31 (PECAM1), KDR (vascular endothelial growth factor receptor 2), CD34, CD146 (melanoma cell adhesion molecule), CD105 (endoglin), vWF (von Willebrand factor), β-catenin, ZO-1 (zona occludens-1) and the absence of hematopoietic lineage-specific markers CD43 and CD45. Scale bars, 50 μm (**e**) Representative microscopic image of an in vitro tube formation assay of hPSC-derived endothelial cells. Scale bar, 200 μm

2 Materials

Reagent preparation (*see* **Notes 1–3**)

1. *Maintenance of hPSCs*:

   mTeSR1, StemCell Technologies.

   hESC-Qualified Matrigel, Corning.

   Gentle cell dissociation Reagent, StemCell Technologies.

2. *Plating step of hPSCs on Day 0*:
   Accutase, Innovative Cell Technologies.
   Y27632 dihydrochloride (ROCK inhibitor), Calbiochem.
   Growth Factor Reduced (GFR) Matrigel, Corning.

   Y-27632 dihydrochloride stock solution (10 mM):
   Dissolve in ultrapure water. Store at −20 °C for up to 1 month.

   Plating Medium:
   mTeSR1 supplemented with 10 μM Y27632.

3. *Priming step on Day 1–4*:
   DMEM/F12, Life Technologies.
   Neurobasal medium, Life Technologies.
   N2 Supplement (100×), Life Technologies.
   Supplement B27 without vitamin A, Life Technologies.
   2-Mercaptoethanol, 50 mM, Life Technologies.
   BMP4, Peprotech.
   CHIR99021, Cayman.
   CP21R7, Roche.

   CHIR99021 stock solution (4.3 μM):
   Dissolve 10 mg of CHIR99021 in 5 mL DMSO.
   Store at −80 °C or −20 °C for up to 1 year.

   CP21R7 stock solution (10 mM):
   Dissolve 2 mg of CP21R in 630 μL DMSO.
   Store at −80 °C or −20 °C for up to 1 year.

   BMP4 stock solution (25 μg/mL):
   Dissolve 10 μg of BMP4 in 4 mL of 510 mM citric acid pH 3.
   Store at −80 °C or −20 °C for up to 1 year.

   N2B27 medium:
   500 mL DMEM/F12 medium +500 mL Neurobasal medium.
   +20 mL B27.
   +10 mL N2.
   +1 mL βMercaptoethanol of 50 mM solution.
   Sterile filtration 0.22 μm, store at 4 °C up to 1 month.

   Priming Medium:
   N2B27 supplemented with 25 μg/mL BMP4 and 1 μM CP21R7 or 8 μM CHIR 99021 (*see* **Note 5**).

4. *Induction step on Day 4–6*:

StemPro34, Life Technologies.

GlutaMAX, Life Technologies.

Forskolin, Tocris.

VEGF165 (VEGFA), Peprotech.

StemPro34 Medium:

500 mL StemPro34 medium.

+5 mL Pen/Strep (1:100).

+5 mL GlutaMAX (1:100).

+StemPro34 Supplement.

Sterile filtration 0.22 μm, store at 4 °C up to 1 month.

VEGF165 stock solution (100 μg/mL):

Dissolve 100 μg of VEGF165 in 1 mL sterile water.

Store at −80 °C or −20 °C for up to 1 year.

Forskolin stock solution (10 mM):

Dissolve 10 mg of forskolin in 2.436 mL DMSO.

Store at −20 °C for up to 1 month.

Induction Medium:

StemPro 34 medium supplemented with 200 ng/mL VEGF and 2 μM forskolin.

5. *Sorting and Expansion step*:

Pre-Separation Filter, Miltenyi Biotec.

CD144 MicroBeads, Miltenyi Biotec.

MACS running buffer, Miltenyi Biotec.

MACS washing buffer, Miltenyi Biotec.

PE mouse anti-human VE-Cadherin/CD144, BD Biosciences.

Alexa Fluor 647 mouse anti-human PDGFRβ/CD 140b, BD Biosciences.

Human fibronectin, Corning.

SB-431542, Tocris.

VascuLife VEGF Medium, Complete Kit, Lifeline Cell Technology.

Fetal bovine serum (FBS), Thermo Fisher Scientific.

VascuLife medium:

500 mL VascuLife medium.

+Complete Supplement Kit.

+8% FBS (adds up to 10% FBS final concentration).

+45 ng/mL VEGF (adds up to 50 ng/mL final concentration).

Sterile filtration 0.22 μm, store at 4 °C up to 1 month.

SB-431542 stock solution (10 mM):

Dissolve in 99.9% ethanol.

Store at −20 °C for up to 1 month.

Expansion Medium:

VascuLife medium supplemented with 50 ng/mL VEGF, 10% Serum, and 10 μM SB-431542.

Fibronectin coated cell culture ware:

Add 1 mL sterile H_2O to 1 mg of fibronectin and incubate at RT for 30 min without shaking. Add 39 mL PBS (w/o Ca and Mg) (=25 μg/mL). Use 6 mL of this solution to coat one T75 flask (=2 μg/cm^2). Incubate for 1 h at RT. Then wash the flask carefully without scrapping the surface once with sterile water. Fibronectin coated culture ware is now ready for use or alternatively be stored at 4 °C for 1 week, only if the fibronectin solution is not aspirated.

3 Methods

3.1 Endothelial Differentiation Method

The text below exemplifies the differentiation procedure for one T175 flask. In case the endothelial cell differentiation is carried out on smaller vessels adjust medium volume accordingly (for instance 3 mL of Priming Medium for one well of a 12 well-plate).

1. *Maintenance of hPSCs*:
   (a) Human pluripotent stem cells are routinely cultured on hESC-Qualified Matrigel in mTeSR1 medium. Cultures are passaged every 3–5 days using Gentle Cell Dissociation Reagent with a daily mTeSR1 medium change. Coat culture ware first by submerging the vial of hESC-Qualified Matrigel on ice in a 4 °C and then by diluting it according to the lot specific dilution factor provided on the Certificate of Analysis.
   (b) An 80% confluent 10 cm plate is required to start differentiation in one T175 (*see* **Note 4**).
2. *Plating step* of hPSCs on Day 0 (Fig. 1):
   (a) Coat one T175 with Growth Factor Reduced (GFR) Matrigel by thawing it on ice and diluting it 1:30.
   (b) Incubate the plates at RT for 1 h. Aspirate the Matrigel, wash once with PBS and add 20 mL.
   (c) Prewarm mTeSR1+ ROCK inhibitor (final conc. 10 μM).

(d) Aspirate growth medium from an 80% confluent 10 cm plate.

(e) Wash with 10 mL prewarmed PBS (Ca^{2+} and Mg^{2+} free).

(f) Add 3 mL prewarmed Accutase and incubate 37 min at 37 °C.

(g) Check cell's detachment under a microscope.

(h) Add 3 mL prewarmed mTeSR1.

(i) Transfer cells to a 15 mL Falcon tube.

(j) Count cells.

(k) Centrifuge the suspension at 1000 rpm ($210 \times g$) for 5 min.

(l) Aspirate supernatant and resuspend the cell pellet in prewarmed Plating Medium (mTeSR1 supplemented with 10 μM Y27632).

(m) Seed single cell suspension at a density of 37,000–47,000 hPSCs per cm^2. Use $2.8–3.5 \times 10^6$ hPSCs in 20 mL per Plating Medium for a T75 flask (*see* **Note 5**).

(n) Incubate cells at 37 °C, 5% CO_2 overnight.

3. *Priming step* on Day 1–3 (Lateral mesoderm induction; Fig. 1):

   (a) Replace medium with prewarmed 50 mL Priming Medium (N2B27 medium supplemented with 8 μM CHIR99021 or 1 μM CP21R7 + 25 ng/mL hBMP4) for a T75 flask 50 mL (*see* **Note 5**).

   (b) Keep the cultures for 2 additional days without medium change!

4. *Induction step* on Day 4 and 5 (conversion into endothelial cells; Fig. 1):

   (a) Replace medium with 40 mL Induction Medium (StemPro34 medium supplemented with 200 ng/mL VEGF and 2 μM forskolin).

   (b) On the next day replace medium with 40 mL of fresh Induction Medium!

5. *Expansion step* on Day 6 (endothelial cell separation and propagation; Fig. 1).

   The following magnetic separation (MACS) protocol describes the enrichment of CD144 positive endothelial cells (Fig. 2a, b) using a semiautomated autoMACS Pro Separator instrument (Miltenyi Biotec). However, differentiated cultures can also readily be sorted by using a MACS Manual Separator (Miltenyi Biotec) or instead by FACS (*see* **Note 6**). Before starting the sorting ensure to have a sufficient amount of fibronectin coated cell culture ware prepared. Final seeding density is 2×10^6 cells in 20 mL Expansion medium (VascuLife supplemented with 50 ng/mL VEGF, 10% Serum and 10 μM SB-431542). Sort CD144 positive endothelial cells and replate enriched CD144 positive cell population onto fibronectin coated tissue culture plates.

(a) Carefully aspirate medium and wash with prewarmed PBS (Ca^{2+} and Mg^{2+} free) without destroying the cell sheet.

(b) Add 5 mL prewarmed Accutase and incubate for 3–5 min at 37 °C until confluent cell layer starts to disintegrate.

(c) Dissociate cell clumps into single cells by gently pipetting up and down a few times to ensure a single cell suspension.

(d) Add 5 mL cold StemPro34 medium and use a Pre-Separation filter to remove cells clumps to transfer cell suspension to a 50 mL Falcon tube.

(e) From this point work fast and keep cells cold using a cooling device (e.g., MACS Chill Rack) and chilled buffers.

(f) Determine cell number and calculate the respective amount of CD144 MicroBeads and MACS running buffer required for your sample (up to 1×10^7 cells in 80 μL cold MACS running buffer +20 μL CD144 MicroBeads). If checking for sorting efficiency by flow cytometry take an aliquot of 3×10^5 cells at this stage and follow instructions for flow cytometer analysis below (*see* **Note 7**).

(g) Centrifuge at 1000 rpm ($210 \times g$) for 5 min.

(h) Aspirate supernatant and loosen cell pellet by gently flicking the tube.

(i) Resuspend cells in an appropriate amount of chilled MACS running buffer and CD144 MicroBeads.

(j) Incubate for 15 min at 2–8 °C.

(k) Remove the unbound beads by adding 5 mL cold MACS running buffer and centrifuge at 1000 rpm ($210 \times g$) for 5 min.

(l) Aspirate supernatant, then loosen cell pellet by gently flicking the tube.

(m) Resuspend the pellet in 0.5 mL MACS running buffer.

(n) Filter the cells by placing a 30 μm Pre-Separation filter (Miltenyi Biotec) on a 15 mL Falcon tube.

(o) Rinse filter once with 0.5 mL cold MACS Buffer to reduce cell loss.

(p) Place the 15 mL falcon in the first rack position of the autoMACS Pro Separator and two empty 15 mL falcons in position 2 and 3 for the depleted and eluted cell fractions.

(q) Choose the positive selection program “Posseld2” (positive selection via two columns) from the instrument’s menu and run program, the sorting will take about 8 min per sample tube (*see* **Note 8**).

(r) Determine yield by counting the eluted cells. If checking for sorting efficiency by flow cytometry (optional) take an aliquot at this stage of 1×10^5 cells from each sorting fraction and follow instructions for flow cytometer analysis below.

(s) Centrifuge at 1000 rpm ($210 \times g$) for 5 min and aspirate supernatant, loosen cell pellet by gently flicking the tube and resuspend cells in prewarmed Expansion Medium.

(t) Seed 2×10^6 cells in 20 mL Expansion Medium onto fibronectin-coated T75 flasks (~26,000 cells/cm^2).

(u) Culture hPSC-derived endothelial cells in Expansion Medium until they reach confluency (usually it takes place between 4 and 6 days). Cells can then be cryopreserved (1.5×10^6 cells/vial) or further expanded for downstream analyses.

(v) Quality assurance tests confirming endothelial cell identity and function are optional but recommended. These assays could comprise but are by no means limited to immune-fluorescent staining for endothelial- (e.g., PECAM1 and vWF) and smooth muscle cell (e.g., PDGFRb/CD140b and αSMA)-specific markers and functional assays such as an in vitro angiogenesis assay (*see* **Notes 9** and **10** and Fig. 2c, d).

Determining efficiencies of sorting and endothelial cells differentiation.

Sorting and endothelial cell conversion efficiency can be determined by using flow-cytometric analysis. Perform the flow cytometry using an endothelial cell-specific anti-CD144-PE antibody (BD Biosciences)

1. For differentiation efficiency incubate 1×10^5 cells with isotype control (IgGAb) (=sample 1), keep 1×10^5 cells as unstained control (=sample 2), and label 1×10^5 cells with CD144-PE antibody (sample 3).
2. For sorting efficiency label 1×10^5 cells of CD144 positive and of CD144 negative fraction with anti-CD144-PE antibody (sample 4 and 5).

3. Centrifuge the five samples in FACS tubes for 5 min at 1000 rpm ($210 \times g$).
4. Resuspend each sample in 90 μL cold MACS buffer.
5. Add 10 μL IgG Isotypic control to sample 1, keep sample 2 untreated and add anti-CD144-PE antibody to sample 3, 4, and 5.
6. Incubate for at least 15 min at 4 °C.
7. Add 1 mL cold MACS running buffer and centrifuge for 5 min at 1000 rpm ($210 \times g$).
8. Remove supernatant and resuspend cell pellets of all samples in 250 μL cold MACS running buffer.
9. Samples are ready for flow cytometry analysis.

4 Notes

1. For sterility opening and reconstitution of all recombinant proteins and other compounds should be performed under a laminar flow hood using an aseptic technique.
2. All DMSO dissolved compounds should be prewarmed for a few minutes in a water bath (37 °C) before adding to the respective cell culture medium.
3. Recombinant proteins are highly unstable. Repeated freeze-thaw cycles would result in a rapid decreasing of their biological activity. Once thawed, aliquots can be kept at 4 °C for up to 1 week.
4. As for all tissue-specific cell differentiations the quality of the used hPSC line for the endothelial cell differentiation method is of utmost importance. We recommend doing a regular quality assurance for the undifferentiated hPSC line in terms of genomic integrity, sterility and pluripotency (for detailed information see also Christensen et al. 2018) [13].
5. This endothelial cell differentiation protocol is routinely used to robustly generate endothelial cells form various human ESC and iPSC lines as well as nonhuman primate iPSCs [14]. However, some hPSC lines may require minor adaptation such as cell seeding density (initial number of hPSCs) or GSK3β inhibitor (CP21R7 and CHIR99021) concentration in order to achieve highest efficiencies. For most PSC lines the concentration of CP21R7 ranges between 1 and 2 μM and for CHIR99021 between 6 and 8 μM. The initial seeding density of hPSCs on day 0 is one of the keys to high differentiation efficiency. Ideally the optimal cell seeding number for each hPSC line is determined in a five-step dilution series ranging from 37,000 to 47,000 cells/cm^2.

6. Immunomagnetic sorting for CD144 positive cells will enrich the EC population up to 95–98% (Fig. 1a, b) but will also result at least in some loss of endothelial cells in the process.
7. If checking for differentiation efficiency by flow cytometer keep 3×10^5 cells at this stage and follow instructions described in the section for determining efficiencies.
8. Program "Possel" (positive selection via one column) can be used to further increase yield if necessary but this could have a negative impact on the purity.
9. For detailed information on possible assays confirming endothelial cell identity and function please review our earlier work published by Patsch and colleagues in 2015 [3].
10. In order to compare hPSC-derived endothelial cells to primary cells such as HUVECs, endothelial cells can be cultured in commonly used endothelial cell maintenance media such as EGM2 medium (Lonza) or VascuLife complete medium without the supplements specifically added to the Expansion Medium.

General Troubleshooting

1. The initial seeding density of hPSCs on day 0 is a key parameter for obtaining high differentiation efficiency. Ideally the optimal cell seeding number for each hPSC line is determined in a five-step dilution series ranging from 37,000 to 47,000 cells/cm^2.
2. Priming Medium on day 4 turns yellow and there are a lot of dead cells floating. This is expected and does not negatively affect our EC differentiations as long as the medium is not turbid. Contrariwise replacing medium with fresh Priming Medium resulted in substantially reduced yields. However, excessive cell death can be observed when cells seeding density is too high.
3. Differentiation efficiency is lower than 20%. This is rather unexpected and could be due to inactive recombinant proteins or wrong medium compositions. Make sure that you prewarm the DMSO stock of GSK3β inhibitors before adding it to the prewarmed Priming Medium. Check if reagents and media were prepared and stored correctly.
4. Differentiation efficiency between 20% and 50% could be due to an inappropriate seeding density or medium volume. Make a serial dilution series for cell seeding density and GSK3β inhibitor concentration to identify the optimal cell line specific condition.
5. Endothelial cells did not attach after sorting. Be sure your fibronectin solution was prepared correctly and that the plates were coated for least 1 h.

6. Endothelial cells are not proliferating. Make sure to replate cells not too sparse, ideally at a seeding density of around 26,000 cells/cm^2 (2×10^6 in VascuLife complete medium supplemented with additional 45 ng/mL VEGF-A, 8% Serum and 10 μM SB-431542). When expanding the cells, do not split the cells at smaller ratio than 1:6.

Acknowledgments

The research leading to these results has received support from the Innovative Medicines Initiative Joint Undertaking under grant agreement number 115439, resources of which are composed of financial contribution from the European Union's Seventh Framework Program (FP7/2007-2013) and EFPIA companies' in kind contribution. This publication reflects only the author's views and neither the IMI JU nor EFPIA nor the European Commission are liable for any use that may be made of the information contained therein.

References

1. Takahashi K, Tanabe K, Ohnuki M, Narita M, Ichisaka T, Tomoda K, Yamanaka S (2007) Induction of pluripotent stem cells from adult human fibroblasts by defined factors. Cell 131 (5):861–872. https://doi.org/10.1016/j.cell.2007.11.019
2. Thomson JA, Itskovitz-Eldor J, Shapiro SS, Waknitz MA, Swiergiel JJ, Marshall VS, Jones JM (1998) Embryonic stem cell lines derived from human blastocysts. Science 282 (5391):1145–1147
3. Patsch C, Challet-Meylan L, Thoma EC, Urich E, Heckel T, O'Sullivan JF, Grainger SJ, Kapp FG, Sun L, Christensen K, Xia Y, Florido MH, He W, Pan W, Prummer M, Warren CR, Jakob-Roetne R, Certa U, Jagasia R, Freskgard PO, Adatto I, Kling D, Huang P, Zon LI, Chaikof EL, Gerszten RE, Graf M, Iacone R, Cowan CA (2015) Generation of vascular endothelial and smooth muscle cells from human pluripotent stem cells. Nat Cell Biol 17(8):994–1003. https://doi.org/10.1038/ncb3205
4. Yamashita J, Itoh H, Hirashima M, Ogawa M, Nishikawa S, Yurugi T, Naito M, Nakao K, Nishikawa S (2000) Flk1-positive cells derived from embryonic stem cells serve as vascular progenitors. Nature 408(6808):92–96. https://doi.org/10.1038/35040568
5. Winnier G, Blessing M, Labosky PA, Hogan BL (1995) Bone morphogenetic protein-4 is required for mesoderm formation and patterning in the mouse. Genes Dev 9 (17):2105–2116
6. Huelsken J, Vogel R, Brinkmann V, Erdmann B, Birchmeier C, Birchmeier W (2000) Requirement for beta-catenin in anterior-posterior axis formation in mice. J Cell Biol 148(3):567–578
7. Kelly OG, Pinson KI, Skarnes WC (2004) The Wnt co-receptors Lrp5 and Lrp6 are essential for gastrulation in mice. Development 131 (12):2803–2815. https://doi.org/10.1242/dev.01137
8. Martyn I, Kanno TY, Ruzo A, Siggia ED, Brivanlou AH (2018) Self-organization of a human organizer by combined Wnt and Nodal signalling. Nature 558 (7708):132–135. https://doi.org/10.1038/s41586-018-0150-y
9. Carmeliet P, Ferreira V, Breier G, Pollefeyt S, Kieckens L, Gertsenstein M, Fahrig M, Vandenhoeck A, Harpal K, Eberhardt C, Declercq C, Pawling J, Moons L, Collen D, Risau W, Nagy A (1996) Abnormal blood vessel development and lethality in embryos lacking a single VEGF allele. Nature 380 (6573):435–439. https://doi.org/10.1038/380435a0
10. Ferrara N, Carver-Moore K, Chen H, Dowd M, Lu L, O'Shea KS, Powell-Braxton L, Hillan KJ, Moore MW (1996) Heterozygous

embryonic lethality induced by targeted inactivation of the VEGF gene. Nature 380 (6573):439–442. https://doi.org/10.1038/380439a0

11. James D, Nam HS, Seandel M, Nolan D, Janovitz T, Tomishima M, Studer L, Lee G, Lyden D, Benezra R, Zaninovic N, Rosenwaks Z, Rabbany SY, Rafii S (2010) Expansion and maintenance of human embryonic stem cell-derived endothelial cells by TGFbeta inhibition is Id1 dependent. Nat Biotechnol 28(2):161–166. https://doi.org/10.1038/nbt.1605
12. Watabe T, Yamashita JK, Mishima K, Miyazono K (2006) TGF-beta signaling in embryonic stem cell-derived endothelial cells. Methods Mol Biol 330:341–351. https://doi.org/10.1385/1-59745-036-7:341
13. Christensen K, Roudnicky F, Patsch C, Burcin M (2018) Requirements for using iPSC-based cell models for assay development in drug discovery. Adv Biochem Eng Biotechnol 163:207–220. https://doi.org/10.1007/10_2017_23
14. Thoma EC, Heckel T, Keller D, Giroud N, Leonard B, Christensen K, Roth A, Bertinetti-Lapatki C, Graf M, Patsch C (2016) Establishment of a translational endothelial cell model using directed differentiation of induced pluripotent stem cells from Cynomolgus monkey. Sci Rep 6:35830. https://doi.org/10.1038/srep35830

Chapter 3

Generating Human iPSC-Derived Astrocytes with Chemically Defined Medium for In Vitro Disease Modeling

Katharina Janssen, Lamiaa Bahnassawy, Claudia Kiefer, Jürgen Korffmann, Georg C. Terstappen, Viktor Lakics, Miroslav Cik, and Peter Reinhardt

Abstract

To better understand and model neurological, in particular neurodegenerative diseases, human induced pluripotent stem cells (hiPSCs) offer a great source for generation of neural cells. We provide a protocol for the differentiation of hiPSc-derived astrocytes in vitro. This protocol not only is chemically defined, that is, it does not use serum, but also allows for the expansion of astrocyte progenitor cells and mature astrocytes. Large batches of hiPSc-derived astrocytes can be stored and used for defined in vitro disease models.

Key words hiPSC, Disease modeling, Neurodegenerative disease, In vitro, Astrocytes

1 Introduction

For decades astrocytes were believed to have a mere supportive function to neurons throughout the brain. Now it is generally accepted that astrocytes play many important roles throughout the central nervous system (CNS) that are critical to normal brain function. Astrocytes are important regulators of brain homeostasis, metabolism, neurogenesis, synaptic transmission, and plasticity. Astrocyte dysfunction has also been shown to be critical in pathophysiological conditions such as neuroinflammation, brain injury, and neurodegenerative disorders such as Alzheimer's disease (AD), Parkinson's disease (PD), or amyotrophic lateral sclerosis (ALS) (for further reading on astrocyte physiopathology refer to [1]).

However, their specific role in disease pathology and progression, as well as whether they are more protective or detrimental remains elusive.

To further understand the disease and the role of certain cell types during disease progression suitable model systems are

Carl-Fredrik Mandenius and James A. Ross (eds.), *Cell-Based Assays Using iPSCs for Drug Development and Testing*, Methods in Molecular Biology, vol. 1994, https://doi.org/10.1007/978-1-4939-9477-9_3,
© Springer Science+Business Media, LLC, part of Springer Nature 2019

necessary that mimic aspects of the disease. Here we demonstrate the generation of astrocytes in vitro at scale from human induced pluripotent stem cell (hiPSC)-derived expandable neural progenitor cells (NPCs). Human neural precursor cells (NPCs) derived from pluripotent stem cells have been shown to be a valuable source for human neurons for disease modeling. As they are already committed to the neural lineage and in several cases suitable for upscaling and freezing their use may reduce the time of neuronal and glial differentiation significantly [2–5].

Most published astrocyte differentiation protocols commencing with hiPSCs are prolonged and require extensive manual handling or growth as aggregates in suspension cultures, making it difficult to handle cells with liquid handling devices and do the appropriate upscaling for drug testing [6–8]. In addition, many of these protocols do not use fully defined conditions during the differentiation process but make use of serum which contains factors known to promote glial differentiation from neural precursor cells [9, 10]. Therefore, we developed a differentiation protocol for the production of expandable astrocytes under chemically defined conditions using NPC starting cultures.

2 Materials

2.1 Cell Culture Reagents

All procedures need to be performed using aseptic techniques in a laminar flow hood by personnel trained in safe aseptic cell culture techniques (Table 1). Media are sterile-filtered after preparation but before addition of growth factors and chemicals. Basal medium (Table 2) can be stored for up to 2 weeks at 4 °C and 1 week after addition of compounds from fresh aliquots to generate the three media used; glial induction medium (Table 3), glial maturation medium (Table 4), and glial proliferation medium (Table 5). Heat medium only to maximum room temperature, also if using a water or heat bath.

2.2 Preparation of Stock Solutions

After preparation, stock solutions are stored at −20 °C in aliquots and used within 6 months.

1. *BMP2, BDNF, NT3, IGF-1, PDGF-AA, bFGF*: Prepare a 100 μg/ml solution in sterile water. Further dilute to 10 μg/ml stock solution with 0.1% BSA/PBS.
2. *T3*: prepare a 1 mg/ml solution. Further dilute to 20 μg/ml stock solution with DMEM/F12.
3. *Purmorphamine*: dissolve in DMSO to a final concentration of 5 mM.
4. *dbcAMP*: Prepare a 100 mM stock solution in sterile water, sterile filter.

Table 1
List of cell culture reagents with suppliers and catalog numbers

Product name	Catalog number	Supplier
DMEM/F-12	11330-032	Thermo
Advanced DMEM/F-12	12634-028	Thermo
Neurobasal™ medium	21103-049	Thermo
B-27 Supplement	17504-044	Thermo
N-2 Supplement	17502-048	Thermo
L-Glutamine	25030-024	Thermo
Penicillin–Streptomycin, Liquid	15140-122	Thermo
BSA 35%	A7979	Sigma
Laminin	L2020-1MG	Sigma Aldrich
Recombinant LIF	LIF1050	Millipore
bFGF/FGF-2	233-FB-025	R&D Systems
dbcAMP	D0627-1G	Sigma-Aldrich
Purmorphamine	4551	Tocris
CNTF	450-13	Peprotech
BMP-2	120-02C	Peprotech
BDNF	450-02	Peprotech
NT3	450-03	Peprotech
IGF-1	100-11	Peprotech
PDGF-AA	100-13	Peprotech
T3	T5516	Sigma
Accutase™ in DPBS	A11105-01	Thermo
D-PBS without Ca^{2+} and Mg^{2+}	10010-015	Thermo
DMSO	D2650	Sigma
Filters for culture medium	5046512 (500 ml) 5046510 (250 ml)	Nalgene
Cell culture plate, 6-well	140,675	Nunc
Cell culture flasks 175 cm^2 75 cm^2 25 cm^2	660175 690175 658175	Greiner
µclear 96-well plate (poly-D-lysine precoated)	655946	Greiner

Table 2
Composition of basal medium

Name	Stock concentration	Final concentration	Volume for 50 ml
Advanced DMEM/F12		1×	24 ml
Neurobasal medium		1×	24 ml
Pen–Strep	100×	1×	500 μl
N-2 Supplement	100×	0.5×	250 μl
B-27 Supplement	50×	0.5×	500 μl
L-Glutamine	200 mM	1 mM	250 μl
BSA, Albumin Solution in DPBS	35%	5 μg/ml	7 μl

Table 3
Composition of glial induction medium

Name	Stock concentration	Final concentration	Volume for 50 ml
Basal medium		1×	50
Purmorphamine	5 mM	1 μM	10 μl
T3	20 μg/ml	40 ng/ml	100 μl
PDGF-AA	10 μg/ml	10 ng/ml	50 μl

Table 4
Composition of glial maturation medium

Name	Stock concentration	Final concentration	Volume for 50 ml
basal medium		1×	50 ml
Purmorphamine	5 mM	1 μM	10 μl
PDGF-AA	10 μg/ml	10 ng/ml	50 μl
IGF	10 μg/ml	10 ng/ml	50 μl
T3	20 μg/ml	40 ng/ml	100 μl
BDNF	10 μg/ml	10 ng/ml	50 μl
NT3	10 μg/ml	10 ng/ml	50 μl
cAMP	100 mM	1 μM	0.5 μl

Table 5
Composition of glial proliferation medium

Name	Stock concentration	Final concentration	Volume for 50 ml
basal medium		1×	50 ml
CNTF	10 µg/ml	5 ng/ml	25 µl
BMP2	10 µg/ml	10 ng/ml	50 µl
FGF-2	10 µg/ml	8 ng/ml	40 µl

2.3 Laminin Coating of Cell Culture Vessels

Add laminin 20 µg/ml in DMEM/F-12 to wells and incubate for at least 2 h or overnight at 37 °C.

1. Discard the laminin and immediately use the wells for cell seeding. Do not allow the laminin to dry.
2. Laminin-coated plates can be stored at 4 °C for 7 days with coating mixture. Prevent drying of laminin during storage by sealing the plates with Parafilm.

2.4 Media

1. Basal medium: (*see* Table 2)
2. Glial induction medium: (*see* Table 3)
3. Glial maturation medium: (*see* Table 4)
4. Glial proliferation medium: (*see* Table 5)

2.5 qRT-PCR reagents

(a) RNeasy Mini Kit (Qiagen) according to manufacturer's instructions.

(b) SuperScript Reverse Transcriptase kit (Thermo) according to manufacturer's instructions.

(c) TaqMan Probes: *GFAP* (total): Hs00909235_gl, *S100beta*: Hs00902901_ml, *CD44*: Hs99999195_ml, *ALDH1L1*: Hs00201836_ml, *GAPDH*: Hs99999905_ml, *ACTB*: Hs99999903_ml.

2.6 Immunofluorescence staining reagents

(a) 4% formaldehyde solution (prepared with PBS from 36% solution supplied by VWR (706445)).

(b) Primary antibodies: CD44 Ab157207, Abcam; GFAP G-9269, Sigma; EAAT1 Ab416, Abcam; S100b Z0311, Dako; ALDH1L1 Ab190298, Abcam.

(c) Secondary antibodies: α-rb-Cy3, (711-165-152), α-ms-Alexa Fluor 488, (715-545-151), α-ch-AlexaFluor488, (703-545-155); all from Jackson Immuno Research.

(d) D-PBS with Ca^{2+} and Mg^{2+} (14040-117, Thermo).

(e) Triton X-100 (T8787, Sigma).

3 Methods

As a starting population we use neural precursor cells (NPCs) generated according to the protocol from [2] employing synergistic inhibition of glycogen synthase kinase 3 (GSK3), transforming growth factor β (TGF-β), and Notch signaling pathways by small molecules. The generated NPCs can be expanded and frozen for several passages according to protocol and further differentiated to neurons or glial cells. During differentiation, medium is changed every 2–3 days, the protocol is weekend-free.

3.1 Generation of Glial-Restricted Progenitors

When NSCs have reached 90% confluency they can be used for astrocyte differentiation (*see* also Fig. 1a for a schematic overview of differentiation).

1. Wash cells with PBS, add Accutase and incubate for 10 min at 37 °C, 5% CO_2.
2. Collect cells in 15 or 50 ml tube in basal medium.
3. Add basal medium to the culture plate to collect remaining cells.
4. Centrifuge at 300 × *g* for 5 min, resuspend cells in basal medium and count the cells.
5. Plate 75,000/cm^2 into a laminin coated cell culture flask/6-Well plate and add Glial induction medium.
6. Passage the cells as soon as they have reached confluency with a ratio of 1:2 up to 1:4. Keep the plating number at 75,000/cm^2 in laminin coated culture vessels.
7. Culture cells in this medium for the next three passages (please *see* **Notes 1** and **2**).

3.2 Maturation and Expansion of Astrocytes

After 3 passages in Glial induction medium the medium is changed to Glial maturation medium.

1. When cells reach confluency split with Accutase into fresh laminin-coated cell culture flasks/plates at a ratio of 1:2 up to 1:4 until passage 10–12. Keep cell seeding densities at 50,000 to 75,000/cm^2 in Glial maturation medium.
2. Continue culturing of cells with above seeding densities, cells can be frozen and stocks of earlier passages saved for later expansions (please *see* **Notes 1** and **3**).
3. After around 12 passages, cells will express an astrocyte gene expression signature marked by the expression of CD44, GFAP, S100b, and ALDH1L1.

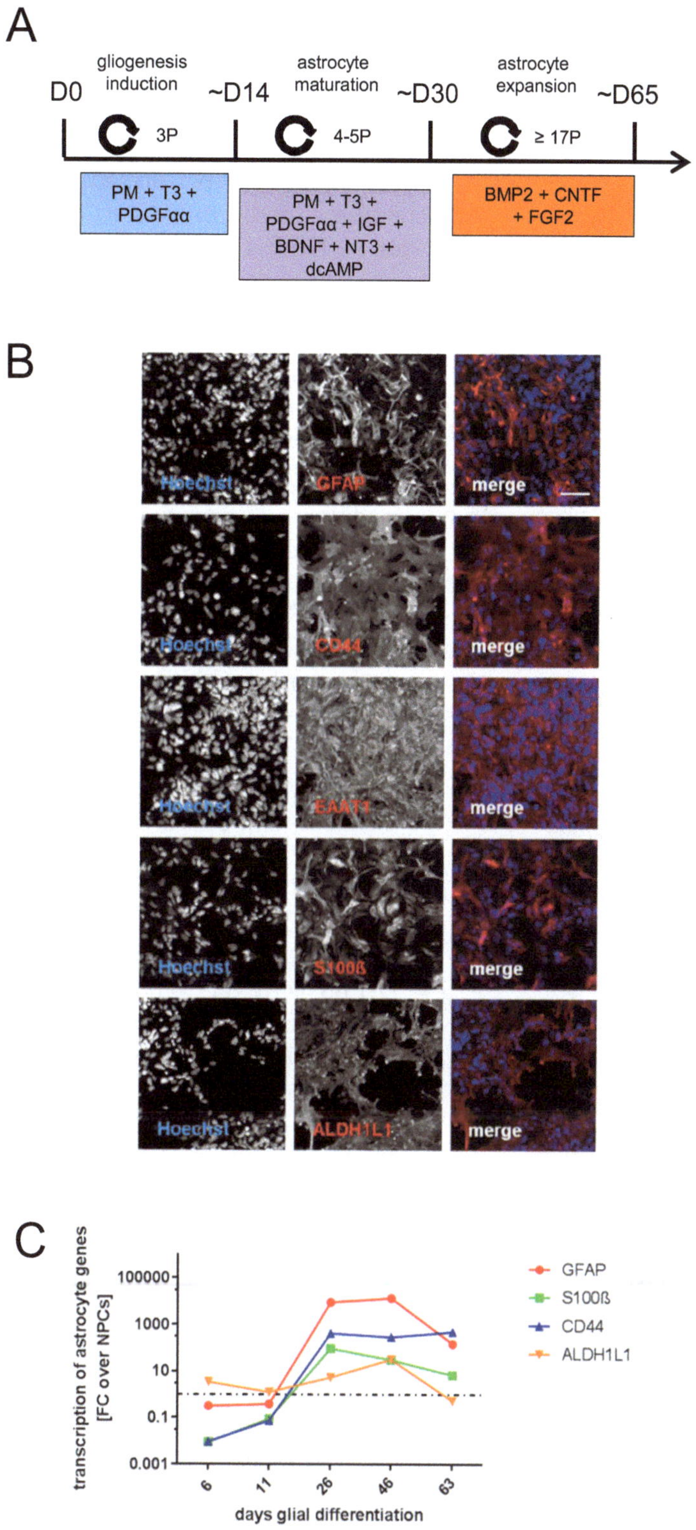

Fig. 1 (**a**) Schematic view of the astrocyte differentiation protocol. (**b**) Immunocytochemistry with antibodies directed against typical astrocyte marker proteins. Images show representative results of healthy control hiPSC-derived astrocytes. Scale bar 100 μm. (**c**) qRT-PCR of astrocyte related genes indicating increasing expression over maturation of astrocytes. The graph shows fold change over expression levels in hiPSC-derived NPCs

3.3 Expansion of Astrocytes

Final astrocyte expansion can be performed using Glial expansion medium, allowing for mass expansion and keeping of large batches, if desired. However, careful observation of astrocyte marker gene expression has to be performed by either qRT-PCR or immunocytochemistry in the process to characterize the astrocyte batches (please *see* **Note 4**).

1. After around 12 passages cells are plated onto fresh laminin-coated culture flasks in Glial expansion medium.
2. When cells reach 80% confluency they can be passaged with Accutase at a 1:3 to 1:6 ratio into laminin-coated flasks/plates. Cells have been successfully expanded until passage 25.
3. Cells can be seeded for further experiments in Glial expansion medium, or, if no further proliferation is desired, into basal medium or other permissive media.

3.4 Characterization of Astrocytes

1. qRT-PCR is performed using TaqMan reagents, using undifferentiated NPCs as calibrator sample. Gene expression is normalized using *GAPDH* and *ACTB* as housekeeping genes and gene expression changes calculated according to the 2^{-ddCt} method (Fig. 1c).
2. Immunofluorescence analysis is performed using standard protocols. Briefly, cells are fixed using 4% formaldehyde in D-PBS with Ca^{2+} and Mg^{2+}, then washed twice with D-PBS with Ca^{2+} and Mg^{2+}, blocked and permeabilized using blocking solution with 3% BSA and 0.1% Triton X-100 in D-PBS with Ca^{2+} and Mg^{2+} for 30 min. Primary antibodies are diluted 1:500 in D-PBS with Ca^{2+} and Mg^{2+} with 0.1% BSA and incubated overnight at 4 °C with agitation. After three wash steps with D-PBS with Ca^{2+} and Mg^{2+}, Primary antibodies diluted 1:500, and if needed nuclear counterstain is applied in D-PBS with Ca^{2+} and Mg^{2+} with 0.1% BSA for 60 min at room temperature and with agitation. After three washing steps cells are stored in D-PBS with Ca^{2+} and Mg^{2+} until imaging analysis.

4 Notes

1. The in vitro-derived astrocytes can be frozen at any passage after switching to Glial maturation medium in basal medium with 10% DMSO. Large batches can be obtained at the end of culture period using Glial maturation medium (around passage 11) or after further expansion using Glial expansion medium.
2. During culture in Glial induction medium (first 3 passages), cells will gradually decrease proliferation and many cells display a neuronal morphology. It is important that the cells appear as nearly post mitotic before switching to Glial maturation medium.

3. After seeding the cells in passage 4 into Glial maturation medium, Glial restricted cells will rapidly increase proliferation and will lead to expanding cultures.
4. In Glial proliferation medium, careful observation of astrocyte marker gene expression has to be performed by either qRT-PCR or immunocytochemistry in the process to characterize the astrocyte batches (Fig. 1b, c).

Acknowledgments

The study leading to these results has received support from the Innovative Medicines Initiative Joint Undertaking under (grant no. 115439), resources of which are composed of financial contribution from the European Union's Seventh Framework Programme (FP7/2007-2013) and EFPIA companies. This publication reflects only the author's views and neither the IMI JU nor EFPIA nor the European Commission are liable for any use that may be made of the information contained therein.

References

1. Rossi D (2015) Astrocyte physiopathology: at the crossroads of intercellular networking, inflammation and cell death. Prog Neurobiol 130:86–120
2. Li W, Sun W, Zhang Y, Wei W, Ambasudhan R, Xia P, Talantova M, Lin T, Kim J, Wang X, Kim WR, Lipton SA, Zhang K, Ding S (2011) Rapid induction and long-term self-renewal of primitive neural precursors from human embryonic stem cells by small molecule inhibitors. Proc Natl Acad Sci U S A 108:8299–8304
3. Heilker R, Traub S, Reinhardt P, Scholer HR, Sterneckert J (2014) iPS cell derived neuronal cells for drug discovery. Trends Pharmacol Sci 35:510–519
4. Corti S, Faravelli I, Cardano M, Conti L (2015) Human pluripotent stem cells as tools for neurodegenerative and neurodevelopmental disease modeling and drug discovery. Expert Opin Drug Discovery 10:615–629
5. Du ZW, Chen H, Liu H, Lu J, Qian K, Huang CL, Zhong X, Fan F, Zhang SC (2015) Generation and expansion of highly pure motor neuron progenitors from human pluripotent stem cells. Nat Commun 6:6626
6. Krencik R, Weick JP, Liu Y, Zhang ZJ, Zhang SC (2011) Specification of transplantable astroglial subtypes from human pluripotent stem cells. Nat Biotechnol 29:528–534
7. Krencik R, Zhang SC (2011) Directed differentiation of functional astroglial subtypes from human pluripotent stem cells. Nat Protoc 6:1710–1717
8. Roybon L, Lamas NJ, Garcia-Diaz A, Yang EJ, Sattler R, Jackson-Lewis V, Kim YA, Kachel CA, Rothstein JD, Przedborski S, Wichterle H, Henderson CE (2013) Human stem cell-derived spinal cord astrocytes with defined mature or reactive phenotypes. Cell Rep 4:1035–1048
9. Kodaira K, Imada M, Goto M, Tomoyasu A, Fukuda T, Kamijo R, Suda T, Higashio K, Katagiri T (2006) Purification and identification of a BMP-like factor from bovine serum. Biochem Biophys Res Commun 345:1224–1231
10. Hu JG, Lu HZ, Wang YX, Bao MS, Zhao BM, Zhou JS (2010) BMP signaling mediates astrocyte differentiation of oligodendrocyte progenitor cells. Tohoku J Exp Med 222:195–200

Chapter 4

Human Induced Pluripotent Stem Cell-Derived Definitive Endoderm Bulk Culture and Hepatic Differentiation

Anwar A. Palakkan, Jyoti Nanda, and James A. Ross

Abstract

We have developed a method to bulk culture definitive endoderm cells generated from human iPSCs which can be stored and differentiated to hepatocytes. Human iPSC-derived definitive endoderm cells were sorted based on the expression of CXCR4. The sorted cells were able to proliferate for extended periods and can be cryopreserved. The definitive endoderm cells were subsequently utilized to generate functional hepatocytes expressing albumin and α-fetoprotein in different multiwell formats. This provides a method to reliably produce more consistent hepatocytes in greater quantities and has enabled the development of high-throughput screening strategies.

Key words Human induced pluripotent stem (hiPS) cells, Hepatic differentiation, Definitive endoderm cells, Bulk culture

1 Introduction

Often, in vivo tests fail to predict human toxicity accurately due to metabolic/genetic differences between humans and other species. Human in vitro models are generally preferable for toxicity screening but the lack of functional maturity of hepatic cell lines and the poor availability of primary cells are major limitations [1]. iPSCs have recently become a more attractive source for differentiated human cells, and protocols for generating various cell types, including hepatocytes, are progressing rapidly. However, current protocols suffer from batch-to-batch variability together with the phenotypic and functional immaturity of the differentiated cells [2].

Hepatic differentiation can be initiated either as 3D embryoid bodies [3] or as a cell monolayer. In monolayer, differentiation can be started with very small stem cell colonies [4], or as single cells [5], on different matrices [6]. Even though hepatic differentiation protocols are improving consistently, they still struggle to produce a pure population of hepatocytes.

Carl-Fredrik Mandenius and James A. Ross (eds.), *Cell-Based Assays Using iPSCs for Drug Development and Testing*, Methods in Molecular Biology, vol. 1994, https://doi.org/10.1007/978-1-4939-9477-9_4,
© Springer Science+Business Media, LLC, part of Springer Nature 2019

Here we describe a method to isolate a pure population of human iPSC-derived definitive endoderm cells from monolayer culture, which can be employed to proliferate definitive endoderm cells or can be cryopreserved. These definitive endoderm cells can be differentiated to hepatocytes and enable the generation of cells with a more consistent phenotype and function.

2 Materials

2.1 Human Induced Pluripotent Stem Cell Culture

1. Human iPSC AD4C1 derived from fibroblasts (Lonza CC-2511, tissue acquisition number 23801; Lonza Group, Ltd., Basel, Switzerland) and characterized within StemBANCC.
2. 1× Dulbecco's Phosphate buffered saline (1× DPBS Ca^{2+}/Mg^{2+} free; cat # 14190; Thermo Fisher Scientific, Waltham, MA).
3. mTeSR™1 (Cat # 05850, Stem cell technologies, Vancouver, Canada).
4. DMEM/F12 (Cat # 21331020, Thermo Fisher Scientific, Waltham, MA).
5. Cellstar™ 6-well plates (cat # 657160; Greiner Bio-One; Frickenhausen, Germany).
6. KnockOut™ DMEM (KO DMEM, Cat # 10829018, Thermo Fisher Scientific, Waltham, MA).
7. Collagenase IV (Collagenase, Cat # 17104019, Thermo Fisher Scientific, Waltham, MA). Collagenase solution: Dissolve required amount of collagenase IV in KO DMEM to get a final concentration of 268 IU/mL. Filter-sterilize and store aliquots of 10 mL at −20 °C. Once thawed, store aliquots at 4 °C and use within 2 weeks.
8. Y27632 (Rock inhibitor, Cat # 72302, Stem cell technologies, Vancouver, Canada). 5 mM stock solution: Dissolve 1 mg of ROCK inhibitor-Y27632 in 624.4 μL of distilled water. Store aliquots of 20 μL at −20 °C for up to 6 months.
9. Matrigel® (cat # 354263; Corning Inc., Corning, NY). Matrigel stock solution: Thaw a vial of Matrigel on ice overnight and dilute 1:1 with ice cold KO DMEM. Store 1 mL aliquots at −20 °C. For coating the plates, to a thawed aliquot add 17 mL of KO DMEM, mix well by pipetting several times and add appropriate volume to the wells (1 mL for 6 well, 0.5 mL for 24 well) [0.25 mg/mL final concentration]. Incubate at room temperature for 1 h or at 37 °C for 30 min before using.
10. KnockOut™ Serum Replacement (KO serum, cat # 10828028; Thermo Fisher Scientific, Waltham, MA). Store 50 mL aliquots at −20 °C.

11. DMSO (Cat #D5879, Sigma-Aldrich, UK).
12. Freezing Medium: Filter-sterilize 90% KO serum and 10% DMSO solution. Store 1 mL aliquots at −20 °C.

2.2 Differentiation of Human iPS Cells to Definitive Endoderm Cells

1. Gentle cell dissociation reagent (Cat # 07174, Stem cell technologies, Vancouver, Canada).
2. STEMdiff Definitive Endoderm Kit (Cat # 05110, Stem cell technologies, Vancouver, Canada).
   (a) "A + B" Medium: Dilute supplement A and supplement B (1:100) in STEMdiff Definitive Endoderm Basal Medium (e.g., add 10 μL of supplement A and 10 μL of supplement B to 980 μL of Basal Medium).
   (b) B medium: Dilute supplement B (1:100) in STEMdiff Definitive Endoderm Basal Medium (e.g., 10 μL of supplement B to 990 μL of Basal Medium).
   (c) "B + R" medium: Dilute supplement B (1:100) in STEMdiff Definitive Endoderm Basal Medium and add Rock inhibitor (final concentration of 10 μM) [e.g., 10 μL of supplement B and 2 μL Rock inhibitor stock (5 mM) to 988 μL of Basal Medium].
3. CD184 (CXCR4) MicroBead Kit (Cat # 130100070, Miltenyi Biotec, Bergisch Gladbach, Germany).
4. Trypsin Inhibitor (Cat # T6522, Sigma-Aldrich, UK). Stock solution (10 mg/mL): Dissolve 125 mg of Trypsin Inhibitor in 2.50 mL of PBS and filter-sterilize. Stock can be stored at 2–8 °C for 3 years. Dilute the stock ten times in medium before use.
5. EDTA solution (0.5 M EDTA, Cat # 03690, Sigma-Aldrich, UK).
6. Trypsin EDTA solution (Cat # R001100, Thermo Fisher Scientific, Waltham, MA).
7. Cell strainer 40 μm (Cat # 431750, Corning, USA).
8. LS columns (Cat # 130042401, Miltenyi Biotec, Germany).
9. PB buffer: DPBS (without Ca^{2+} and Mg^{2+}) with 0.5% BSA.
10. PEB buffer: DPBS (without Ca^{2+} and Mg^{2+}) with 0.5% BSA and 0.1 M EDTA.

2.3 Differentiation to Hepatocytes

1. Hepatocyte Growth Factor (cat No: 100-39, PeproTech, UK). HGF Stock (10 μg/mL): Dissolve 100 μg of HGF in 200 μL of water and incubate for 1 h at room temperature. Transfer this solution to 9.8 mL of 0.1% BSA (in PBS), mix well and aliquots can be stored at −80 °C.
2. Oncostatin M (cat No: 300-10T, PeproTech, UK). HGF Stock (20 μg/mL): Dissolve 100 μg of OSM in 200 μL of water. Transfer this solution to 4.8 mL of 0.1% BSA (in PBS), mix well and aliquots can be stored at −80 °C.

3. SR-DMSO Medium: KO-DMEM (Cat # 10829, Thermo Fisher Scientific, Waltham, MA) with 10% KO-Serum (Cat # 10828, Thermo Fisher Scientific, Waltham, MA), 1× GlutaMAX (Cat # 35050, Thermo Fisher Scientific, Waltham, MA), 1× nonessential amino acids (Cat # 11140, Thermo Fisher Scientific, Waltham, MA), 0.1 mM 2-mercapto ethanol (Cat # 31350, Thermo Fisher Scientific, Waltham, MA), 1% DMSO (Cat # D5879, Thermo Sigma-Aldrich, UK) and 1% Penicillin/Streptomycin (Cat # 15140, Thermo Fisher Scientific, Waltham, MA). SR-DMSO should be sterile filtered using a 0.22 μm filter unit under vacuum and aliquots are stored at −20 °C or at 4 °C for 1 month.
4. HZM Medium: Base medium: HepatoZYME-SFM (Cat # 17705, Thermo Fisher Scientific, Waltham, MA) with 1× GlutaMAX (Cat # 35050, Thermo Fisher Scientific, Waltham, MA), 1% Penicillin/Streptomycin (Cat # 15140, Thermo Fisher Scientific, Waltham, MA), and 10 μM Hydrocortisone 21-hemisuccinate (Cat # H4881, Sigma-Aldrich, UK). Aliquots of base medium are stored at −20 °C or at 4 °C for 1 month. Complete HZM medium should be prepared on the day of use by adding hepatocyte growth factor (10 ng/mL) and oncostatin M (20 ng/mL) by diluting the stock (1:1000).

2.4 Immunostaining

1. 4′,6-Diamidin-2-phenylindol (DAPI; cat # D1306; Thermo Fisher Scientific), 10.8 μM stock solution: Dissolve 10 mg of DAPI in 2 mL of distilled water. Store aliquots of 50 μL at −20 °C. For working solution (300 nM) dilute 50 μL stock to 1.8 mL of PBS.
2. DRAQ5™ (cat # DR50050, Biostatus, UK). Nuclei was stained by incubating with 5 μM DRAQ5 for 5 min at room temperature.
3. Fixing solution: 4% (w/v) paraformaldehyde (cat # P6148, Sigma-Aldrich) dissolved in PBS.
4. Methanol (cat # 322415, Sigma-Aldrich).

3 Method

3.1 Human Induced Pluripotent Stem Cell Culture and Maintenance

3.1.1 Thawing Vial of Pluripotent Stem Cells

1. Prepare required amount of thawing medium (2 mL/6 well) by adding 2 μL of Rock inhibitor stock solution per mL of mTeSR™1.
2. Prepare Matrigel-coated 6-well plate. Incubate Matrigel-containing plates at room temperature for an hour or at 37 °C for 30 min before using. Remove the Matrigel and add 1.5 mL of thawing medium per well.
3. Take a vial of iPSC from liquid nitrogen and keep in dry ice until use.

4. Using forceps immerse the vial in a water bath without submerging the cap. Swirl the vial gently. When only an ice crystal remains, remove the vial from the water bath and wipe away any water from the outside of the vial. Spray with ethanol and leave to air-dry in laminar flow.
5. Using a wide-bore tip transfer cells gently into a sterile 15 mL conical tube.
6. Slowly, add 10 mL of mTeSR™1 medium dropwise to cells. While adding, gently move the tube back and forth to mix cells.
7. Add 1 mL of mTeSR™1 to the vial and collect remaining cells. Centrifuge cells at 200 × *g* for 5 min.
8. Aspirate and discard the supernatant. Resuspend the cell pellet in 0.5 mL of thawing medium by gently pipetting a few times using a wide bore tip.
9. Slowly add 0.5 mL of cell suspension dropwise into each well. Place plate in the incubator and gently shake the plate back and forth and side to side to distribute the cells evenly.

3.1.2 Culture Maintenance

1. Exchange medium daily with fresh mTeSR™1 (3 mL/well) until well are 80% confluent. Remove any differentiation by pruning after marking the differentiated area.
2. Cells need to be passaged when wells reach 80% confluence. Wash wells with DPBS (Ca^{2+}/Mg^{2+} free) and incubate with collagenase solution for 3 min at 37 °C. Carefully aspirate collagenase solution and flush the loosened differentiated cells with DPBS (Ca^{2+} and Mg^{2+} free).
3. Scrape the intact colonies (add 2 mL medium/well) and transfer to a conical 15 mL tube. Add 1 mL of medium to the wells and collect remaining cells.
4. Pipet up and down slowly (use 10 mL pipette) to make even-sized small colonies.
5. Seed 1 mL of cell suspension to Matrigel-coated 6-well plate (split ratio 1:3).
6. Shake the wells to get uniform seeding and incubate at 37 °C.
7. Exchange medium daily with fresh mTeSR™1 until next passaging step at 80% confluence.
8. NANOG and TRA-160 expression can be checked frequently to confirm pluripotency of cells (Fig. 1).

3.1.3 Cryopreservation

1. For cryopreservation, wells at 80% confluent are treated with collagenase solution for 3 min at 37 °C.
2. Break the confluent cells to make medium-sized colonies (small colonies are not preferred for freezing) and collect them in a 50 mL conical tube.

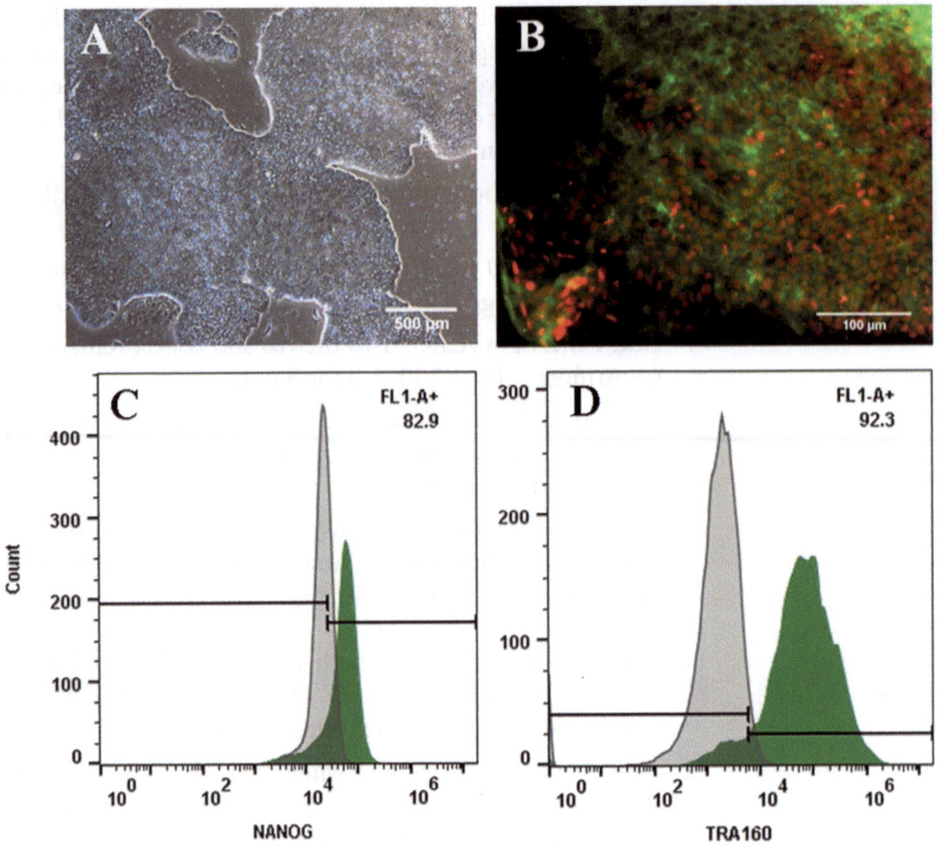

Fig. 1 Human iPSC morphology and characterization. (**a**) Compact morphology of human iPSCs cultured on Matrigel. (**b**) E-cadherin expression by iPSC colonies. Flow cytometry analysis of (**c**) NANOG and (**d**) TRA-160 expression. Green and gray peaks represent protein of interest and isotype control respectively. Inset value (FL1-A$^+$) represents percentage of positive cells

3. Centrifuge at 200 × *g* for 5 min. Remove the supernatant and resuspend cells in 1 mL of freezing medium (1 mL/well) and transfer to the labelled cryovials (use wide-bore tips).
4. Store cryovials for 24 h at −80 °C in a freezing container, and then transfer to the liquid nitrogen storage tank for long time storage.

3.2 Differentiation of iPS Cells to Definitive Endoderm Cells

Endoderm differentiation is promoted using the STEMdiff Definitive Endoderm Kit as per the manufacturer instructions. hIPSC at 80% confluent were used for differentiation.

3.2.1 Preparing Cells for the Induction

1. Warm a sufficient volume of mTeSR™1 and DMEM/F12 to room temperature.
2. Prepare single-cell passaging medium by adding Y-27632 ROCK inhibitor to mTeSR™1 to obtain a final concentration of 10 μM ROCK inhibitor.

3. Rinse the wells to be passaged with DPBS (Ca^{2+}/Mg^{2+} free).
4. Add 1 mL of "gentle cell dissociation reagent" to each well and incubate at 37 °C for 10 min.
5. Dislodge cells by pipetting and transfer them to a tube containing an equal volume of DMEM/F12.
6. To collect any remaining cells in the well, wash with 1 mL of DMEM/F12 and transfer them to the tube.
7. Centrifuge the tube at 300 × *g* for 5 min.
8. Resuspend cells in 1 mL of Single-Cell passaging medium and count the number of live cells using a hemocytometer.
9. Plate the cells at a density of $2.1 \times 10^5/cm^2$ (i.e., 2×10^6 cells per 6 wells) on Matrigel-coated plates and incubate at 37 °C for 24 h (considered as day 0). (Note: Adjust density if necessary, so that the cells are approximately 90–100% confluent on Day 1.)

3.2.2 Inducing Endoderm Differentiation

1. Warm sufficient volumes of DMEM/F12 and STEMdiff™ Definitive Endoderm Basal Medium to 37 °C.
2. On Day 1, Prepare "Medium A + B" by diluting supplement A and supplement B (1:100) in Basal Medium (e.g., add 10 μL of supplement A and 10 μL of supplement B to 980 μL of Basal Medium).
3. Aspirate medium from the cells and wash with 1 mL DMEM/F12.
4. Add 2 mL of "Medium A + B" to each well and incubate at 37 °C for 24 h.
5. On Day 2, prepare "Medium B" by diluting supplement B (1:100) in Basal Medium (e.g., add 10 μL of supplement B to 990 μL of Basal Medium). (Note: Prepare sufficient "Medium B" to be used on Days 2, 3, and 4 (i.e., 6 mL per well.)
6. Aspirate medium from the well and add 2 mL of "Medium B" and incubate at 37 °C for 24 h.
7. On Day 3 and 4 replenish the "Medium B."

3.2.3 Sorting Definitive Endoderm Cells

On Day 5, definitive endoderm cells need to be sorted based on the expression of CXCR4.

1. Wash wells with DPBS (Ca^{2+}/Mg^{2+} free) and add 1 mL trypsin-EDTA/6 well and incubate for 30 s to 1 min at 37 °C.
2. Remove trypsin and add 1 mL of trypsin inhibitor.
3. Collect the cell suspension in DMEM/F12 and pass through a 40 μm cell strainer.

4. Centrifuge cells at 300 × *g* for 5 min and resuspend cells in 100 μL of PEB buffer.
5. Estimate number of cells using a hemocytometer (approximate cell yield is 1×10^6 cells/6 well).

Labeling Cells

Volumes for the procedure below are for up to 10^7 total cells. Scale the volume if more cells are present.

1. Add 10 μL CD184 (CXCR4)-APC and incubate for 10 min at 4–8 °C.
2. Wash by adding 2 mL of PEB buffer and centrifuge at 300 × *g* for 5 min.
3. Resuspend cell pellet in 80 μL of PEB buffer.
4. Add 20 μL Anti-APC microbeads and incubate for 15 min.
5. Wash by adding 2 mL of PEB buffer and centrifuge at 300 × *g* for 5 min.
6. Resuspend cell pellet in 500 μL of PB buffer.

Magnetic Separation

1. Place LS column in the magnetic field of MACS separator.
2. Prepare LS column by rinsing with 3 mL of PEB buffer.
3. Add cell suspension slowly on top of the column.
4. Wash three times with 3 mL of PEB buffer.
5. Remove column from separator and place on a 15 mL centrifuge tube.
6. Add 5 mL of PB buffer to the column and flush out the cells.
7. Centrifuge the collected cells (300 × *g* for 5 min) and resuspend in 500 μL of "B + R" medium.

Magnetic Separation

1. Seed the cells at a cell density of 2×10^4 cells/cm^2 in "B + R" medium on Matrigel-coated plates.
2. Next day change the medium to "Medium B," and thereafter replenish medium every 2 days.
3. Split cells every 6–7 days using trypsin–EDTA and plate at 2×10^4 cells/cm^2 ~(1:5 or 1:6 split ratio). Cells can be cultured for up to ten passages or cryopreserved (Fig. 2).

Freezing Definitive Endoderm Cells

1. Trypsinize cells after washing with PBS (1 mL of trypsin-EDTA/6 well, 30 s at 37 °C).
2. Remove trypsin and add 1 mL of trypsin inhibitor.
3. Collect the cell suspension in DMEM/F12.
4. Centrifuge cells at 300 × *g* for 5 min.
5. Resuspend cell pellet in freezing medium and store aliquots of 1×10^6 cells at −80 °C or liquid nitrogen.

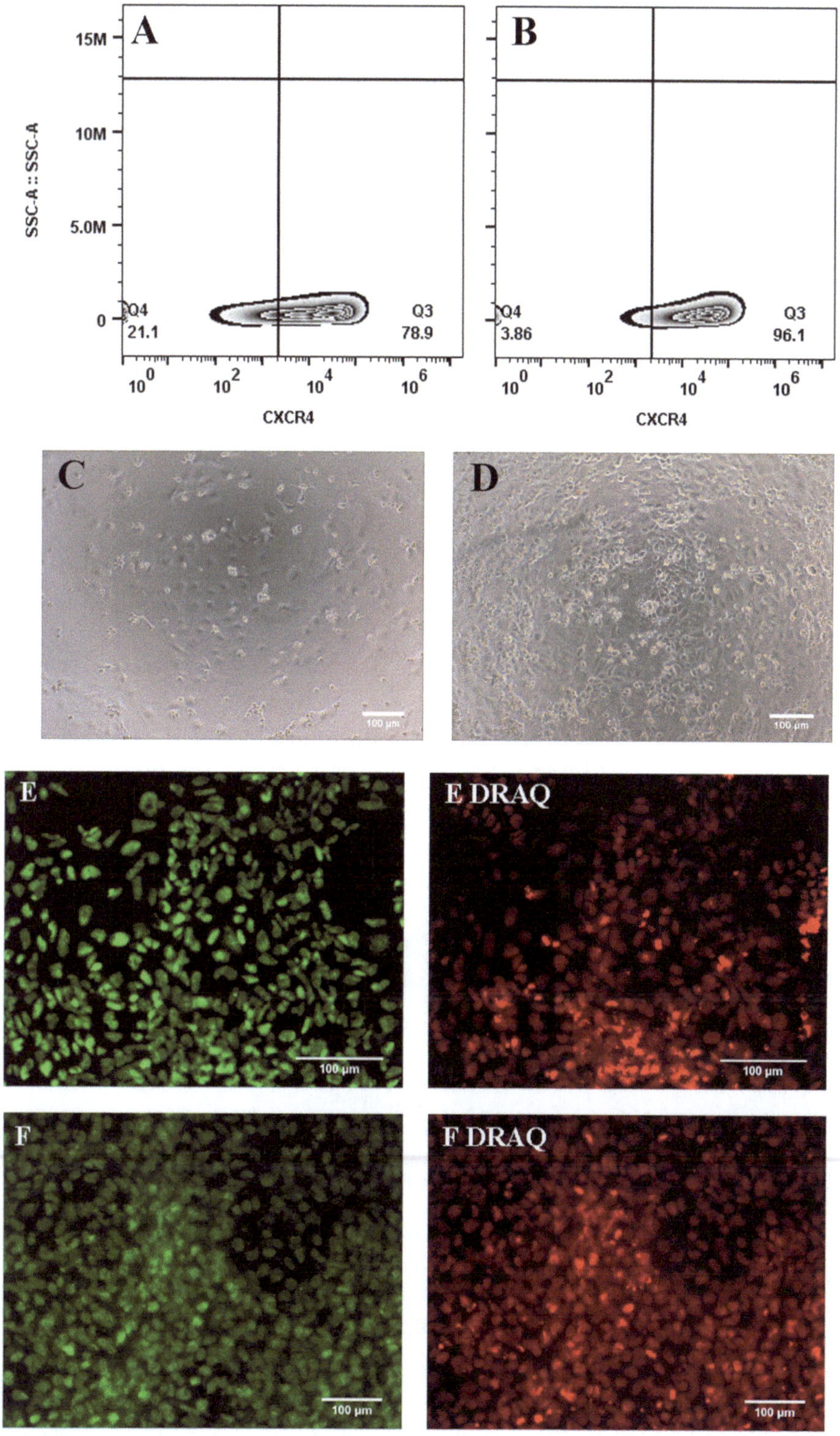

Fig. 2 MACS sorting of endoderm cells. Flow cytometry analysis of definitive endoderm cells (**a**) before and (**b**) after MACS sorting expression. Inset value (Q3) represents percentage of positive cells. Morphology of cultured definitive endoderm cells (**c**) next day after sorting and at (**d**) passage 6. Definitive endoderm cells at passage 6 showing the expression of (**e**) SOX17 and (**f**) FOXA2

6. While thawing, seed the cells on Matrigel-coated wells plated at a density of 3×10^4 cells/cm^2 in "B + R" medium. Note: Adjust seeding density if it is necessary to have confluent cells the next day.

3.2.4 Differentiation to the Hepatic Lineage

1. For differentiation to the hepatic lineage, cells need to be cultured in SR DMSO medium for 4 days (4 mL/6 well plate), with a medium change every alternate day.
2. To mature the hepatoblasts, cells need to be cultured in HZM medium (4 mL/6 well) for a week with a medium change every alternate day.

3.3 Immunofluorescence Staining

After differentiation perform immunofluorescence staining on a parallel culture to check the expression of albumin and α-fetoprotein (Fig. 3).

1. Wash cells three times with PBS and add 4% warm paraformaldehyde.
2. Incubate cells for 10 min at 37 °C.
3. Wash cells three times with 500 μL/well of PBS.
4. Aspirate PBS and permeabilize with ice-cold methanol for 20 min on ice.
5. Wash cell twice with PBS and add 100 μL of primary antibodies (Table 1) in 2% serum and 1% BSA solution.
6. Incubate cells for 1 h at room temperature.

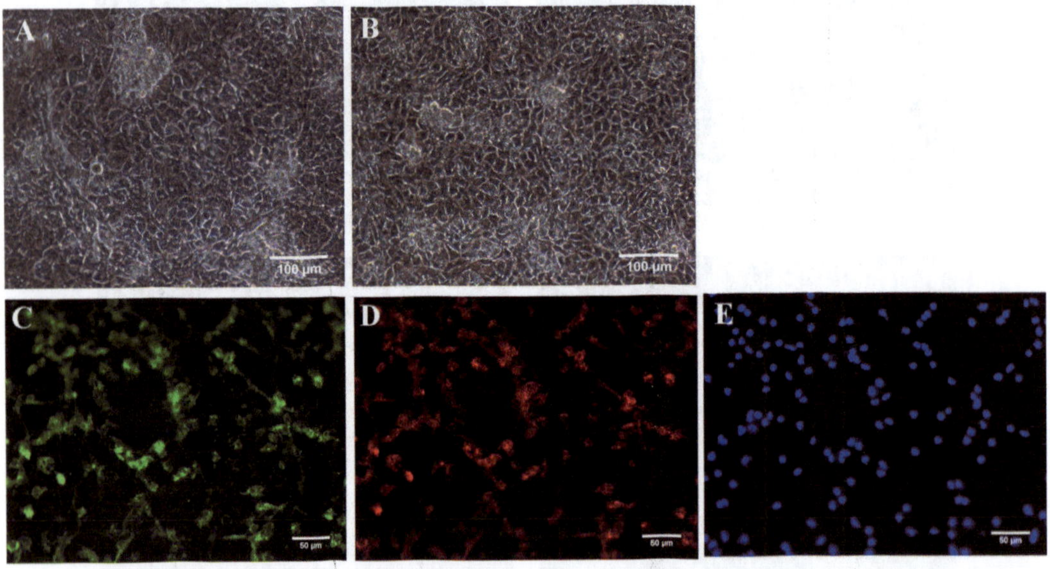

Fig. 3 Morphology of differentiated hepatocytes. Morphology of differentiated hepatocytes on (**a**) day 9 and (**b**) Day 16. Expression of (**c**) albumin and (**d**) α-fetoprotein by differentiated hepatocytes. (**e**) shows the nuclei counterstained with DAPI

Table 1
List of antibodies

Antigen	Dilution	Supplier	Cat #
Primary antibodies Primary antibodies are listed according to recognized antigen, abbreviation (abbr.), dilution, supplier, and catalog number (cat #)			
SOX17	1:2000	R&D, USA	AF1924
CXCR4 (CD184), APC conjugated	1:10	Miltenyi Biotec, Germany	130-100-070
NANOG	1:400	Cell Signalling Technologies, USA	3580
TRA-1-60, FITC conjugated	1:50	Millipore, UK	FCMAB115F
FOXA2	1:3000	Abcam, UK	ab40874
E cadherin	1:100	BD, Belgium	BD610181
Alpha fetoprotein (AFP)	1:500	Sigma, UK	A8452
Albumin (ALB)	1:2000	Sigma, UK	A6684

Target species	Fluorochrome	Dilution	Supplier	Cat #
Secondary antibodies Secondary antibodies are listed according to target species, fluorochrome, dilution, supplier, and catalog number (cat #)				
Donkey anti-goat IgG (H + L)	488	1:250	Thermo Fisher Scientific	A-11055
F(ab′)2 goat anti-Mouse IgG (H + L)	488	1:250	Thermo Fisher Scientific	A-11017
F(ab′)2 goat anti-Rabbit IgG (H + L)	488	1:250	Cell signalling	4412
Donkey anti-mouse (H + L) highly cross-adsorbed secondary antibody	594	1:250	Thermo Fisher Scientific	A21203

7. Wash cells three times with PBS and add 100 μL of species-specific secondary Alexa Fluor™ antibodies diluted in 2% serum and 1% BSA solution.
8. Incubate cells for 30 min at room temperature in the dark.
9. Wash cells three times with PBS and add 100 μL of DAPI solution or DRAQ5.
10. Incubate cells for 5 min at RT in the dark.
11. Wash cells three times with PBS.
12. Perform imaging using a fluorescence microscope (e.g., Leica DMi8, Germany).

3.4 Flow Cytometry Analysis

Routinely perform analysis of the expression of NANOG and TRA-160 by human iPSCs by flow cytometry.

1. Trypsinize the cells.
2. Pass through a cell strainer, centrifuge, and count.
3. Add 250 μL of 4% paraformaldehyde and Incubate cells for 10 min at 37 °C.
4. Centrifuge cells at 300 × *g* for 5 min and wash cells with PBS (500 μL/tube).
5. Centrifuge cells at 300 × *g* for 5 min and permeabilize with Triton X for 10 min.
6. Centrifuge cells at 300 × *g* for 5 min and wash cells with PBS (500 μL/tube).
7. Add 100 μL of primary antibodies (Table 1) in 2% serum and 1% BSA solution.
8. Incubate cells for 1 h at room temperature.
9. Centrifuge cells at 300 × *g* for 5 min and wash cells with PBS (500 μL/tube).
10. Add 100 μL of species-specific secondary Alexa Fluor™ antibodies diluted in 2% serum and 1% BSA solution.
11. Incubate cells for 30 min at room temperature in the dark.
12. Centrifuge cells at 300 × *g* for 5 min and wash cells with PBS (500 μL/tube).
13. Load the samples on a flow cytometer (e.g., BD Accuri C6) and perform the analysis.

Acknowledgments

The study leading to these results has received support from the Innovative Medicines Initiative Joint Undertaking under (grant no. 115439), resources of which are composed of financial contribution from the European Union's Seventh Framework Programme (FP7/2007-2013) and EFPIA companies. This publication reflects only the author's views and neither the IMI JU nor EFPIA nor the European Commission are liable for any use that may be made of the information contained therein.

References

1. Palakkan AA, Hay DC, Anil Kumar PR, Kumary TV, Ross JA (2013) Liver tissue engineering and cell sources: issues and challenges. Liver Int 33 (5):666–676. https://doi.org/10.1111/liv.12134
2. Palakkan AA, Nanda J, Ross JA (2017) Pluripotent stem cells to hepatocytes, the journey so far. Biomed Rep 6(4):367–373. https://doi.org/10.3892/br.2017.867
3. Pettinato G, Ramanathan R, Fisher RA, Mangino MJ, Zhang N, Wen X (2016) Scalable differentiation of human iPSCs in a multicellular spheroid-based 3D culture into hepatocyte-like cells through direct Wnt/β-catenin pathway

inhibition. Sci Rep 6:32888. https://doi.org/10.1038/srep32888

4. Hay DC, Fletcher J, Payne C, Terrace JD, Gallagher RC, Snoeys J, Black JR, Wojtacha D, Samuel K, Hannoun Z, Pryde A, Filippi C, Currie IS, Forbes SJ, Ross JA, Newsome PN, Iredale JP (2008) Highly efficient differentiation of hESCs to functional hepatic endoderm requires ActivinA and Wnt3a signaling. Proc Natl Acad Sci U S A 105(34):12301–12306. https://doi.org/10.1073/pnas.0806522105
5. Liu J, Brzeszczynska J, Samuel K, Black J, Palakkan A, Anderson RA, Gallagher R, Ross JA (2015) Efficient episomal reprogramming of blood mononuclear cells and differentiation to hepatocytes with functional drug metabolism. Exp Cell Res 338(2):203–213. https://doi.org/10.1016/j.yexcr.2015.08.004
6. Cameron K, Tan R, Schmidt-Heck W, Campos G, Lyall MJ, Wang Y, Lucendo-Villarin B, Szkolnicka D, Bates N, Kimber SJ, Hengstler JG, Godoy P, Forbes SJ, Hay DC (2015) Recombinant laminins drive the differentiation and self-organization of hESC-derived hepatocytes. Stem Cell Reports 5(6):1250–1262. https://doi.org/10.1016/j.stemcr.2015.10.016

Chapter 5

Production of Cardiomyocytes from Human Pluripotent Stem Cells by Bioreactor Technologies

Caroline Halloin, Michelle Coffee, Felix Manstein, and Robert Zweigerdt

Abstract

Cardiomyocytes from human pluripotent stem cells (hPSCs) have the ability to advance specificity of in vitro assays for drug discovery and safety pharmacology. They may also provide a superior cell source for envisioned cell therapies to repair damaged hearts. All applications will require the production of cardiomyocytes (CMs) by robust upscalable bioprocesses via industry-compliant technologies. This paper describes a detailed procedure for producing hPSC-CMs in stirred tank bioreactors in 100 ml process scale. The strategy combines both hPSC expansion in suspension culture and, directly followed by, cardiogenic differentiation using small molecule-Wnt pathway modulators. We also provide a protocol describing how to plan and expand the pluripotent stem cells to enable parallel inoculation of 4× 150 ml parallel bioreactor differentiations, potentially producing more than 240×10^6 cardiomyocytes in 22 days.

Key words Bioreactor, Cardiac differentiation, Process development, Human pluripotent stem cells, Cardiomyocytes

1 Introduction

The development of more physiological drug discovery assays [1, 2], organ-specific disease models [3, 4], and cell therapies for treating heart diseases [5, 6] requires large amounts of human cardiomyocytes (CMs). It has been estimated that more than one billion cardiomyocytes are irreversibly lost during a myocardial infarction but could potentially be replaced through the use of human pluripotent stem cell (hPSC)-derived cardiomyocyte-like progenies (hPSC-CMs) [7]. Due to their expandability and multi-lineage differentiation potential in vitro, hPSCs present excellent starting material to produce the required amount of CMs. We and others have shown that CM production can be scaled up into stirred tank bioreactors by cultivation and differentiation of hPSC aggregates in suspension culture [8–10]. The use of stirred tank bioreactors has several advantages, including the possibility of linear

Carl-Fredrik Mandenius and James A. Ross (eds.), *Cell-Based Assays Using iPSCs for Drug Development and Testing*, Methods in Molecular Biology, vol. 1994, https://doi.org/10.1007/978-1-4939-9477-9_5,
© Springer Science+Business Media, LLC, part of Springer Nature 2019

scale-up of process dimensions and the extensive knowledge that is available on bioreactor technology from industry. Modulation and control of specific process parameters such as pH, dissolved oxygen (DO), and metabolites are also enabled by advanced stirred tank reactor systems and allows for a more homogeneous culture environment. Following process inoculation with a single-cell suspension, aggregates of hPSCs can be formed in stirred tank bioreactors in the presence of the Rho-associated coiled-coil forming protein serine/threonine (ROCK inhibitor, RI) Y27623, which transiently supports single-cell survival ahead of aggregation [11–13]. Directed differentiation into specific lineages such as endothelial cells [14] or CMs [8] in suspension culture can then be performed through the stepwise application of small molecules such as the glycogen synthase kinase 3 beta (GSK3ß) inhibitor CHIR99201 (CHIR) and inhibitor of WNT production-2 (IWP-2), established modulators of Wnt-pathway activity [15–17]. It has been shown that linage specific differentiation of hPSC is a complex, multiparametric process that requires highly defined and controlled conditions to ensure efficiency and reproducibility [18]. In addition to the importance of using exact concentration and timing of chemical molecules, it has been shown that the cell density during the first step of differentiation can have a profound impact on the differentiation efficiency [18]. We therefore propose a differentiation protocol where the cell density—prior to the addition of differentiation inducing molecule CHIR—is closely defined and monitored to ensure a more reliable process outcome. Inoculating the process with an established density of 0.33 million hPSCs/ml, we generated ~0.5 million differentiated cells/ml at ~80% CMs purity, thus producing ~40 million CMs in 100 ml process scale. Expanding on this previously established strategy [8] we now show how this platform technology can be expanded to enable 4× parallel bioreactor differentiations at 150 ml working scale. This enables, on the one hand, further parallel optimization of process conditions and, on the other hand, the production of ~250×10^6 hPSC-CMs in one differentiation approach.

2 Materials

2.1 Bioreactor

1. DASbox Mini bioreactor system for cell culture applications (Eppendorf, cat. no. 76DX04CC) including 4× DASbox Mini bioreactor vessel for cell culture applications (Eppendorf, cat. no. 76DS0250ODSS).
2. Sampling valves (Eppendorf, cat. no. 78200077).
3. DASGIP Dip Tube Pg 13.5, 220 mm (Eppendorf, cat. no. 76DGDT220).

4. DASbox exhaust system (Eppendorf, cat. no. 76DXOFF).
5. DASbox exhaust condenser, Peltier (Eppendorf, cat. no. 76DXCOND).
6. DASbox overhead drive (DASGIP Eppendorf, cat. no. 76DXOHD).
7. Impeller: 8-blade, 60° pitch (available from Eppendorf upon request).
8. Polytetrafluoroethylene (PTFE) membrane inline vent filter; pore size, 0.2 µm (mdi Membrane Technologies, cat. no. ITFX0801BBXX109).
9. EasyFerm Plus 120 pH sensor (Hamilton, cat. no. 238643).
10. OxyFerm FDA 225 (Hamilton, cat. no. 237452).
11. DASGIP compression fitting for outer diameter (o.d.) 12 mm w/Pg 13.5 male thread (Eppendorf, cat. no. 76DGCF12).
12. Mohr pinchcock clamp (Carl Roth, cat. no. KY01.1).
13. Female Luer-lock connectors (Carl Roth, cat. no. CT62.1).
14. Male Luer-lock connectors (Carl Roth, cat. no. CT58.1).
15. Silicone tubing inner diameter (i.d.) 4.8 mm (Watson-Marlow, cat. no. 913.A048.024).
16. Disposable syringes Omnifix 5 ml with Luer-lock (B. Braun Melsungen, cat. no. 4617053V).

2.2 Perfusion System

1. Dosage Line C-Flex, with 2× Luer-lock fittings; i.d. 0.8 mm, length 2 m, male/male (Eppendorf, cat. no. 78510310)
2. Pump Head Tubing, Bioprene i.d. 0.5/W 1.05 mm, female/female (Eppendorf, cat. no. 78510118).
3. DASGIP Head Gear for addition bottles with GL45 neck, C-Flex; female Luer-Lock (Eppendorf, cat. no. 78510311).
4. Laboratory glass bottle, 100 ml, GL 45 neck and Laboratory glass bottle, 500 ml, GL 45 neck.

2.3 Monolayer Expansion

1. T25 flasks (TPP, cat. no. 90026).
2. T75 and T175 flask (Greiner Bio-One, cat. nos. 658175 and 660175).
3. Multi-Flask, 3-layer with Vented Cap (Corning Life Sciences Art. 353143) (optional).
4. Centrifuge tube support cushions for 500 ml centrifuge tubes (Corning #431124) and Centrifuge tubes 500 ml (Corning life science, CLS431123-36EA) (optional).
5. Vi-CELL™ XR from Beckman Coulter (optional).

2.4 Cell Staining and Flow Cytometry

1. 1.5 ml microcentrifuge tubes (Sarstedt).
2. 15 ml Greiner centrifuge tubes (Sigma-Aldrich).
3. Eppendorf® Thermomixer Compact (Sigma-Aldrich).
4. 96-Well V-bottom plate (Carl Roth).
5. BD Accuri™ C6 (BD Biosciences).
6. FlowJo V10.1 software (FlowJo LLC, Ashland, OR, USA).

2.5 Culture Medium, Enzymes, and Chemicals

1. Medium for Stem Cell cultivation and expansion: mTeSR1 (or E8).
2. Medium during differentiation: 500 ml RPMI 1640 supplemented with either 10 ml B27-insulin (RB^- medium) or 10 ml B27 with insulin (RB^+ medium).
3. DMEM/F12.
4. Geltrex basement membrane matrix (Invitrogen).
5. Trypan blue solution (0.4% (wt/vol)).
6. Solution to dissociate monolayer: Accutase (Life Technologies).
7. Sigmacote (Sigma-Aldrich): use 1 ml to coat one bioreactor vessel.
8. Low-Ca^{2+} solution: 120 mM NaCl, 5.4 mM KCl, 5 mM $MgSO_4$, 5 mM sodium pyruvate, 20 mM glucose, 20 mM taurine, 10 mM HEPES, and 30 μM $CaCl_2$ adjusted to pH 6.9 by the addition of NaOH.
9. Solution to dissociate aggregates: dissolve 500 mg collagenase B (Roche) in 500 ml low-Ca^{2+} solution. Filter to sterilize (0.22 μm), aliquot and store at −20 °C for up to 6 months (*see* **Note 1**).
10. 10 mM Rho kinase inhibitor (RI) Y27632 stock solution: 20 mg of Y27632 in 5.88 ml pure water. Filter to sterilize. Aliquot and store at −20 °C for up to 3 months (*see* **Note 2**).
11. CHIR99021 stock solution: 10 mg CHIR99021 in 1.79 ml DMSO. Aliquot and store at −20 °C for up to a year (*see* **Note 3**).
12. IWP2 stock solution: 10 mg in 2.17 ml DMSO, aliquot and store at −20 °C for up to a year.
13. Perform® (Schülke).
14. Optional: Penicillin/Streptomycin during preculture in the bioreactor.
15. 70% ethanol.
16. 90% ice-cold methanol: dilute 100% methanol with PBS w/o to produce 90% methanol and store methanol at −20 °C.

17. Flow Buffer 1 solution: 0.5% BSA, 3 mM ethylenediaminetetraacetic acid (EDTA, Sigma-Aldrich), in 500 ml phosphate buffered saline without Ca^{2+}/Mg^{2+} (PBS w/o).
18. Flow Buffer 2 solution: 0.5% BSA, 0.1% Triton X-100 (Sigma-Aldrich), 3 mM EDTA, in 500 ml PBS w/o.
19. Primary antibodies: Cardiac Troponin T (cTnT, Invitrogen, dilution 1:200, species—mouse, class—IgG) and sarcomeric α-actinin (SA, Sigma-Aldrich, dilution 1:800, species—mouse, class—IgG), mouse IgG (Thermo Scientific, dilution 1:100).
20. Secondary antibody: Cy5 (Jackson ImmunoResearch, dilution 1:200, species—donkey, anti-mouse IgG).

3 Methods

3.1 Preparation and Assembly of the Bioreactor

1. Prepare bioreactor at least 2 days before inoculation of the bioreactor. For precise timeline of the preparations and differentiation protocol, *see* Fig. 1.
2. Siliconize the bioreactor glass vessel. Repeatedly pipet 1 ml of Sigmacote solution to the sides and bottom of the vessel. Keep the vessel at room temperature (RT) overnight (*see* **Note 4**).
3. Examine head-plate and DO sensors for damage and replace damaged O-rings if necessary. Make sure a sampling port, a waste port, and a feed port are present as well as an exhaust and inlet air filter in the head plate (*see* **Notes 5** and **6**).
4. Assemble perfusion circuit and calibrate pumps according to Eppendorf-DASbox manual (*see* **Note 7**).
5. One day before inoculation of the bioreactors: Rinse bioreactor vessel thoroughly with water to remove any trace of Sigmacote and cover the bottom of the vessel with pure water to prevent pH probes to dry out during steam autoclaving.
6. Attach impeller to the lowest possible level of the bioreactor shaft.
7. Calibrate pH probes, screw pH probes and DO probes to the headplate of the bioreactor (*see* **Notes 8** and **9**).
8. The perfusion system can be connected to the bioreactor before sterilization. Open head-plate screw cap and autoclave the bioreactor for 20 min at 120 °C. Close the head-plate screw cap immediately afterward before placing the bioreactor under the flow hood.
9. Replace water with 100 ml PBS w/o and place the bioreactor into the DASbox station. Connect the temperature and DO probe, the gas supply, the agitation and the exhaust air

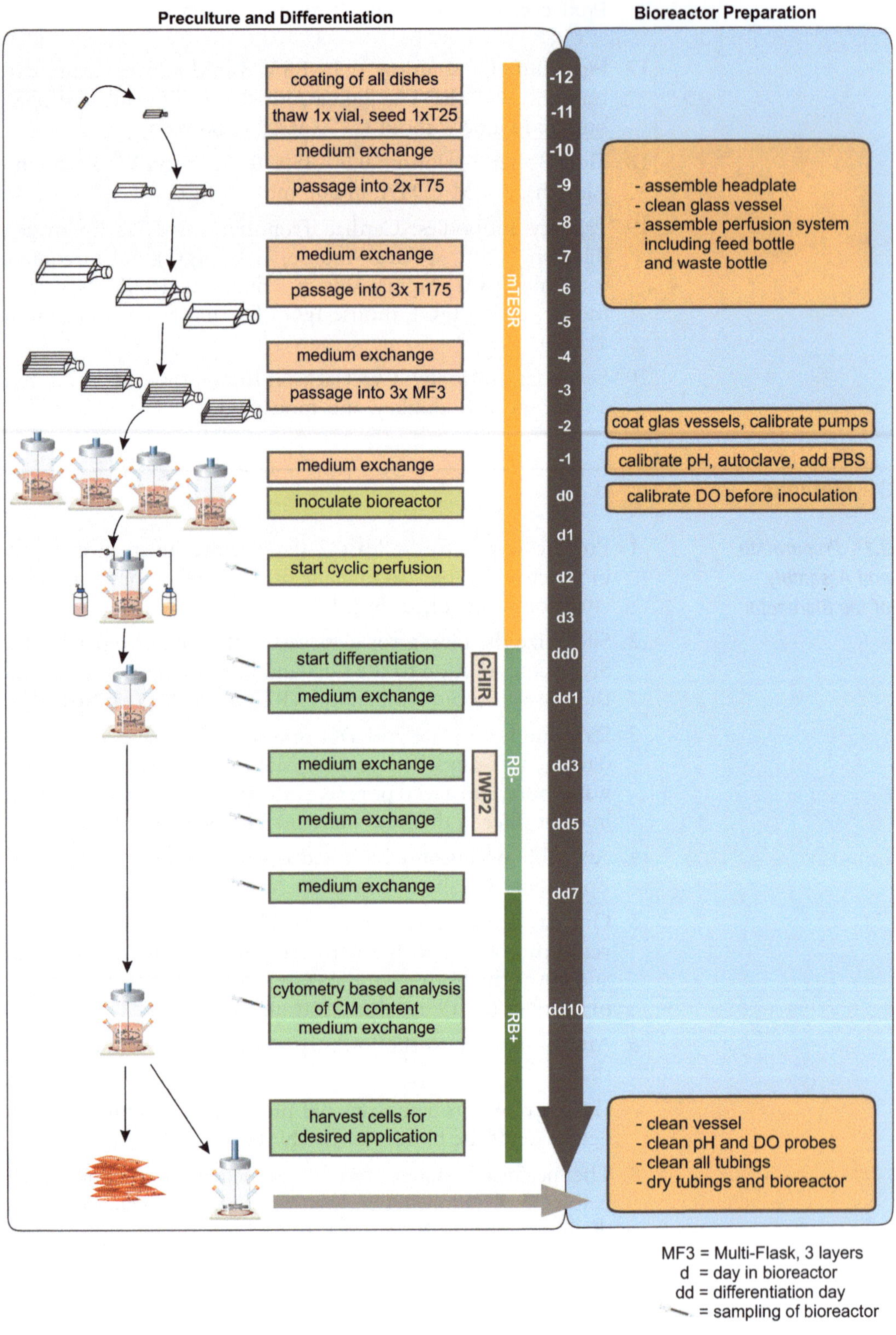

Fig. 1 Timetable and handling procedures for hPSC expansion and cardiomyogenic differentiation in four parallel bioreactors at 150 ml process scale. The necessary steps for bioreactor preparation and passaging rhythm are depicted on both sides of the timeline. The timeline is shown as preculture days (−12 to −1), bioreactor expansion days (d0 to d3), and bioreactor differentiation days (dd0 to dd10)

Table 1
Checklist for connection of the bioreactor to the station

	Connections and settings					
Software settings	70 rpm ccw		37 °C			3 sl/h air headspace CO_2 headspace Oxygen headspace 5% CO_2
Connect to bioreactor:	Overhead drive	pH probe	Temperature	Gas supply headspace	DO probe	Condenser
Switch on in software:	Agitation		Temperature		Gassing	

condenser. Start the DO calibration procedure: switch on agitation, gassing (21% O_2), and temperature (37 °C) in the software (*see* **Note 10**).

10. Before detaching the cells on day 0: calibrate DO in software (offset 100%), then remove the PBS under the flow hood and replace it by 120 ml mTeSR + 10 μM Y27632 (addition of Penicillin/Streptomycin optional). Place the bioreactor vessel into the bioreactor station and connect temperature, overhead drive and gas supply. Start stirring (60 rpm with E8, 70 rpm with mTeSR. Prepare the settings needed for the cultivation of cells in the software (settings are shown in Table 1) (*see* **Notes 11** and **12**).

3.2 Expansion of Cells as a Monolayer and Inoculation of Four Bioreactors

1. Obtain a stock of frozen hPSCs, 3 × 10^6 cells in one vial (frozen from monolayer cultures) (*see* **Notes 13** and **14**).
2. Start 12 days prior to the planned bioreactor inoculation (see timeline for scale up and bioreactor preparation and differentiation depicted in Fig. 1) (*see* **Note 15**).
3. Coat 1× T25, 2× T75, 3× T175, and 4× Multi-Flasks with Geltrex: thaw 1375 μl Geltrex into 550 ml DMEM/F12 and mix. Add 5 ml into T25, 15 ml into T75, 35 ml into T175 and 100 ml into a Multi-Flask, respectively and incubate at 37 °C for at least 1 h (*see* **Note 16**).
4. 11 days prior to bioreactor inoculation, thaw one vial in 37 °C water bath until a small ice cube is left. Remove the cell suspension sterilely and dilute it ten times in 4 °C cold mTeSR + 10 μM Y27632. Centrifuge tubes at 300 × *g* for 3 min. Discard supernatant and gently resuspend in 5 ml mTeSR + 10 μM Y27632. Remove DMEM/F12 from coated T25 and transfer cell suspension into T25 flask (*see* **Note 17**).
5. Full medium exchange (ME) 1 day after thawing and seeding cells.

6. Passage the cells into 2× T75 flasks 2 days after thawing the cells: Aspirate the medium, wash with 5 ml of PBS and remove PBS. Add 1 ml of Accutase solution and incubate the plate for 3–4 min at 37 °C. Add 5 ml cold DMEM/F12 (4 °C). Pipet medium gently over the surface to detach cells and transfer suspension into a centrifugation tube to centrifuge at 300 × *g* for 3 min. Discard supernatant, resuspend in 1 ml mTeSR + 10 μM Y27632 and count cells by using trypan blue cell counting or applying the ViCell (dilute 100 μl sample in 400 μl PBS ($f = 5$) to count with the ViCell device). Remove medium in coated T75 flask and add 15 ml mTeSR + 10 μM Y27632 in both T75. Seed one T75 with 40,000 cells/cm^2 (3 × 10^6 cells per T75) (*see* **Note 18**).
7. Full ME 2 days after seeding the cells.
8. Passage cells after a total of 3 days into 2–3× T175 using the same procedure described in Subheading 3.7.
9. ME 2 days after passage of the cells.
10. Same procedure is used to further cultivate the cells and to inoculate up to 4 bioreactors 3 days later (*see* Fig. 1 and Table 2) (*see* **Note 19**).
11. Bioreactor inoculation: after preparing the medium and the cultivation settings in the software for each bioreactor, cells can be detached from all four Multi-Flasks as described in Subheading 3.2.6 for passaging the cells. (Necessary volume can be found in Table 2.) Resuspend in 50 ml mTeSR + 10 μM Y27632. Take three separate samples to count the cells according to standard counting procedures or via the ViCell. Calculate the average cell density and pipet cell suspension into the bioreactor in order to seed 75 × 10^6 cells in one bioreactor. Add medium (mTeSR + 10 μM Y27632) to reach a total volume of 150 ml. Final cell density: 0.5 × 10^6 cells/ml (*see* **Note 20**).

Table 2
Specifications for monolayer expansion in flasks

Flask	Culture volume (ml)	PBS for washing (ml)	Accutase (ml)	DMEM/F12 to dilute Accutase (ml)	Geltrex volume (ml)	Cells (1 × 10^6)
T25	5	3	1	5	5	1
T75	15	9	3	15	15	3
T175	35	21	7	35	35	7
Multi-Flask 3 layers	100	60	21	100	100	21

12. Place the bioreactor back into the DASBox station and connect all probes, gas supply, overhead drive, and the condenser. Start the inoculation time clock in the software (*see* **Note 21**).
13. Make sure everything is connected and switched on properly—*see* checklist in Table 1.
14. Take a first sample after around 30 min of inoculation to count and confirm the inoculation cell density. Sample procedure is explained below. No enzymatic dissociation is necessary at this time point. Sample can be placed into ViCell without any dilutions (*see* **Note 22**).

3.3 Sampling of Bioreactor

1. Sampling of bioreactor: remove cap from bioreactor sample port and spray with 70% ethanol.
2. Connect a sterile Luer-lock syringe to the valve and open the Mohr pinchcock clamp of the sampling tube while slowly starting to draw suspension with the syringe. Draw ~1 ml of suspension into the syringe and close the clamp immediately. After the syringe has been unplugged, the 1 ml volume can be discarded and another 2 ml can be pulled into the syringe. Place the 2 ml into the well of a low attachment 6-well plate. Clean the sample port with 70% ethanol before closing the cap (*see* **Notes 23** and **24**).
3. Microscopic assessment of the aggregates is now possible before gathering aggregates in the middle of the well by gentle shaking. Transfer aggregates and medium into centrifuge tubes. Cells can be centrifuged at 300 × *g* for 3 min. Supernatant can be frozen at −20 °C to be used for metabolic analysis such as glucose and lactate measurement.
4. Wash pellet with 1 ml PBS w/o, centrifuge at 300 × *g* for another 3 min before adding 500 μl Accutase or collagenase B. Incubate at 37 °C for at least 5 min when using collagenase B and 3 min for Accutase. Aggregates should then fall apart and cells can be centrifuged at 300 × *g* for 3 min. Discard the supernatant and resuspend cells in 1 ml PBS to count using the ViCell. Cells can then be used for additional analysis as needed.
5. To monitor the cultivation of hPSCs, samples can be taken every second day. During the first days of differentiation, daily sampling is advisable. It is then enough to sample on day 5, 7, and finally on day 10 for endpoint analysis (*see* **Note 25**).

3.4 Expansion Phase in Bioreactor Using Cyclic Perfusion

1. Examine whether a drop in pH and DO can be observed 48 h after inoculation and take a sample to ensure aggregates were formed.

2. If aggregates are present, pipet 150 ml of mTeSR medium into the feed bottle in a sterile way (*see* **Note 26**).
3. Prime the feed tubing by switching the pump on, and press "shot" for 5 ml at 40 ml/min.
4. Load and start the script for cyclic perfusion provided by Eppendorf upon request (*see* **Note 27**).
5. Start the cyclic perfusion 48 h after inoculation (*see* **Note 28**).
6. Refill feed bottle every day with fresh medium.

3.5 Differentiation

1. Start differentiation 4 days after inoculation of the bioreactor.
2. Take a sample to count the cells as described above.
3. Prepare 620 ml of RB^- differentiation medium with 7.5 μM CHIR through addition of 387.5 μl CHIR stock solution (*see* **Note 29**).
4. Turn off the agitation, gassing and temperature (software) and detach all probes, overhead drive, gas supply and temperature sensors to place the bioreactor below the flow hood. Pipet the suspension into the 3× 50 ml centrifuge tubes and note down the exact volume in order to calculate the number of cells in the bioreactor (*see* **Note 30**).
5. Centrifuge the 3× 50 ml centrifuge tubes for 3 min at 300 × *g*. Resuspend and pool all three pellets to create a cell suspension of 10×10^6 cells/ml.
6. Pipet 145 ml of the prepared differentiation medium + CHIR into one bioreactor and add 5 ml of the resuspended cell suspension to reach a cell density of 0.33×10^6 cells/ml in one bioreactor. Place the bioreactor back into the DASbox station and reconnect all probes, start gassing, agitation and temperature in the software (*see* **Note 31**).
7. Precisely 24 h after start of differentiation: transfer cell suspension to 3× 50 ml centrifuge tubes, centrifuge at 300 × *g* for 1 min, discard the supernatant and add fresh RB^- differentiation medium (*see* **Note 32**).
8. After 48 h (a total of 72 h after the start of the differentiation): remove the medium and replace it with differentiation medium + 5 μM IWP-2 through a centrifugation step described above. Prepare the medium in advance by addition of 305 μl IWP stock solution to 610 ml RB^- for 4 bioreactors at 150 ml.
9. After 48 h (120 h after initiation of the differentiation), replace the medium with differentiation medium RB^- through centrifugation.
10. After 48 h (7 days after initiation of the differentiation) the medium can be exchanged with differentiation medium RB^+. From now on the medium can be changed every 2–3 days (*see* **Note 33**).

11. Aggregates can be analyzed after 10 days of differentiation to evaluate cardiomyocyte content. Aggregates can stay in bioreactor for more than 20 days but need to be fed every third day.

3.6 Quantitative Analysis of Differentiation Efficiency by Flow Cytometry

1. Take a sample from the bioreactor as described above. Transfer aggregates to a 15 ml centrifuge tube.
2. Centrifuge the aggregates at 300 × *g* for 3 min and discard the supernatant.
3. Wash the aggregates with 1 ml PBS w/o, centrifuge again at 300 × *g* for 3 min and discard the supernatant.
4. Transfer the aggregates to a 1.5 ml microcentrifuge tube and dissociate the aggregates by resuspending the pellet in 500 μl collagenase B. Incubate it in a Thermomixer at 37 °C and 1000 rpm for 20–30 min.
5. Once the aggregates are completely dissociated, transfer the solution to a 15 ml centrifuge tube again and centrifuge at 300 × *g* for 3 min. Discard the supernatant. Cells can be resuspended in 1 ml PBS w/o or cell culture medium.
6. Count the cells by means of trypan blue cell counting or using a ViCell device.
7. For staining, transfer cells to a 96-well V-bottom plate—about 150,000 cells per well. Centrifuge at 300 × *g* for 3 min. Discard the supernatant (*see* **Note 34**).
8. Fix the cells by adding 100 μl cold 90% methanol per well, for 15 min and keep the cells on ice or at 4 °C (*see* **Note 35**).
9. Centrifuge the plate at 300 × *g* for 3 min and discard the supernatant. Wash the cells twice with Flow Buffer 1 (100 μl per well). Centrifuge again at 300 × *g* for 3 min and discard supernatant each time.
10. Add the primary antibodies. Antibodies that can be used to identify the cardiomyocyte population include cTnT, SA and mouse IgG as a negative control/isotype. Antibodies are diluted in Flow Buffer 2 and 50 μl is added per well. Cells are incubated with the primary antibodies for 1 h at RT.
11. Centrifuge the plate at 300 × *g* for 3 min and discard the supernatant. Wash the cells once with Flow Buffer 2 and centrifuge again (*see* **Note 36**).
12. Add the secondary antibody. Dilute Cy5 in Flow Buffer 2. Add 50 μl per well and incubate the cells with the secondary antibody for 30 min in the dark at RT (*see* **Note 37**).
13. Centrifuge the plate at 300 × *g* for 3 min and discard the supernatant. Wash the cells twice with Flow Buffer 2. Centrifuge after each wash and discard the supernatant.
14. Resuspend the cells in 100 μl Flow Buffer 1.

15. Analyze the cells with a flow cytometer such as the BD Accuri™ C6 (BD Biosciences).
16. The flow cytometry results can be analyzed with FlowJo V10.1 software (FlowJo LLC, Ashland, OR, USA).

3.7 Cleaning of Bioreactor

1. Under the flow hood: Remove cell suspension. Cells can be processed as required.
2. Add 150 ml of 10 g/l perform disinfection solution, place the bioreactor back in the bioreactor station, set the stirring at 150 rpm and temperature at 37 °C and incubate for 30 min. Rinse the sample port, feeding and waste tubes with perform disinfection solution.
3. Rinse the bioreactor and sample port, feeding and waste tubes with water.
4. Remove the pH and DO probes to wash them with distilled water outside of the flow hood.
5. Clean sample port, feeding and waste tubes with 70% ethanol before rinsing the tubes several times with distilled water.
6. Clean glass vessel with 70% ethanol and rinse with water.

4 Notes

1. Collagenase B causes skin, eye and respiratory irritation. Avoid inhalation and skin contact. Wear protective gloves and eye protection.
2. Y27632 is toxic. Avoid skin contact, ingestion, and inhalation. Wear protective gloves and eye protection when preparing the stock solution.
3. CHIR99021 is toxic. Avoid skin contact, ingestion and inhalation. Wear protective gloves and eye protection when you are preparing the stock solution.
4. Sigmacote is harmful if swallowed or inhaled. Wear protective gloves and work under the fume hood.
5. Examine the membrane of DO sensors for damage and check whether the volume of electrolyte solution is sufficient.
6. The waste port should neither be too high nor too deep into the medium to ensure that no aggregates are pulled into the waste. If you plan to work with 100 ml bioreactor volume, place the waste port at a filling level of ~90 ml. When working with 150 ml, place the waste port at a level of ~130 ml. The waste bottle should be at least a 1 l bottle.
7. Calibrate by pumping 30 ml water at 30 ml/h into empty centrifuge tubes.

8. When connecting the DO probes, the lowest point of the probe should end slightly above the impeller to ensure a more homogenous aggregation. If the probe is too low, the impeller can be damaged by the probe.
9. Screw caps on sensors to protect the connections and wrap vent filters with aluminum foil before autoclaving.
10. There are two possible gas supply tubings in a DASbox station: head space/overlay and submersed. During DO calibration, the submersed gas supply tube has to be connected.
11. The mTeSR volume prepared in the bioreactor depends on the planned working volume. When aiming to work with 150 ml bioreactors, an initial addition of 120 ml is advisable. When aiming at 100 ml working volume, no more than 80 ml medium should be prepared in each bioreactor.
12. Impeller speed is dependent on cell line and medium and should be tested beforehand. In general, 70 rpm works nicely when using mTESR medium and 60 rpm with E8 medium.
13. The cell lines should be regularly checked for authenticity, possible karyotype abnormalities and mycoplasma infections.
14. Alternatively it is possible to start with hPSCs maintained on irradiation-inactivated murine embryonic fibroblasts (MEFs).
15. Expansion of pluripotent cells is calculated for an inoculation of 4 bioreactors with 150 ml working volume. Expansion time can be shortened when working with less medium or bioreactors. Rule of thumb: 1× Multi-Flasks produces enough cells for one 150 ml bioreactor. 2× T175 are enough for one bioreactor with a 100 ml working volume.
16. Make sure that the medium is distributed evenly into all three layers of Multi-Flasks as described in the manual of the Multi-Flasks. Coated flasks can be kept in the fridge for a period of up to 4 weeks.
17. Coated T25, T75, T175, and Multi-Flasks should be at RT before use.
18. Do not leave Accutase longer than 6 min on the cells! Prolonged incubation with Accutase can be detrimental to the cells.
19. Beware when seeding Multi-Flasks to homogenously distribute the cells throughout the three layers. Medium can be added and emptied by pouring carefully.
20. Optional use of a 500 ml centrifuge tube. It improves the speed and efficiency during detachment and inoculation of four bioreactors. Do not resuspend in more than 50 ml and measure the exact volume after resuspension in order to know the exact number of cells in the suspension.

21. Do not forget the temperature sensor when detaching and reconnecting all connections during the inoculation of the bioreactor. The resulting overheating of the cell suspension can be detrimental to the cells, ultimately leading to cell death.
22. The cell density in the bioreactor should lie between 0.4 and 0.6×10^6 cells/ml. If the cell density deviates too much from 0.5×10^6 cells/ml, it might be necessary to correct it by addition and/or removal of culture medium and cell suspension from the bioreactor.
23. Aggregates tend to stick to the walls of the syringe. It is therefore advisable to rinse the syringe. To this reason, the six-well plate can be tilted and after waiting ~30 s for the aggregates to settle into the corner, some of the medium can be drawn without pulling any aggregates into the syringe. Tap the syringe to detach aggregates attached to the wall and immediately push the medium back into the same well.
24. Check whether sampling of the bioreactors may have removed tubings for gassing or temperature probes.
25. Be aware that each sampling removes a total of 3 ml out of the bioreactor. The number of samples should be limited when working with a 100 ml bioreactor. The waste port is located at a height of around 80 ml. Not more than 20 ml should therefore be removed.
26. If a flow hood is close to the DASbox station, attach a longer tubing to the feed bottle. Feed bottles can then be refilled under the flow hood. If the sterile hood is too far from the DASbox station, clamp the tubing before detaching it from feed bottle by unscrewing the Luer-lock and place the ending of the tubing in 70% ethanol. Then place the feed bottle below the flow hood and (re)fill it with 150 ml medium.
27. You can adapt the script according to your working volume in the bioreactor. The script has been written in order to change 7 ml every second hour. It is possible to increase the volume and frequency.
28. Make sure the pumps are functioning properly and that the pump direction is correct.
29. CHIR is light sensitive. Work without direct light.
30. Make sure to remove all aggregates from bioreactor. Do not leave any on the bottom of the vessel.
31. Do not forget to replace the temperature sensor.
32. The removal of CHIR should be performed exactly $24 \text{ h} \pm 15$ min. This is crucial for the differentiation outcome.
33. It is not necessary to perform a full medium exchange from day 7 onward. The bioreactor can be placed under the flow hood

for about 5–10 min to let aggregates settle. Then gently pipet medium out of the bioreactor and replace the same amount of medium.

34. If cells were resuspended in PBS w/o, no additional washing is required. Otherwise, wash the cells with PBS w/o before proceeding.
35. Methanol is harmful and should be used under a fume hood.
36. To discard the supernatant, the plate can be turned upside down in a fluid movement. Dry the plate on a piece of paper.
37. Secondary antibodies are light sensitive and should also be kept in the dark during preparation of the dilutions.

Acknowledgments

This work was supported by StemBANCC (support from the Innovative Medicines Initiative joint undertaking under grant 115439-2, whose resources are composed of financial contribution from the European Union [FP7/2007-2013] and EFPIA companies' in-kind contribution). RZ received funding from the German Research Foundation (DFG; including the Cluster of Excellence REBIRTH EXC 62/2 and grants ZW64/4-1, KFO311, and ZW64/7-1), the German Ministry for Education and Science (BMBF; grants: 13N14086, 01EK1601A, and 01EK1602A) and by the European Union (TECHNOBEAT; grant 66724).

References

1. Christoffersson J, Meier F, Kempf H, Schwanke K, Coffee M, Beilmann M, Zweigerdt R, Mandenius CF (2018) A cardiac cell outgrowth assay for evaluating drug compounds using a cardiac spheroid-on-a-chip device. Bioengineering 5(2). https://doi.org/10.3390/bioengineering5020036
2. Christoffersson J, Bergstrom G, Schwanke K, Kempf H, Zweigerdt R, Mandenius CF (2016) A microfluidic bioreactor for toxicity testing of stem cell derived 3D cardiac bodies. Methods Mol Biol 1502:159–168. https://doi.org/10.1007/7651_2016_340
3. Iorga B, Schwanke K, Weber N, Wendland M, Greten S, Piep B, Dos Remedios CG, Martin U, Zweigerdt R, Kraft T, Brenner B (2017) Differences in contractile function of myofibrils within human embryonic stem cell-derived cardiomyocytes vs. adult ventricular myofibrils are related to distinct sarcomeric protein isoforms. Front Physiol 8:1111. https://doi.org/10.3389/fphys.2017.01111
4. Weber N, Schwanke K, Greten S, Wendland M, Iorga B, Fischer M, Geers-Knorr C, Hegermann J, Wrede C, Fiedler J, Kempf H, Franke A, Piep B, Pfanne A, Thum T, Martin U, Brenner B, Zweigerdt R, Kraft T (2016) Stiff matrix induces switch to pure beta-cardiac myosin heavy chain expression in human ESC-derived cardiomyocytes. Basic Res Cardiol 111(6):68. https://doi.org/10.1007/s00395-016-0587-9
5. Templin C, Zweigerdt R, Schwanke K, Olmer R, Ghadri JR, Emmert MY, Muller E, Kuest SM, Cohrs S, Schibli R, Kronen P, Hilbe M, Reinisch A, Strunk D, Haverich A, Hoerstrup S, Luscher TF, Kaufmann PA, Landmesser U, Martin U (2012) Transplantation and tracking of human-induced pluripotent stem cells in a pig model of myocardial infarction: assessment of cell survival, engraftment, and distribution by hybrid single photon emission computed tomography/computed tomography of sodium iodide symporter transgene expression. Circulation 126(4):430–439.

https://doi.org/10.1161/CIRCULATIONAHA.111.087684

6. Rojas SV, Kensah G, Rotaermel A, Baraki H, Kutschka I, Zweigerdt R, Martin U, Haverich A, Gruh I, Martens A (2017) Transplantation of purified iPSC-derived cardiomyocytes in myocardial infarction. PLoS One 12 (5):e0173222. https://doi.org/10.1371/journal.pone.0173222
7. Zweigerdt R (2009) Large scale production of stem cells and their derivatives. Adv Biochem Eng Biotechnol 114:201–235. https://doi.org/10.1007/10_2008_27
8. Kempf H, Olmer R, Kropp C, Ruckert M, Jara-Avaca M, Robles-Diaz D, Franke A, Elliott DA, Wojciechowski D, Fischer M, Roa Lara A, Kensah G, Gruh I, Haverich A, Martin U, Zweigerdt R (2014) Controlling expansion and cardiomyogenic differentiation of human pluripotent stem cells in scalable suspension culture. Stem Cell Reports 3(6):1132–1146. https://doi.org/10.1016/j.stemcr.2014.09.017
9. Fonoudi H, Ansari H, Abbasalizadeh S, Larijani MR, Kiani S, Hashemizadeh S, Zarchi AS, Bosman A, Blue GM, Pahlavan S, Perry M, Orr Y, Mayorchak Y, Vandenberg J, Talkhabi M, Winlaw DS, Harvey RP, Aghdami N, Baharvand H (2015) A universal and robust integrated platform for the scalable production of human cardiomyocytes from pluripotent stem cells. Stem Cells Transl Med 4(12):1482–1494. https://doi.org/10.5966/sctm.2014-0275
10. Chen VC, Ye J, Shukla P, Hua G, Chen D, Lin Z, Liu JC, Chai J, Gold J, Wu J, Hsu D, Couture LA (2015) Development of a scalable suspension culture for cardiac differentiation from human pluripotent stem cells. Stem Cell Res 15(2):365–375. https://doi.org/10.1016/j.scr.2015.08.002
11. Watanabe K, Ueno M, Kamiya D, Nishiyama A, Matsumura M, Wataya T, Takahashi JB, Nishikawa S, Nishikawa S, Muguruma K, Sasai Y (2007) A ROCK inhibitor permits survival of dissociated human embryonic stem cells. Nat Biotechnol 25 (6):681–686. https://doi.org/10.1038/nbt1310
12. Kropp C, Kempf H, Halloin C, Robles-Diaz D, Franke A, Scheper T, Kinast K, Knorpp T, Joos TO, Haverich A, Martin U, Zweigerdt R, Olmer R (2016) Impact of feeding strategies on the scalable expansion of human pluripotent stem cells in single-use stirred tank bioreactors. Stem Cells Transl Med 5(10):1289–1301. https://doi.org/10.5966/sctm.2015-0253
13. Olmer R, Haase A, Merkert S, Cui W, Palecek J, Ran C, Kirschning A, Scheper T, Glage S, Miller K, Curnow EC, Hayes ES, Martin U (2010) Long term expansion of undifferentiated human iPS and ES cells in suspension culture using a defined medium. Stem Cell Res 5(1):51–64. https://doi.org/10.1016/j.scr.2010.03.005
14. Olmer R, Engels L, Usman A, Menke S, Malik MNH, Pessler F, Gohring G, Bornhorst D, Bolten S, Abdelilah-Seyfried S, Scheper T, Kempf H, Zweigerdt R, Martin U (2018) Differentiation of human pluripotent stem cells into functional endothelial cells in scalable suspension culture. Stem Cell Reports 10 (5):1657–1672. https://doi.org/10.1016/j.stemcr.2018.03.017
15. Lian X, Hsiao C, Wilson G, Zhu K, Hazeltine LB, Azarin SM, Raval KK, Zhang J, Kamp TJ, Palecek SP (2012) Robust cardiomyocyte differentiation from human pluripotent stem cells via temporal modulation of canonical Wnt signaling. Proc Natl Acad Sci U S A 109(27): E1848–E1857. https://doi.org/10.1073/pnas.1200250109
16. Gonzalez R, Lee JW, Schultz PG (2011) Stepwise chemically induced cardiomyocyte specification of human embryonic stem cells. Angew Chem Int Ed Engl 50(47):11181–11185
17. Burridge PW, Matsa E, Shukla P, Lin ZC, Churko JM, Ebert AD, Lan F, Diecke S, Huber B, Mordwinkin NM, Plews JR, Abilez OJ, Cui B, Gold JD, Wu JC (2014) Chemically defined generation of human cardiomyocytes. Nat Methods 11(8):855–860. https://doi.org/10.1038/nmeth.2999
18. Kempf H, Olmer R, Haase A, Franke A, Bolesani E, Schwanke K, Robles-Diaz D, Coffee M, Gohring G, Drager G, Potz O, Joos T, Martinez-Hackert E, Haverich A, Buettner FF, Martin U, Zweigerdt R (2016) Bulk cell density and Wnt/TGFbeta signalling regulate mesendodermal patterning of human pluripotent stem cells. Nat Commun 7:13602. https://doi.org/10.1038/ncomms13602

Chapter 6

Improved Protocol for Cardiac Differentiation and Maturation of Pluripotent Stem Cells

Imen Jebeniani, Shunli Ding, and Michel Pucéat

Abstract

Pluripotent stem cells feature the capacity to differentiate into any somatic cell types including cardiomyocytes. We report a cost-effective and simple protocol for the differentiation of specific ventricular cardiomyocytes. These cells are elongated, do not spontaneously beat, and do not feature any Ca^{2+}-transient, an index of their stage of maturation toward adult cardiac cells. They represent a suitable model to screen both the efficiency and toxicology of drugs.

Key words Cardiomyocytes, Pluripotent stem cells, Heart, Metabolism

1 Introduction

Induced pluripotent stem cells are capable of differentiating toward any somatic cell type including cardiomyocytes [1]. Patient-specific cells provide a potentially faithful model of cardiac diseases [2]. The cells can also be used to test the efficacy as well as any toxic effects of molecules including therapeutic drugs [3, 4]. Many 2D protocols have thus been proposed to drive pluripotent cells toward a myocardial cell [5]. However, most of these protocols drive iPS cells toward embryonic cardiomyocytes reminiscent of the early stages of development. They are spontaneously beating with an index of unstable resting membrane potential and have a likely absence of a functional sarcoplasmic reticulum and/or T-tubules which are key intracellular structures to ensure a proper excitation–contraction coupling [6]. This raises questions as to the physiological relevance of such cells in drug efficacy or toxicology assays [7].

Recently, using a bioengineering 3D culture approach, two laboratories reported that cardiac differentiation of iPSCs can indeed be pushed forward an "adult-like" stage [8, 9]. However, these protocols are time-consuming, labor-intensive and require some skills in bioengineering.

Carl-Fredrik Mandenius and James A. Ross (eds.), *Cell-Based Assays Using iPSCs for Drug Development and Testing*, Methods in Molecular Biology, vol. 1994, https://doi.org/10.1007/978-1-4939-9477-9_6,

© Springer Science+Business Media, LLC, part of Springer Nature 2019

In order to drive iPS cells toward functional, more mature and ventricular-specific cardiomyocytes, a protocol of cardiomyocyte differentiation should mimic the step-by-step developmental pathway [10]. Here, we describe a simple 2D protocol of differentiation using modulators of the Wnt pathway, together with the morphogen BMP2, to drive cells toward mesoderm and then Sonic Hedgehog (SHH) to drive MESP1+ mesodermal cells toward the fate of cardiac progenitors and early cardiomyocytes. We reasoned that when in vivo cardiomyocytes stop proliferating and start to mature, when epicardium-derived fibroblasts start populating the ventricle, they switch metabolism from glycolysis to fatty acid oxidation [11] as they get more and more exposed to oxygen [12]. Herein, we report a low-cost protocol using a metabolic medium in order to differentiate and mature ventricular-specific cardiomyocytes.

2 Materials

2.1 Cell Culture

Prepare all stocks of molecules as a 1000 or 10,000× stock concentrations.

1. CHIR9901 (STEMCELL technologies): Prepare a 10 mg/ml stock in DMSO, make aliquots and store at −20 °C.
2. BMP2 (ThermoFisher, Life Technologies, PHC7145): Prepare a 100 μg/ml stock in acetic acid 10 mM, make aliquots and store at −20 °C.
3. IWP4 (STEMCELL Technologies): Make up a 2.5 mM stock solution in DMSO; make aliquots and store at −20 °C.
4. SHH (Peprotech): Make a 100 μg/ml stock in PBS added with BSA 0.5%, make aliquots and store at −20 °C.
5. RPMI medium 1640 (Thermofisher, France, 21875091/ RPMI medium 1640).
6. DMEM low glucose (Thermofisher, France, 10567014).
7. Matrigel (Corning 354277/Matrigel hESC-Qualif): To coat plates, use 1/100 Matrigel in cold DMEM according to manufacturer's instructions. Add 10,000 mitomycin-arrested MEF/cm^2 for the differentiation plates.
8. Fibronectin (Sigma-Aldrich F1141)/use the solution at 1/100 to coat the plates.
9. Knock-out serum replacement (KO-SR, Thermofisher, France): Use 5% in RPMI medium 1640 added with nonessential amino acids (Thermofisher, France, 11140035/NEAA 100×) and mercaptoethanol (10^{-7} M).
10. Linoleic and oleic acids bound to BSA (Sigma-Aldrich, I2771-5ML/ITS+3 Liquid Media Supplement (100×).

11. Accutase (Thermofisher, France, A11105-01/StemPro® Accutase® Cell Dissociation Reagent).
12. Thiazovivin (Sigma-Aldrich, SML1045).

2.2 Mitotic Arrest of Mouse Embryonic Fibroblasts

Mouse Embryonic Fibroblasts (MEF) isolated from E14.5 mouse embryos are cultured in DMEM supplemented with 10% fetal calf serum. Mitomycin (10 μg/ml) is added to confluent plates for 2 h. Then cells are washed twice with PBS and isolated after 5 min trypsin–EDTA treatment. Mitomycin-inactivated cells can be frozen in DMEM/50%FCS/10% DMSO and store as aliquots at −80 °C for 3 months.

2.3 Cell Permeation

Fluo4-AM cell permeant (Thermofisher, F14201). Resuspend Fluo4-AM with DMSO and 10% pluronic acid to get a stock concentration of 1 mg/ml; use 10 mg/ml to load cells.

2.4 Antibodies

1. Anti-Mesp1 (Abcam ab129387).
2. Anti-Brachyury (R&D AF2085).
3. Anti-Nkx-2.5 (R&D AF2444).
4. Anti MLCv (Clone F109. 3E0; Enzo Life Sciences).

3 Methods

Carry out all procedures at 37 °C unless otherwise specified. IPS cells should be maintained under rigorous culture conditions in antibiotic-free medium. Absence of mycoplasma should be checked on a regular basis. Expression of pluripotency markers (OCT4, NANOG, SOX2) should also be checked by immunofluorescence to visualize homogeneity of the culture every ten passages.

Cells are gown on Matrigel-coated plates in Essential 8 (Thermofisher) or Stem Macs IPS brew (Miltenyi) medium. We favor iPS brew medium for mesodermal and cardiac differentiation.

3.1 Cell Differentiation

1. The day prior to differentiation, prepare a dish coated with Matrigel and 10^5 mitomycin-inactivated MEF/cm^2 (*see* Subheading 2.2 above) (*see* **Note 1**).
2. Add thiazovivin (1/10,000) to the cells 1 h before splitting cells. Switch cells to new Matrigel-coated plates with 10^5 MEF/cm^2 using Accutase for 4 min at RT to split cells and plate cells to reach 50% confluency for the next few days with small colonies [3].
3. Day 1: Wash cells with RPMI-KOSR once and add (1 ml per well in a 6 wells/plate or 8 ml/10 cm plate) RMPI-KOSR added with 10 μM CHIR9901 for 14 h (overnight).

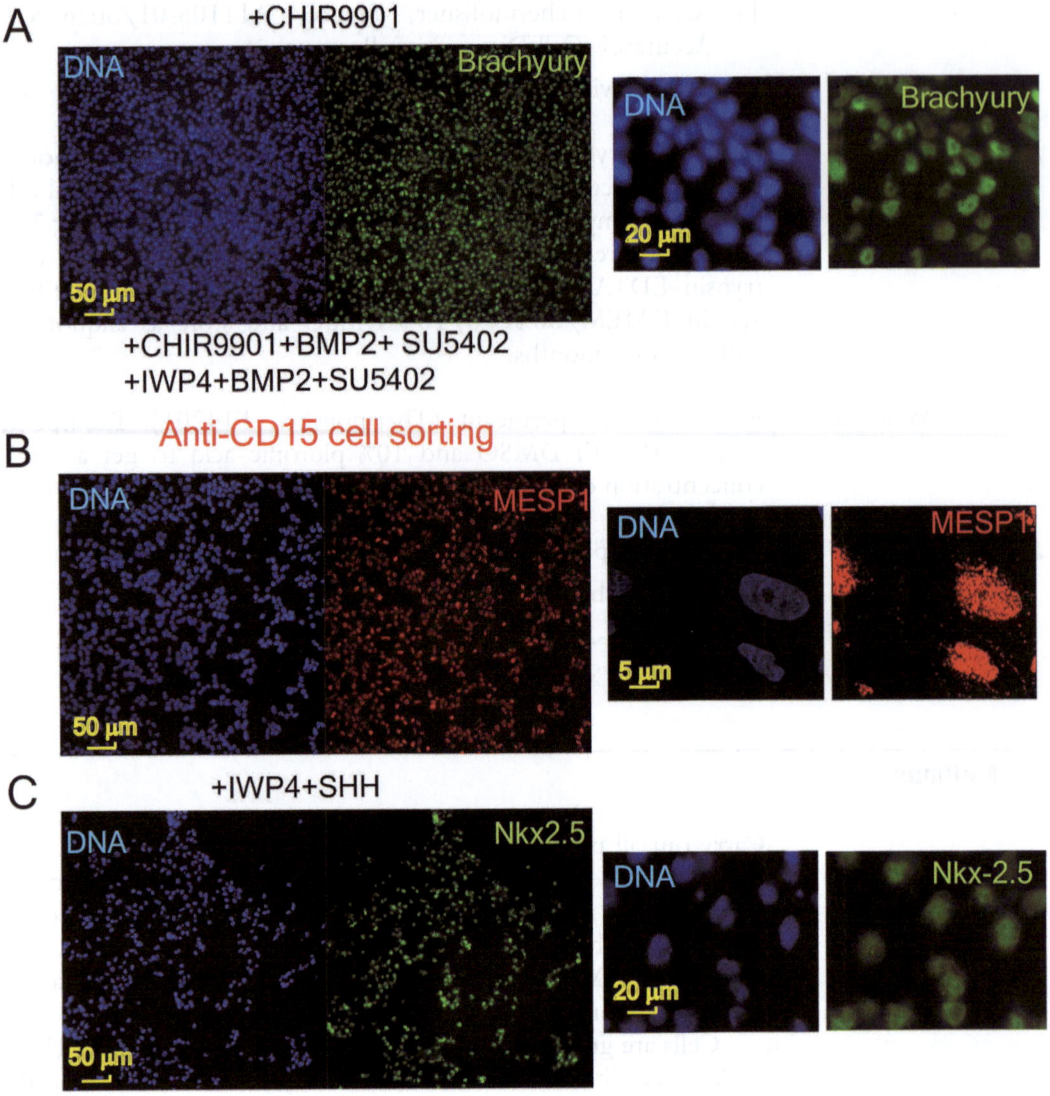

Fig. 1 (**a**) Anti-brachyury immunostaining of cells treated with CHIR9:901, CHIR9901 + BMP2 + SU5402, and IWP4 + BMP2 + SU5402. (**b**) After anti-CD15 sorting cells were stained with anti-MESP1. (**c**) After 1 day IWP4 + SHH treatment, cells were stained with anti-Nkx2.5 antibody

4. Day 2: Replace medium with 1 ml RPMI-KOSR added with 5 μM CHIR9901 and 10 ng/ml BMP2.

   At this stage, most cells should express Brachyury (Fig. 1a).

5. Day 3: Sort mesodermal cells using anti-CD15 antibody-conjugated beads (easy step STEM CELL technologies kit) according to manufacturer's protocol. Plate CD15+ cells on Matrigel-coated plates. Depending upon the iPS cell line, 40–60% cells should be CD15+.

   At this step, most cells should express MesP1 (Fig. 1b) (*see* **Note 2**).

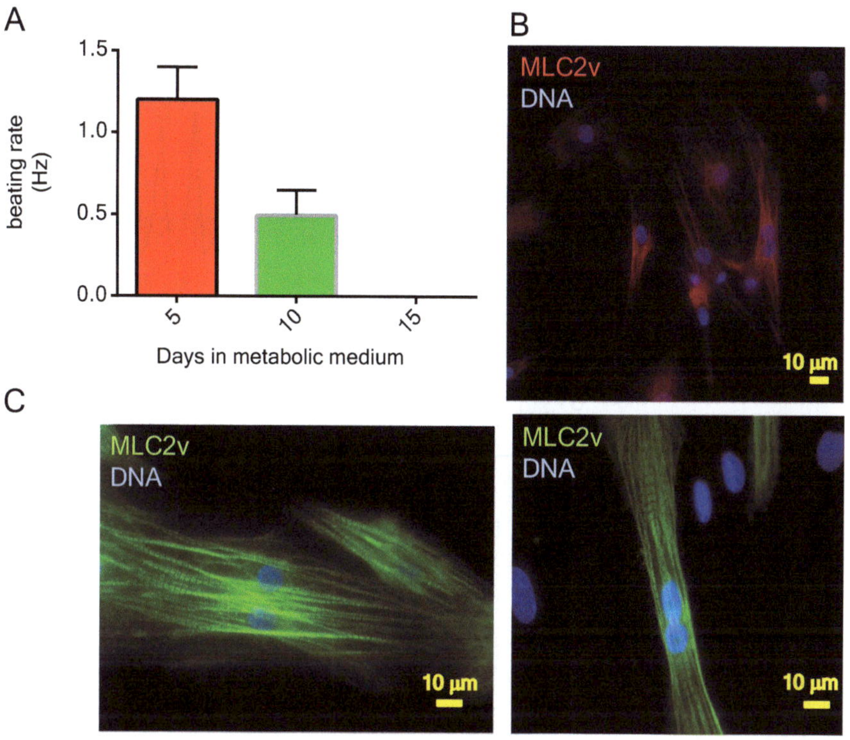

Fig. 2 (**a**) Beating was scored in three experiments at days 5, 10, and 15 after addition of the metabolic medium. (**b**) Anti-MLC2v immunostaining of day 15 postmaturation cardiomyocytes. (**c**) High magnification images of day 15 postmaturation cardiomyocytes

6. Day 4: Replace medium by 1 ml RPMI-KOSR added with 2.5 μM IWP4 and 10 ng/ml SHH.
7. Day 5: Replace medium by 1 ml RPMI-KOSR.

   At this stage, cells express Nkx-2.5 (Fig. 1c).
8. Day 7: Replace medium with DMEM-low glucose 1.5% BSA every 3 days until beating cells are observed in the dish (Days 10–15) (*see* **Note 3**).
9. When cells beat, split the cells using TrypLE express enzyme and replate them on fibronectin-coated dishes. First, wash the cells with PBS + Ca^{2+} + Mg^{2+}, then add TrypLE for 5 min at 37 °C and add 1/3 DMEM-low glucose 1.5% BSA to stop enzymatic activity. Spin down the cells (300 × *g*, 4 min).

   Plate cells in metabolic medium: DMEM-low glucose and ITS+3 Liquid Media Supplement (1/100).
10. Recordings of Ca^{2+} transients: The beating rate is monitored using Ca^{2+} transients rate as an index of beating frequency. Cardiomyocytes are loaded with 10 μg/ml Fluo-4 AM for 15 min at 37 °C, washed with culture medium. Images are acquired using a ZEISS observer equipped with a fast acquisition Hamamatsu CCD driven by Zen physiology software.

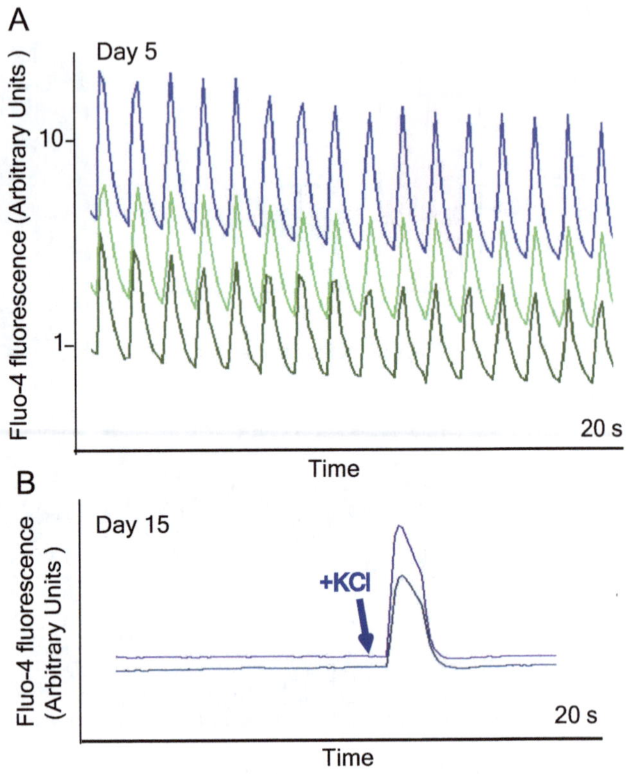

Fig. 3 Recordings of Ca^{2+} transients in Fluo-4-loaded cardiomyocytes at day 5 (**a**) and 15 (**b**) of culture in metabolic medium

Cells should stop beating within 10–15 days (Fig. 2a) and should express myosin light chain 2 (MLC2v) an index of their ventricular phenotype (Fig. 2b, c).

11. To check excitability of cells, add a puff of 15 mM KCl with a micropipette to trigger an action-potential and in turn a long-shaped Ca^{2+} transient (Fig. 3).

4 Notes

1. MEF prevent cell death during differentiation and allow differentiating cells to express membrane SSEA-1 (CD15) [13].
2. Single cell-seq analysis of CD15+ cells confirmed high expression of mesendodermal and mesodermal genes such as *MESP1*, *EOMES*, *Wnt3*, *MIXL1*, *GSC*, and *SALL4* [14].
3. Cells should never become more than 70% confluent before splitting them [13].

References

1. Yoshida Y, Yamanaka S (2017) Induced pluripotent stem cells 10 years later: for cardiac applications. Circ Res 120(12):1958–1968
2. Giacomelli E, Mummery CL, Bellin M (2017) Human heart disease: lessons from human pluripotent stem cell-derived cardiomyocytes. Cell Mol Life Sci 74(20):3711–3739
3. Sinnecker D, Laugwitz KL, Moretti A (2014) Induced pluripotent stem cell-derived cardiomyocytes for drug development and toxicity testing. Pharmacol Ther 143(2):246–252
4. Blin G, Neri T, Stefanovic S, Pucéat M (2010) Human embryonic and induced pluripotent stem cells in basic and clinical research in cardiology. Curr Stem Cell Res Ther 5(3):215–226
5. Mummery CL, Zhang J, Ng ES, Elliott DA, Elefanty AG, Kamp TJ (2012) Differentiation of human embryonic stem cells and induced pluripotent stem cells to cardiomyocytes: a methods overview. Circ Res 111(3):344–358
6. Kane C, Terracciano CM (2015) Induced pluripotent stem cell-derived cardiac myocytes to understand and test calcium handling: pie in the sky? J Mol Cell Cardiol 89(Pt B):376–378
7. Sala L, Bellin M, Mummery CL (2017) Integrating cardiomyocytes from human pluripotent stem cells in safety pharmacology: has the time come? Br J Pharmacol 174 (21):3749–3765
8. Tiburcy M, Hudson JE, Balfanz P, Schlick S, Meyer T, Chang Liao ML, Levent E, Raad F, Zeidler S, Wingender E, Riegler J, Wang M, Gold JD, Kehat I, Wettwer E, Ravens U, Dierickx P, van Laake LW, Goumans MJ, Khadjeh S, Toischer K, Hasenfuss G, Couture LA, Unger A, Linke WA, Araki T, Neel B, Keller G, Gepstein L, Wu JC, Zimmermann WH (2017) Defined engineered human myocardium with advanced maturation for applications in heart failure modeling and repair. Circulation 135(19):1832–1847
9. Ronaldson-Bouchard K, Ma SP, Yeager K, Chen T, Song L, Sirabella D, Morikawa K, Teles D, Yazawa M, Vunjak-Novakovic G (2018) Advanced maturation of human cardiac tissue grown from pluripotent stem cells. Nature 556(7700):239–243
10. Van Vliet P, Wu SM, Zaffran S, Pucéat M (2012) Early cardiac development: a view from stem cells to embryos. Cardiovasc Res 96(3):352–362
11. Lopaschuk GD, Jaswal JS (2010) Energy metabolic phenotype of the cardiomyocyte during development, differentiation, and postnatal maturation. J Cardiovasc Pharmacol 56 (2):130–140
12. Menendez-Montes I, Escobar B, Palacios B, Gómez MJ, Izquierdo-Garcia JL, Flores L, Jiménez-Borreguero LJ, Aragones J, Ruiz-Cabello J, Torres M, Martin-Puig S (2016) Myocardial VHL-HIF signaling controls an embryonic metabolic switch essential for cardiac maturation. Dev Cell 39(6):724–739
13. Leschik J, Stefanovic S, Brinon B, Pucéat M (2008) Cardiac commitment of primate embryonic stem cells. Nat Protoc 3 (9):1381–1387
14. Neri T, Hiriart E, van Vliet3 PP, Faurel E, Norris RA, Farhat B, Jagla B, Lefrancois J, Sugi Y, Moore-Morris T, Zaffran S, Faustino RS, Zambon AC, Desvignes J-P, Salgado D, Levine RA, de la Pompa JL, Terzic A, Evans SM, Markwald R, Pucéat M (2019) Human pre-valvular endocardial cells derived from pluripotent stem cells recapitulate cardiac pathophysiological valvulogenesis. Nat Commun 10(1):1929–1943

Chapter 7

Human Pluripotent Stem Cell Expansion in Stirred Tank Bioreactors

Felix Manstein, Caroline Halloin, and Robert Zweigerdt

Abstract

This chapter describes a detailed protocol on human pluripotent stem cells (hPSCs) cultivation as matrix-free cell-only aggregates in defined and xeno-free culture medium in stirred tank bioreactors (STBRs). Starting with a frozen stock pre-expanded on conventional culture dishes (2D), the ultimate process is performed in 150 mL culture scale in stirred tank bioreactors (3D) and is designed to produce up to 500 million pluripotent hPSC within 7 days. The culture strategy includes perfusion-based cell feeding facilitating process control, automation, and higher cell yields. Ultimately, this detailed protocol describes an important step for generating a defined starting cell population for directed lineage differentiation and subsequently fueling human cell-based assays and regenerative medicine approaches.

Key words Human pluripotent stem cells (hPSCs), Stirred tank bioreactors (STBRs), Suspension culture, Perfusion feeding, Process development, Automation

1 Introduction

The envisioned use of human pluripotent stem cells (hPSCs) ranges from developing advanced drug discovery and safety pharmacology assays, innovative in vitro disease modelling and ultimately base-repair of cells in diseased organs [1]. For this to occur, an efficient, economically viable supply of hPSCs in large quantities and high quality is required [2]. At typical laboratory conditions hPSCs are cultured as surface- or matrix-adherent colonies or cell layers in dishes or flasks (two-dimensional cultivation; 2D) [3]. These culture platforms, however, lack the possibility to monitor culture conditions properly, which is required for developing efficient mass cell production [4]. Instrumented stirred-tank bioreactors (STBRs) have the potential to overcome these limitations. Besides systematic process development, STBR systems also support linear

Felix Manstein and Caroline Halloin contributed equally to this work.

Carl-Fredrik Mandenius and James A. Ross (eds.), *Cell-Based Assays Using iPSCs for Drug Development and Testing*, Methods in Molecular Biology, vol. 1994, https://doi.org/10.1007/978-1-4939-9477-9_7,
© Springer Science+Business Media, LLC, part of Springer Nature 2019

process upscaling by applying equivalent bioreactor design of incremental dimensions. Importantly, STBRs ensure homogeneous distribution of cells, nutrients and gases and enable continuous online monitoring and control of key process parameters such as pH and dissolved oxygen (DO) [4]. Given that hPSCs conventionally require matrix adherence, two parallel approaches have been developed to enable transition from 2D toward cultivation in 3D suspension culture. This was achieved either by cultivation on microcarriers, thereby providing an attachment surface and matrix equivalent to 2D conditions [5], or by the formation of matrix-free "cell-only aggregates" [6]. Following inoculation of single cells most recent studies, applying the cell-only approach, achieved densities of up to a ~2.85×10^6 hPSCs per mL at the 100 mL bioreactor scale, which was enabled by using the perfusion feeding strategy [6]. This feeding strategy is based on a continuous influx of fresh medium and parallel efflux of used medium at the same rate. For cell retention (i.e., to avoid cell washout by perfusion feeding) various devices can be used, but porous stainless steel or glass filters are the most established in the stem cell field. The main advantage of perfusion feeding, as compared to conventional repeated batch feeding, includes the implementation of a more homogenous concentration of nutrients such as glucose or waste metabolites such as lactate in the culture medium [6]. Not only hPSC expansion but also lineage-specific differentiation into cardiomyocytes or endothelial cells can be efficiently performed by applying STBR technology [7, 8].

Starting from one frozen vial of 3×10^6 hPSC, the protocol described is designed to generate up to two billion cells within 19 days, by utilizing four parallel bioreactors conducting a 150 mL process scale in each vessel. Therefore, the strategy provides a useful starting point for further modulation of hPSC culture conditions and can be utilized for the induction of directed differentiation of hPSC into desired functional progenies.

2 Materials

2.1 Bioreactor

1. DASbox Mini bioreactor system for cell culture applications (Eppendorf, cat. no. 76DX04CC) including 4× DASbox Mini bioreactor vessel for cell culture applications (Eppendorf, cat. no. 76DS0250ODSS).
2. Sampling valves (Eppendorf, cat. no. 78200077).
3. DASGIP Dip Tube Pg 13.5, 220 mm (Eppendorf, cat. no. 76DGDT220).
4. DASbox exhaust system (Eppendorf, cat. no. 76DXOFF).

5. DASbox exhaust condenser, Peltier (Eppendorf, cat. no. 76DXCOND).
6. DASbox overhead drive (DASGIP Eppendorf, cat. no. 76DXOHD).
7. Impeller: 8-blade, 60° pitch; prototype available from Eppendorf on request.
8. Polytetrafluoroethylene (PTFE) membrane inline vent filter; pore size, 0.2 μm (mdi Membrane Technologies, cat. no. ITFX0801BBXX109).
9. EasyFerm Plus 120 pH sensor (Hamilton, cat. no. 238643).
10. OxyFerm FDA 225 (Hamilton, cat. no. 237452).
11. DASGIP compression fitting, for outer diameter (o.d.) 12 mm, w/Pg 13.5 male thread; Eppendorf, cat. no. 76DGCF12).
12. Mohr pinchcock clamp (Carl Roth, cat. no. KY01.1).
13. Female Luer-lock connectors (Carl Roth, cat. no. CT62.1).
14. Male Luer-lock connectors (Carl Roth, cat. no. CT58.1).
15. Silicone tubing, inner diameter (i.d.) 4.8 mm (Watson-Marlow, cat. no. 913.A048.024).
16. Disposable syringes, Omnifix 5 mL with Luer-lock (B. Braun Melsungen, cat. no. 4617053 V).

2.2 Perfusion System

1. Dosage Line C-Flex, with 2× Luer-lock fittings; i.d. 0.8 mm, length 2 m, male/male (Eppendorf, cat. no. 78510310).
2. Pump Head Tubing, Bioprene i.d. 0.5/W 1.05 mm, female/female (Eppendorf, cat. no. 78510118).
3. DASGIP Head Gear for addition bottles with GL45 neck, C-Flex; female Luer-Lock (Eppendorf, cat. no. 78510311).
4. Laboratory glass bottle, 100 mL, GL 45 neck and Laboratory glass bottle, 500 mL, GL 45 neck, sintered glass frit, AD 6 × 15 mm, Por.3, pore size16–40 μm, glass tube AD4 × 0.8 mm, L35 mm (Eppendorf, cat. no. 78903230).

2.3 Monolayer Expansion

1. T25 flasks (TPP, cat. no. 90026).
2. T75 and T175 flask (Greiner Bio-One, cat. nos. 658175 and 660175).
3. Multiflask, three-layer with vented cap (Corning Life Sciences, Art.353143).
4. Centrifuge tube support cushions for 500 mL centrifuge tubes (Corning, #431124) (optional).
5. Centrifuge tubes 500 mL (Corning Life Sciences, CLS431123-36EA) (optional).

6. Conical tubes (15 and 50 mL; Greiner Bio-One, cat. nos. 188271 and 227261).
7. Serological pipettes (5, 10, 25, and 50 mL; cat. nos. 86.1253.001, 86.1254.001, 86.1685.001, and 86.1689.001).
8. Vi-CELL™ XR from Beckman Coulter (optional).

2.4 Culture Medium, Enzymes, and Chemicals

Medium for Stem Cell cultivation and expansion is the E8 preparation according to Chen et al. [9].

1. DMEM/F-12, HEPES (Thermo Fisher Scientific, cat. no. 11330057).
2. Sodium selenite (Sigma-Aldrich, cat. no. S5261).
3. Ascorbic acid 2-phosphate (Sigma-Aldrich, cat. no. A8960).
4. $NaHCO_3$ (Sigma-Aldrich, cat. no. S6041).
5. Insulin (Sigma-Aldrich, cat. no. I9278).
6. Human recombinant Transferrin (Sigma-Aldrich, cat. no. T3705).
7. Recombinant Human FGF-basic (Peprotech, cat. no. 100-18B).
8. TGFβ1 (human recombinant—CHO) (Peprotech, cat. no. 100-21C).
9. Recombinant human Albumin solution (Sigma-Aldrich, cat. no. A6608).
10. Geltrex™ LDEV-Free, hESC-Qualified, Reduced Growth Factor Basement Membrane Matrix (Thermo Fisher Scientific, cat. no. A1413202). To prepare, thaw the Geltrex® vial overnight on ice in the fridge. Spray the thawed vial with 70% ethanol, wipe-dry, and put under laminar air flow. Keep the vial on ice while aliquoting 125 μL per 0.5 mL tube and keep on ice before refreezing aliquots at −80 °C for long-term storage and avoid repeated freeze thaw cycles.
11. Trypan blue solution (0.4%) (Sigma-Aldrich cat. no. T8154-100ML).
12. Solution to dissociate monolayer: StemPro™ Accutase™ Cell Dissociation Reagent (Thermo Fischer Scientific, cat. no. A1110501).
13. Sigmacote (Sigma-Aldrich, cat. no. SL2) use 1 mL to coat 1 bioreactor vessel.
14. 10 mM Rho kinase inhibitor (RI) Y27632 stock solution: 20 mg of Y27632 in 5.88 mL pure water. Filter to sterilize. Aliquot and store at −20 °C for up to 3 months (*see* **Note 1**).
15. Perform® disinfection reagent (Schülke & Mayr GmbH).

16. Terg-a-zyme® enzyme detergent (Sigma-Aldrich, cat. no. Z273287-1EA).
17. PBS 10× w/o Mg^{2+} and Ca^{2+} (Thermo Fischer Scientific, cat. no. 70011051). To prepare, add 50 mL of PBS 10× w/o Mg^{2+} and Ca^{2+} to 450 mL of pure water and autoclave it afterward. The final solution can be stored at room temperature for up to 24 months.
18. Technical buffer, pH 4.01 (WTW, cat. no. 108800).
19. Technical buffer, pH 7.00 (WTW, cat. no. 108802).
20. KCl (Merck Millipore, cat. no. 1.04936.025).
21. Penicillin/Streptomycin (Biochrom, cat. no. A2213) during cultivation in the bioreactor (optional).

3 Methods

3.1 Preparation and Assembly of the Bioreactor

Prepare bioreactor at least 2 days before inoculation of the bioreactor.

1. Siliconize the bioreactor glass vessel under the flow hood. Repeatedly pipet 1 mL of Sigmacote solution to the sides and bottom of the vessel. Keep the vessel at RT overnight (*see* **Note 2**).
2. Examine headplate and DO sensors for damage and replace damaged O-rings if necessary. Make sure a sampling, a waste and a feed port is present as well as an exhaust and inlet air filter in the headplate (*see* **Note 3**).
3. Assemble perfusion circuit. Note that the feed tubes need to be long enough that feed bottles can be refilled under the flow hood, while still attached to the bioreactor.
4. For pump calibration, load the pump head tubing of feed and waste lines into the respective pump heads. Place the tube inlet sides of the dosage lines into a bottle with pure water and prime the lines. Weigh empty canonical tubes with caps. Place one tube outlet of feed or waste line into one tube each. Start dispensation with a defined flow rate that is going to be used during cultivation (e.g., 6.3 mL per h for a cultivation volume of 150 mL). Stop pumping after a defined dispensing time (e.g., at least 60 min). Weigh the filled canonical tubes with caps. With the help of the DASware software by comparing theoretic and actual dosed amount per time unit.

   1 day before inoculation of the bioreactors:
5. Rinse bioreactor vessel thoroughly with water to remove any trace of Sigmacote and cover the bottom of the vessel with pure water to prevent pH probes to dry out during steam autoclaving.

6. Attach impeller to the lowest possible level of the bioreactor shaft.
7. Attach the sintered glass frit with a small piece of tubing to the dip tube and position it directly above the impeller. It will function as a cell retention filter while perfusion. Waste tube need to be attached to the filter.
8. For pH probe calibration, remove pH probe from the storage solution (3 M KCl). Check that the pH probe's diaphragm is not soiled or damaged. Attach the probe to the cable of the bioreactor station. Rinse the sensor with pure water and carefully pat it dry. Immerse the pH and temperature sensor into pH 7.00 calibration buffer. Make sure that the pH sensor diaphragm is completely submerged. Wait until the pH and temperature readings are stable, and subsequently reset the zero value at the bioreactor control unit (calibrate offset). Repeat the previous steps with a pH 4.01 calibration buffer. Wait until the pH and temperature readings are stable, and calibrate the slope. Screw pH probes and DO probes to the headplate of the bioreactor (*see* **Notes 4** and **5**).
9. The perfusion system can be connected to the bioreactor before sterilization. Open headplate screw cap and autoclave the bioreactor for 20 min at 120 °C. Close the headplate screw cap immediately afterward before placing the bioreactor under the flow hood.
10. Replace water with 100 mL PBS w/o and place the bioreactor into the DASbox station. Attach the temperature and DO probe, the gas supply, the agitation, and the exhaust air condenser. Start the DO calibration procedure: switch on agitation, gassing (21% O_2) and temperature (37 °C) set in the software (*see* **Note 6**).
    Before detaching the cells on day 0:
11. Calibrate DO in software (offset 100%), then remove the PBS under the flow hood and replace it by 120 mL E8 + 10 μM Y27632 (addition of Penicillin/Streptomycin optional). Place the bioreactor vessel into the bioreactor station and connect temperature, overhead drive, and gas supply. Start stirring at 60 rpm. Prepare the settings needed for the cultivation of cells in the software (settings in Table 1) (*see* **Note 7**).

3.2 Expansion of Cells as a Monolayer and Inoculation of 4 Bioreactors

1. Obtain a stock of frozen hPSCs, three million cells in one vial (frozen from monolayer cultures) (*see* **Notes 8** and **9**).
   Start 13 days prior to planned bioreactor inoculation (see ***Note 10****).*
2. Coat 1× T25, 2× T75, 3× T175, and four multiflasks with Geltrex: thaw 1375 μL Geltrex into 550 mL DMEM/F12 and mix. Add 5 mL into T25, 15 mL into T75, 35 mL into T175

Table 1
Checklist when inoculating a bioreactor

	Stirring	Temperature	Gassing	pH	DO	Perfusion
Software settings	60 rpm ccw	37 °C	3 sL/h, 21% DO, 5% CO_2 air, O_2, CO_2, N_2 = headspace			
Attach to bioreactor:	Overhead drive	Temperature	Gas headspace Condenser	pH probe	DO probe	
Switch on in software:	Agitation	Temperature	Gassing			

and 100 mL into a multiflask, respectively and incubate at 37 °C for at least 1 h (*see* **Note 11**).

12 days prior to bioreactor inoculation:

3. Thaw one vial in 37 °C water bath until a small ice cube is left. Remove the cell suspension sterilely and dilute it ten times in 4 °C cold E8 + 10 μM Y27632. Centrifuge tubes at 300 × *g* for 3 min. Discard supernatant and gently resuspend in 5 mL E8 + 10 μM Y27632. Remove DMEM/F12 from coated T25 and transfer cell suspension into T25 (*see* **Note 12**).
4. Full medium exchange (ME) 2 days after thawing and seeding cells.

   9 days prior to bioreactor inoculation:
5. Passage the cells into 2× T75 3 days after thawing the cells: Aspirate the medium, wash with 5 mL of PBS, and remove PBS. Add 1 mL of Accutase solution and incubate the plate for 3–4 min at 37 °C. Add 5 mL cold DMEM/F12 (4 °C). Pipet medium gently over the surface to detach cells and transfer suspension into a conical tube to centrifuge at 300 × *g* for 3 min. Discard supernatant, resuspend in 1 mL E8 + 10 μM Y27632 and count cells by using trypan blue cell counting or applying the ViCell (dilute 100 μL sample in 400 μL PBS ($f = 5$) to count with the ViCell device). Remove medium in coated T75 flask and add 15 mL E8 + 10 μM Y27632 in both T75. Seed one T75 with 40,000 cells/cm^2 (3 × 10^6 cells per T75) (*see* **Note 13**).
6. Full ME 2 days after seeding the cells.

   6 days prior to bioreactor inoculation:
7. Passage cells after a total of 3 days into 3× T175 using the same procedure described in **step 5**.
8. ME 2 days after passage of the cells.

   3 days prior to bioreactor inoculation:

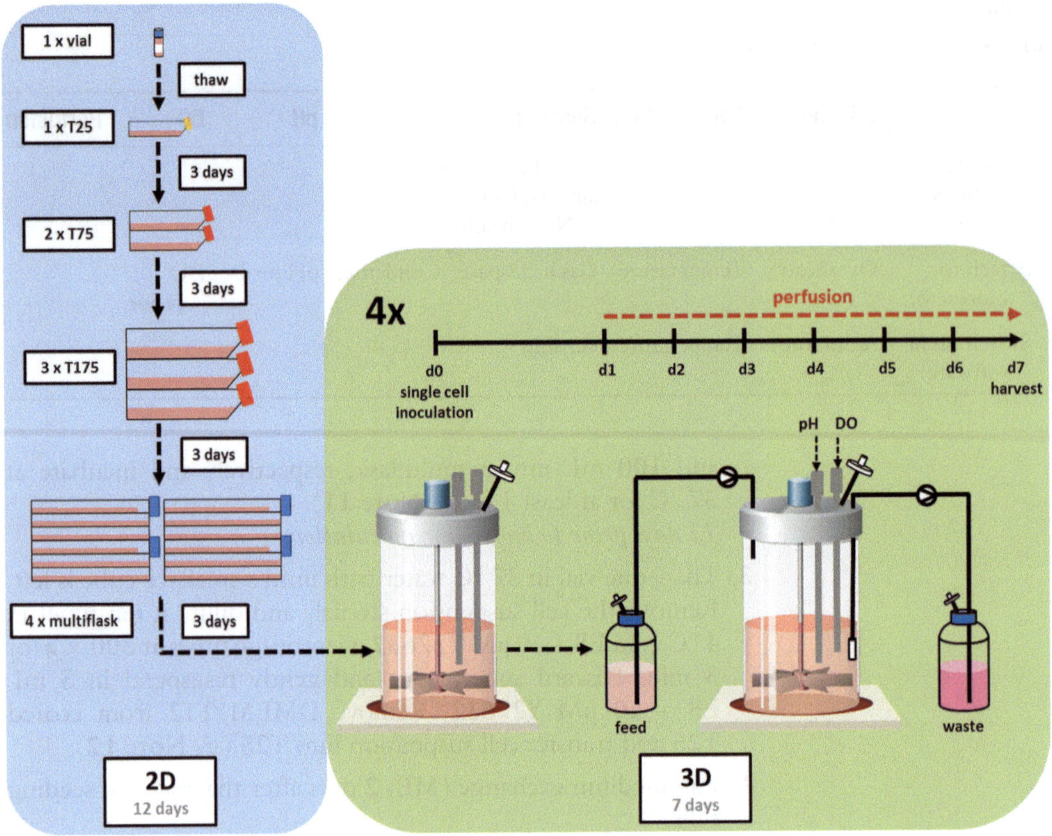

Fig. 1 Schematic illustration of the 12 day hPSC expansion in 2D (blue) followed by 7-day suspension-based expansion (green) in 4 parallel bioreactors at 150 mL process scale. Cells are thawed and sequentially expanded on T-flasks in 2D monolayers. Afterward, cells are detached from monolayer cultures and seeded as single-cell suspensions on day 0 to stirred tank bioreactors. Probes are installed into the bioreactor for measurement and potentially control of pH and dissolved oxygen (DO). During the first 24 h cultures were maintained without any medium exchange to form aggregates. On culture day 1 automated perfusion medium change is started (6.3 mL/h), resulting in one medium exchange per day. On culture day 7 cells can be harvested, analyzed, and further used

9. Same procedure is used to further cultivate the cells in multiflasks and to inoculate up to 4 bioreactors 3 days later (Fig. 1 and Table 2) (*see* **Note 14**).

   Bioreactor inoculation:

10. After preparing the medium and the cultivation settings in the software for each bioreactor, cells can be detached from all four multiflasks as described in **step 5** for passaging the cells. (Necessary volumes can be found in Table 2.) Resuspend in 50 mL E8 + 10 μM Y27632. Take three separate samples to count the cells according to standard counting procedures or via the ViCell. Calculate the average cell density and add cell suspension to the bioreactor in order to seed 75×10^6 cells in one

Table 2
Specifications for monolayer expansion flasks

Flask	Culture volume (mL)	PBS for washing (mL)	Accutase (mL)	DMEM/F12 to dilute Accutase (mL)	Geltrex in DMEM/F12 (mL)	Cells (10^6)
T25	5	3	1	5	5	1
T75	15	9	3	15	15	3
T175	35	21	7	35	35	7
Multiflask three layers	100	60	21	100	100	21

bioreactor. Add medium (E8 + 10 μM Y27632) to reach a total volume of 150 mL. Final cell density: 0.5×10^6 cells/mL (*see* **Note 15**).

11. Place the bioreactor back into the DASBox station and connect all probes, gas supply, overhead drive, and the condenser. Start the inoculation time clock in the software.
12. Make sure everything is attached and switched on properly—*see* checklist in Table 1.
13. Take a first sample after around 30 min of inoculation to count and confirm the inoculation cell density. Sample procedure is explained below. No enzymatic dissociation is necessary at this time point. Sample can be placed into ViCell without any dilutions (*see* **Note 16**).

3.3 Sampling of Bioreactor

1. Sampling of bioreactor: remove cap from bioreactor sample port and spray with 70% ethanol.
2. Connect a sterile Luer-lock syringe to the valve and open the Mohr pinchcock clamp of the sampling tube while slowly starting to draw suspension with the syringe. Draw ~1 mL of suspension into the syringe and close the clamp immediately. After the syringe has been unplugged, the 1 mL volume can be discarded and another 2.5 mL can be pulled into the syringe. Place the 2.5 mL into a well of a low attachment 6 well plate. Clean the sample port with 70% ethanol before closing the cap (*see* **Notes 17** and **18**).
3. Microscopic assessment of the aggregates is now possible before gathering aggregates in the middle of the well by gentle shaking. Transfer 2 mL of aggregate solution into centrifuge tubes. Cells can be centrifuged at $300 \times g$ for 3 min. Supernatant can be frozen at −20 °C to be used for metabolic analysis such as glucose and lactate measurement.

Table 3
Checklist when starting the perfusion after 24 h

	Stirring	Temperature	Gassing	pH	DO	Perfusion
Software settings	60 rpm ccw	37 °C	3 sL/h, 21% DO, 5% CO_2 air, O_2, CO_2, N_2 = headspace			6.3 mL/h
Attach to bioreactor:	Overhead drive	Temperature	Gas headspace Condenser	pH probe	DO probe	
Switch on in software:	Agitation	Temperature	Gassing			Waste/feed pump

4. Wash pellet with 1 mL PBS w/o, centrifuge at 300 × *g* for another 3 min before adding 500 μL Accutase. Incubate the mixture at 37 °C for 3–4 min until aggregates are completely dissociated into single cells. Gently shake the mixture every 1 min. Centrifuge at 300 × *g* for 3 min. Discard the supernatant and resuspend cells in 2 mL PBS to count using the ViCell. Cells can then be used for additional analysis as needed.
5. To monitor the cultivation of hPSCs, samples should be taken every 24 h.

3.4 Expansion of hPSCs in Bioreactor Using Perfusion Mode

1. After the first 23 h of cultivation pipet 150 mL of E8 medium into the feed bottle in a sterile way (*see* **Note 19**).
2. Prime the feed tubing by switching the pump on, and press "shot" for 5 mL at 40 mL/min.
3. Start the perfusion 24 h after inoculation by starting the feed and waste pump with a pump speed of 6.3 mL/h. Make a mark at the bioreactor vessel to check whether the culture volume stays constant for 24 h (*see* **Note 20**).
4. Make sure everything is attached and switched on properly—*see* checklist in Table 3.
5. After 24 h of perfusion check the culture volume and adapt pumps if necessary (*see* **Note 21**).
6. Refill the feed bottle with fresh medium and do so every 24 h.

3.5 Cleaning of Bioreactor

1. Under the flow hood: Remove cell suspension. Cells can be processed as required.
2. Add 150 mL of 10 g/L Perform disinfection solution place the bioreactor back in the bioreactor station, set the stirring at 150 rpm and temperature at 37 °C and incubate for 30 min. Rinse the sample port, feeding and waste tubes with perform disinfection solution.

3. Rinse the bioreactor and sample port, feeding and waste tubes with water.
4. Remove the pH and DO probes to wash them with distilled water outside of the flow hood.
5. Clean sample port, feeding, and waste tubes with 70% ethanol before rinsing the tubes several times with distilled water.
6. Clean glass vessel with 70% ethanol and rinse.
7. Incubate the sintered glass frit in Terg-a-zyme® enzyme detergent for 24 h at 37 °C, and afterward rinse with water.

4 Notes

1. Y27632 is toxic. Avoid skin contact, ingestion, and inhalation. Wear protective gloves and eye protection when preparing the stock solution.
2. Sigmacote is harmful if swallowed or inhaled. Wear protective gloves and work under the fume hood.
3. Examine the membrane of DO sensors for damage and check whether the volume of electrolyte solution is sufficient.
4. When attaching the DO probes, the lowest point of the probe should end slightly above the impeller to ensure a more homogenous aggregation. If the probe is too low, the impeller can be damaged by the probe.
5. Screw caps on sensors to protect the connections and wrap vent filters with aluminum foil before autoclaving.
6. There are two possible gas supply tubings in a DASbox station: head space/overlay and submersed. During DO calibration, the submersed gas supply tube has to be used!
7. The E8 volume prepared in the bioreactor depends on the planned working volume. When aiming to work with 150 mL bioreactors, an initial addition of 120 mL is advisable. When aiming at 100 mL working volume, no more than 80 mL medium should be prepared in each bioreactor.
8. The cell lines should be regularly checked for authenticity, possible karyotype abnormalities and mycoplasma infections.
9. Alternatively it is possible to start with hPSCs maintained on irradiation-inactivated murine embryonic fibroblasts (MEFs).
10. Expansion of pluripotent cells is calculated for an inoculation of 4 bioreactors with 150 mL working volume. Expansion time can be shortened when working with less medium or bioreactors. Rule of thumb: 1× multiflask produces enough cells for one 150 mL bioreactor. 2× T175 are enough for one bioreactor with a 100 mL working volume.

11. Make sure that the medium is distributed evenly into all three layers of multiflask as described in the manual of the multiflasks. Coated flasks can be kept in the fridge for a period of up to 4 weeks.
12. Coated T25, T75, T175 and multiflask should not be at RT before use.
13. Do not leave Accutase longer than 6 min on the cells! Prolonged incubation with Accutase can be detrimental to the cells.
14. Beware when seeding multiflasks to homogenously distribute the cells throughout the three layers. Medium can be added and emptied by pouring. Be careful not to touch waste bottles when pouring medium out to ensure sterility of the multiflasks.
15. Optional use of a 500 mL centrifuge tube. It improves the speed and efficiency during detachment and inoculation of 4 bioreactors. Do not resuspend in more than 50 mL and measure the exact volume after resuspension in order to know the exact number of cells in the suspension.
16. The cell density in the bioreactor should lie between 0.45 and 0.55×10^6 cells/mL. If the cell density deviates too much from 0.5×10^6 cells/mL, it might be necessary to correct it by addition and/or removal of culture medium and cell suspension from the bioreactor.
17. Check whether sampling of the bioreactors may have removed tubings for gassing or temperature probes.
18. Be aware that each sampling removes a total of 3.5 mL out of the bioreactor. Therefore, the number of samples should be limited.
19. Feed bottles can then be refilled under the flow hood. If the sterile hood is too far from the DASbox station, clamp the tubing before detaching it from feed bottle by unscrewing the Luer-lock and place the ending of the tubing in 70% ethanol. Then place the feed bottle below the flow hood and (re)fill it with 150 mL medium.
20. Make sure the pumps are functioning properly and that the pump direction is correct.
21. In case of a low culture volume, check the feed bottle to evaluate whether the feed is too slow or the waste too fast. Refill the volume by shot feeding and adapt the pump speed. For the case of a high culture volume enhance the speed of the waste pump and perfume a shoot with 40 mL/h until the right volume is reached and adapt the pump speed.

Acknowledgments

This work was supported by StemBANCC (support from the Innovative Medicines Initiative joint undertaking under grant 115439-2, whose resources are composed of financial contribution from the European Union [FP7/2007-2013] and EFPIA companies' in-kind contribution). RZ received funding from the German Research Foundation (DFG; including the Cluster of Excellence REBIRTH EXC 62/2 and grants ZW64/4-1, KFO311, and ZW64/7-1), the German Ministry for Education and Science (BMBF; grants: 13 N14086, 01EK1601A, and 01EK1602A) and by the European Union (TECHNOBEAT; grant 66724).

References

1. Inoue H, Nagata N, Kurokawa H, Yamanaka S (2014) iPS cells: a game changer for future medicine. EMBO J 33:409–417
2. Kempf H, Andree B, Zweigerdt R (2016) Largescale production of human pluripotent stem cell derived cardiomyocytes. Adv Drug Deliv Rev 96:18–30
3. Serra M, Brito C, Correia C, Alves PM (2012) Process engineering of human pluripotent stem cells for clinical application. Trends Biotechnol 30:350–359
4. Zweigerdt R (2009) Large scale production of stem cells and their derivatives. Adv Biochem Eng Biotechnol 114:201–235
5. Serra M, Brito C, Sousa MF, Jensen J, Tostões R, Clemente J et al (2010) Improving expansion of pluripotent human embryonic stem cells in perfused bioreactors through oxygen control. J Biotechnol 148:208–215
6. Kropp C, Kempf H, Halloin C, Robles-Diaz D, Franke A, Scheper T et al (2016) Impact of feeding strategies on the scalable expansion of human pluripotent stem cells in single-use stirred tank bioreactors. Stem Cells Transl Med 5:1289–1301
7. Kempf H, Olmer R, Kropp C, Ruckert M, Jara-Avaca M, Robles-Diaz D et al (2014) Controlling expansion and cardiomyogenic differentiation of human pluripotent stem cells in scalable suspension culture. Stem Cell Reports 6:1132–1146
8. Olmer R, Engels L, Usman A, Menke S, Malik MNH, Pessler F et al (2018) Differentiation of human pluripotent stem cells into functional endothelial cells in scalable suspension culture. Stem Cell Reports 5:1657–1672
9. Chen G, Gulbranson DR, Hou Z, Bolin JM, Ruotti V, Probasco MD et al (2011) Chemically defined conditions for human iPSC derivation and culture. Nat Methods 8:424–429

Chapter 8

A Protocol for One-Step Differentiation of Human Induced Pluripotent Stem Cells into Mature Podocytes

Cormac Murphy, Elisabeth Feifel, Paul Jennings, Gerhard Gstraunthaler, and Anja Wilmes

Abstract

Within the glomerulus, podocytes are highly specialized visceral epithelial cells that are part of the glomerular filtration barrier. Human podocyte cell culture is rather challenging for primary or immortalized cells, due to the nonproliferative state of the cells. In addition, rapid dedifferentiation is often observed. Hence, iPSC-derived podocytes offer an exciting alternative to culture podocyte-like cells from different donors over prolonged time. Here we report a simple and rapid one-step protocol that drives iPSC into podocyte-like cells in 10 days.

Key words Kidney, Podocyte, Differentiation, Human induced pluripotent cells, Pericytes

1 Introduction

Podocytes, specialized pericytes with distinctive foot-like projections, are essential components of the glomerular filtration barrier (GFB) of the nephron [1]. They contribute to both, the structure and function of the GFB and also to the production of factors that assist with the growth and maintenance of components of the GFB [2]. They have also developed a reputation as the "weak link" of the glomerulus [3] and the loss or injury of podocytes can result in a breakdown of the GFB and leakage of excessive amounts of protein in the urine, also known as proteinuria [4]. Podocyte loss is an important factor that may contribute to chronic and end-stage kidney disease. While injury to podocytes has been linked to multiple causes such as genetic conditions [5], diabetes [6], and exposure to toxic compounds [7], the mechanisms of toxicity remain poorly understood [8]. This is in part due to the difficulty of culturing mature terminally differentiated podocytes in in vitro cultures [9–11]. There is one conditionally immortalized human

Carl-Fredrik Mandenius and James A. Ross (eds.), *Cell-Based Assays Using iPSCs for Drug Development and Testing*, Methods in Molecular Biology, vol. 1994, https://doi.org/10.1007/978-1-4939-9477-9_8,
© Springer Science+Business Media, LLC, part of Springer Nature 2019

podocyte cell line available, but these cells have shown phenotypic and behavioral variations that differ from in vivo podocytes [12].

Induced pluripotent stem cells (iPSC), with their capacity for differentiation and self-renewal [10] offer a potential long-lasting source of patient-derived cells. These can be differentiated into podocyte-like cells with phenotypic characteristics that closely resemble in vivo counterparts. Recently, several differentiation protocols to drive iPSCs into podocytes have been developed [13–17]. Many of these protocols are quite complex and long and use high amounts of fetal bovine serum (FBS). Here, we report a simple and fast protocol [18], using a one-step method with low concentrations of FBS (1.25%), to differentiate iPSC into mature functional podocytes over the course of 10 days.

2 Materials

Ensure that all experiments are carried out under good laboratory practice (GLP) conditions. All cell culture work has to be under sterile conditions.

2.1 iPS Cell Culture Media

1. For the cultivation of iPSC cultures mTeSR™1 cell culture medium and Geltrex® coating (A1413302, ThermoFisher) is used. Store mTeSR™1 basal medium (StemCell Technologies, #85851) at 4 °C and mTeSR™1 5× Supplement (#85852) at −20 °C.
2. Complete medium is prepared by thawing 100 ml of mTeSR™1 5× supplement and adding to 400 ml of mTeSR™1 basal medium. Mix thoroughly.
3. Complete medium can be stored for up to 14 days at 4 °C or be aliquoted and stored at −20 °C for up to 6 months.
4. Geltrex® coating is prepared according to manufacturer's instruction at 0.08 mg/ml. Coated plates are prewarmed to 37 °C before use.
5. iPSC cultures are passaged twice a week with Versene (17-711, Lonza) at a split ratio of 1:3–1:8.

2.2 iPSC Lines

1. Human iPSCs were obtained from the IMI-funded StemBANCC project (http://stembancc.org) [19]. The following iPSC lines were tested and differentiated into podocyte-like cells: SBAD3 clone 1, SBAD2 clone 1, SBNEO clone 1, SFC086-03-01, SFC800-03-03, and SFC018-03-01.

2.3 Podocyte Differentiation Cell Culture Media

1. Geltrex® coated plates (A1413302, ThermoFisher) are required as described above.
2. For the differentiation, iPSCs are split with Accutase® (GIBCO, A11105-01).

3. The differentiation medium consists of: DMEM/HAM-F-12 (GIBCO cat. No. 31330-038), 1.25% FBS (GIBCO, 10270), 100 μM nonessential amino acids (NEAA, 11140-050), 15 ng/ml BMP7 (PHC9544), 10 ng/ml activin A (PHC9564, Thermo Fisher Scientific), 100 nM retinoic acid (R2625, Sigma), and 100 U/ml penicillin, 0.1 mg/ml streptomycin (P4333, Sigma). DMEM/Ham-F12 supplemented with NEAA 100 μM, 1.25% FBS, and 100 U/ml penicillin, 0.1 mg/ml streptomycin can be mixed in advance and kept at 4 °C for up to 2 months.
4. Growth factors (BMP4, activin A, and retinoic acid) should be prepared freshly for every feeding cycle. These should be reconstituted according to the manufacturer's recommendations and frozen as stock solutions in aliquots (*see* **Note 1**).
5. For initial seeding, 5 μM Rho-associated kinase (ROCK) inhibitor Y-27632 (Abcam, cat. No. ab120129) is added for 24 h.

2.4 Immunofluorescent Microscopy

1. The following antibodies for immunocytochemistry are recommended: anti-synaptopodin (Abcam ab224491, 1:75) and WT-1 (R&D Systems, AF5729, 1:100), for 30 min. Secondary antibodies Alexa Fluor 568 (Goat: A-11056, Rabbit: A-10040, Thermofisher, 1:250).
2. For F-actin staining, phalloidin is used (ActinGreen™ 488, GeneCopoeia, Rockville, MD).

3 Methods

3.1 iPSC Cultivation

1. Warm up only the required amount of mTeSR™1 cell culture medium for each feeding cycle at 37 °C. Aspirate old medium from cultures and apply fresh medium. iPSC cultures should be fed every day unless cultures are maintained on coverslips as previously described [20].

3.2 Podocyte Differentiation

1. Prepare podocyte differentiation medium (PDM), by adding the growth factors (15 ng/ml BMP7, 10 ng/ml activin A, and 100 nM retinoic acid) to the podocyte culture medium (PCM) (DMEM/Ham-F12, NEAA 100 μM, 1.25% FBS, and 100 U/ml penicillin, 0.1 mg/ml streptomycin).
2. Have a Geltrex®-coated plate ready at 37 °C. The following volumes are based on differentiation in 6-well plates and should be adjusted if other well-formats are used.
3. Aspirate medium from one 6 well of iPSCs that are at a maximum of 65% confluency (as shown in Fig. 1) and wash once with 1 ml PBS (*see* **Note 2**).

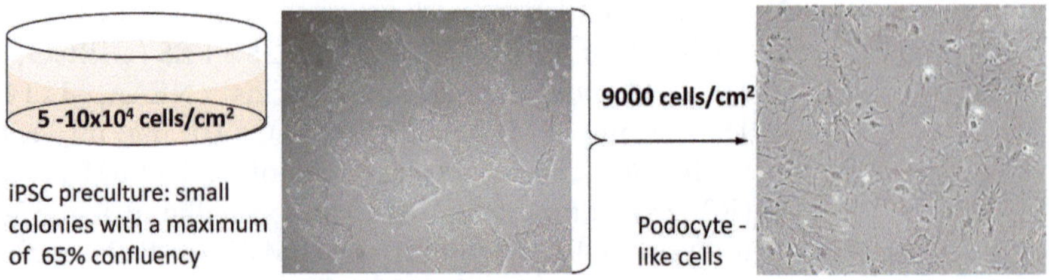

Fig. 1 Optimal seeding densities of a typical differentiation of iPSCs into podocyte-like cells

4. Add 400 μl of Accutase®, incubate for approximately 3–4 min then carefully drain before the cells detach (*see* **Note 3**).
5. Gently suspend cells using 1 ml of the PDM, pipet up and down only two times (*see* **Note 4**).
6. Count cells quickly, calculate for 9000 cells/cm^2 in half of the final culture volume (e.g., 1 ml per 6 well) (*see* **Notes 5** and **6**).
7. Add Rock inhibitor (to a final concentration of 5 μM) to the resuspended cells, volume head over head twice, and spread on the prepared Geltrex®-coated plates.
8. Perform back and forward and left and right gentle shaking movements instead of rotation before placing the plates into the incubator to avoid uneven concentric distribution of cells.
9. After 3–5 h, when most cells have visibly attached, add the remaining medium carefully to make up the final culture volume (e.g., topped up to 2 ml in a 6 well).
10. After 24 h (on day 1) change to PDM without Rock inhibitor.
11. Feed cells on days 3, 6, and 9 with PDM.
12. From day 10 onward, change to PCM without the growth factors and feed every 2–3 days. Podocyte-like cells can be kept in culture for 3–4 weeks (*see* **Note 7**).

3.3 Characterization of Podocyte Markers

1. Characteristic foot processes are a marker of podocytes and can be observed in phase contrast microscopy (Fig. 2).
2. For immunofluorescence staining of podocytes markers (Fig. 3), cells should be differentiated over 10 days on sterile glass coverslips, coated with Geltrex®.
3. On day 10, wash podocyte-like cells three times with PBS and fix them by incubating with 4% paraformaldehyde for 10 min at room temperature.
4. Wash the cells three times with PBS. At this point the fixed cells can be covered with PBS and kept at 2–8 °C.
5. Incubate fixed cells for 30 min in blocking buffer (PBS with 5% BSA, 1% Triton X-100).

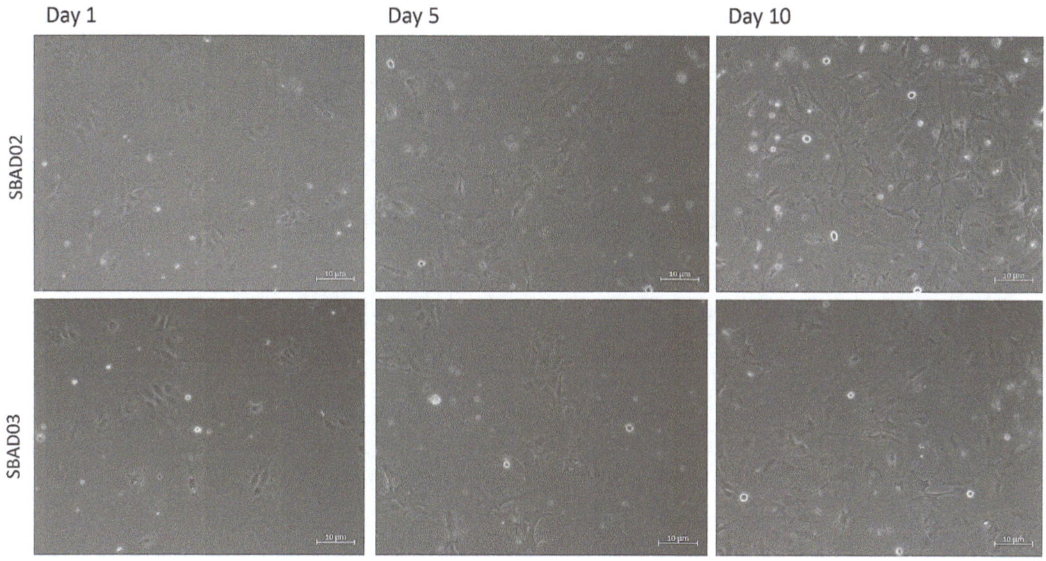

Fig. 2 Differentiation of iPSCs from the iPSC lines SBAD2 and SBAD3 into podocyte-like cells over the course of 10 days

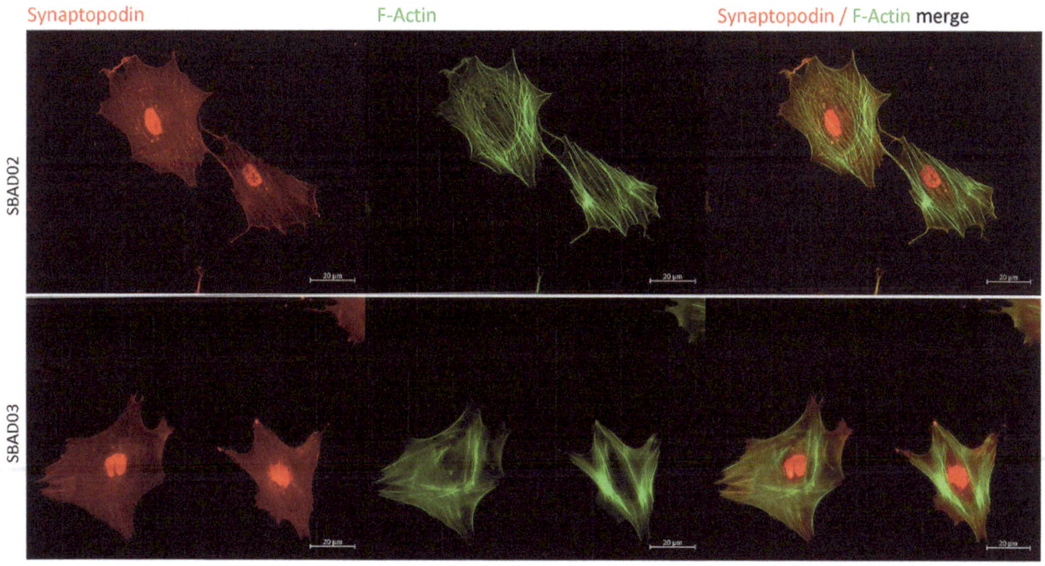

Fig. 3 (**a**) Immunostaining of podocyte-like cells derived from the iPSC lines SBAD2 and SBAD3 with synaptopodin and F-actin staining. iPSC were differentiated on glass coverslips, fixed and stained for synaptopodin and F-actin. Red and green colors were applied postcapture. (**b**) Immunostaining of podocyte-like cells derived from the iPSC lines SBAD2 and SBAD3 with WT1 and F-actin staining. iPSC were differentiated on glass coverslips, fixed and stained for WT1 and F-actin. Red and green colors were applied postcapture

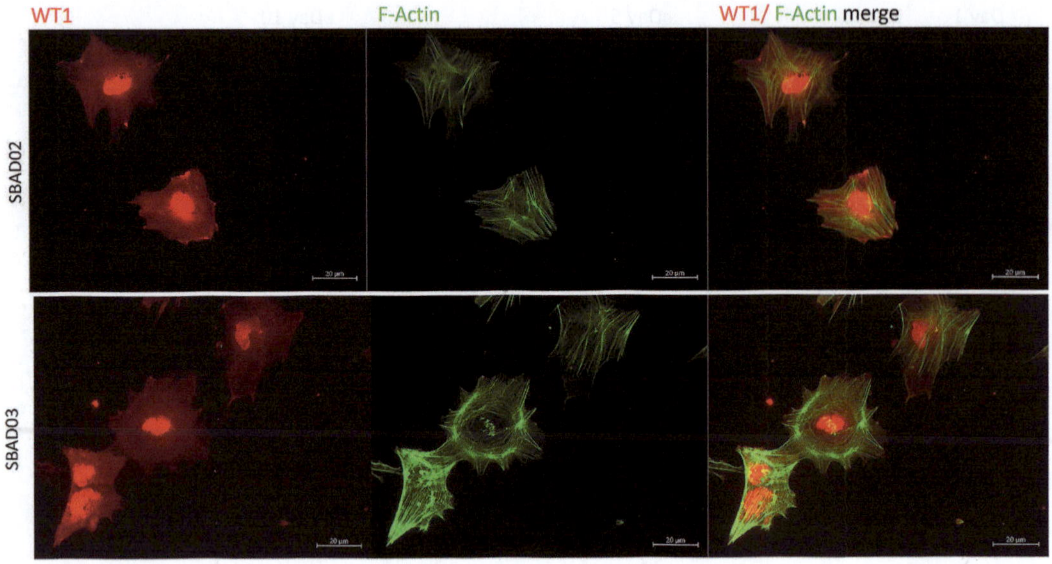

Fig. 3 (continued)

6. Remove the blocking buffer and incubate cells with primary antibodies diluted in 10% blocking buffer (diluted in PBS) at RT for 30 min. The following antibodies were tested and are recommended: anti-synaptopodin (Abcam ab224491) at a dilution of 1:75, and WT1 at a dilution of 1:100 (R&D Systems, AF5729). After incubation, wash three times with PBS.
7. Incubate the cells with the appropriate secondary antibody Alexa Fluor 568 (Thermofisher; Goat: A-11056, Rabbit: A-10040) at a dilution of 1:250 and add 1 drop of Actin-Green™ 488 stain (GeneCopoeia, Rockville, MD) in 10% blocking buffer (diluted in PBS). Wash three times in PBS and mount to coverslips.

4 Notes

1. Avoid freeze-thaw cycles of growth factors (BMP7, activin A, and retinoic acid).
2. The confluency of the iPSCs from the original plate at the time of passaging is very important for the quality and yield of podocyte-like cells.
3. Acctuase® should be left on the cells long enough that they clearly appear to be detaching from each other but removed before they detach from the plate.
4. The iPSCs should be resuspended gently, disrupting them as little as possible. This will allow for better cell attachment.

5. 9000 cells/cm^2 works well for 10 cm^2 and 6 well plates, but should be tested and adjusted for smaller formats
6. For the initial seeding volume, using half of the final amount of medium is recommended to aid attachment and to prevent initial clustering. If using 96-well formats or smaller, it is advised to seed the cells in the full amount of medium.
7. From day 10 of the differentiation, the podocytes are mature and nonproliferative.

References

1. Greka A, Mundel P (2012) Cell biology and pathology of podocytes. Annu Rev Physiol 74:299–323
2. Scott RP, Quaggin SE (2015) The cell biology of renal filtration. J Cell Biol 209:199–210
3. Delézay O, He Z, Hodin S et al (2017) Glomerular filtration drug injury: in vitro evaluation of functional and morphological podocyte perturbations. Exp Cell Res 361:300–307
4. Shankland SJ (2006) The podocyte's response to injury: role in proteinuria and glomerulosclerosis. Kidney Int 69:2131–2147
5. Chugh SS (2007) Transcriptional regulation of podocyte disease. Transl Res 149:237–242
6. Li JJ, Kwak SJ, Jung DS et al (2007) Podocyte biology in diabetic nephropathy. Kidney Int 72:s36–s42
7. Hirschberg R (2012) Renal complications from bisphosphonate treatment. Curr Opin Support Palliat Care 6:342–347
8. Haley KE, Kronenberg NM, Liehm P et al (2018) Podocyte injury elicits loss and recovery of cellular forces. Sci Adv 4(6):eaap8030
9. Ni L, Saleem M, Mathieson PW (2012) Podocyte culture: tricks of the trade. Nephrology (Carlton) 17:525–531
10. Shankland SJ, Pippin JW, Reiser J et al (2007) Podocytes in culture: past, present, and future. Kidney Int 72:26–36
11. Krtil J, Pláteník J, Kazderová M et al (2007) Culture methods of glomerular podocytes. Kidney Blood Press Res 30:162–174
12. Saleem MA, O'Hare MJ, Reiser J et al (2002) A conditionally immortalized human podocyte cell line demonstrating nephrin and podocin expression. J Am Soc Nephrol 13:630–638
13. Sharmin S, Taguchi A, Kaku Y et al (2016) Human induced pluripotent stem cell-derived Podocytes mature into vascularized glomeruli upon experimental transplantation. J Am Soc Nephrol 27:1778–1791
14. Starokozhko V, Hemmingsen M, Larsen L et al (2018) Differentiation of human induced pluripotent stem cell under flow conditions to mature hepatocytes for liver tissue engineering. J Tissue Eng Regen Med 12:1273–1284
15. Musah S, Mammoto A, Ferrante TC et al (2017) Mature induced-pluripotent-stem-cell-derived human podocytes reconstitute kidney glomerular-capillary-wall function on a chip. Nat Biomed Eng 1:0069
16. Song B, Smink AM, Jones CV et al (2012) The directed differentiation of human iPS cells into kidney podocytes. PLoS One 7:e46453
17. Ciampi O, Iacone R, Longaretti L et al (2016) Generation of functional podocytes from human induced pluripotent stem cells. Stem Cell Res 17:130–139
18. Rauch C, Feifel E, Kern G et al (2018) Differentiation of human iPSCs into functional podocytes. PLoS One 13(9):e0203869
19. Morrison M, Klein C, Clemann N et al (2015) StemBANCC: governing access to material and data in a large stem cell research consortium. Stem Cell Rev 5:681–687
20. Wilmes A, Rauch C, Carta G et al (2017) Towards optimisation of induced pluripotent cell culture: extracellular acidification results in growth arrest of iPSC prior to nutrient exhaustion. Toxicol In Vitro 45:445–454

[illegible]

Chapter 9

Differentiation of hiPS Cells into Definitive Endoderm for High-Throughput Screening

Teresa Bluhmki, Stefanie Traub, Eva Schruf, James Garnett, Florian Gantner, Daniel Bischoff, and Ralf Heilker

Abstract

In drug discovery, there is an increasing demand for more physiological in vitro models that recapitulate the disease situation in patients. Human induced pluripotent stem (hiPS) cell-derived model cells could serve this purpose. To date, several directed differentiation approaches have been described to generate definitive endoderm (DE) from hiPS cells, but protocols suitable for drug development and high-throughput screening (HTS) have not been reported yet. In this work, a large-scale expansion of hiPS cells for high-throughput adaption is presented and an optimized stepwise differentiation of hiPS cells into DE cells is described. The produced DE cells were demonstrated to express classical DE markers on the gene expression and protein level. The here described DE cells are multipotent progenitors and act as starting points for a broad spectrum of endodermal model cells in HTS and other areas of drug discovery.

Key words hiPS cells, Definitive endoderm, Multipotent progenitor cell, Drug discovery, High-throughput-screening

1 Introduction

In the past, the random testing of large compound collections in HTS assays profited from the robustness and the typically large signal windows of target-overexpressing model cells. However, such artificial assays lack the physiological context of the disease tissue, so that more physiological approaches have experienced a broad renaissance in recent years. In parallel, the technical basis for more physiological high-throughput formats has improved over the past decade through the development of model systems based upon induced pluripotent stem (iPS) cells [1]. For instance, hiPS cell-derived neural progenitors were expanded in sufficient numbers to meet the needs of an HTS campaign on neuronal model cells [2, 3]. Likewise, hiPS cells could serve as a basis for HTS campaigns using model cells derived from the endodermal germ

Carl-Fredrik Mandenius and James A. Ross (eds.), *Cell-Based Assays Using iPSCs for Drug Development and Testing*, Methods in Molecular Biology, vol. 1994, https://doi.org/10.1007/978-1-4939-9477-9_9,
© Springer Science+Business Media, LLC, part of Springer Nature 2019

layer. In the latter context, this work describes the large-scale expansion of hiPS cells for high-throughput applications and the subsequent differentiation of the hiPS cells into DE cells.

During human embryonic development the gastrulation results in the production of all three principal germ layers: mesoderm, ectoderm, and endoderm [4]. Subsequent development and patterning of the DE leads to the formation of various major organs of the human body including the lungs, liver, intestines, thyroid gland, and pancreas [5, 6]. In consequence, an upscaled production of DE cells out of hiPS cells could serve as a basis for subsequent differentiation toward model cells representing the above described tissues [7]. Efficient production of DE cells in vitro needs highly regulated and coordinated interactions of different transcription factors and molecular signals [8]. The underlying signaling pathways have led to the introduction of key components such as activin A [9, 10], Wnt, and BMP [11–13] into the published DE differentiation protocols [8, 11]. The accordingly generated DE cells can be characterized by the expression of specific DE markers such as SOX17, CXCR4, EpCAM [14–16], and by the loss of their initial pluripotency markers such as OCT4. The here described protocol (Figs. 1 and 2) to produce large numbers of hiPS cell-derived DE cells constitutes a promising basis for the development of highly physiological in vitro models in the context of HTS. Apart from HTS applications, the directed differentiation of these DE cells into a variety of different cell types of the human body opens a powerful tool for regenerative medicine, cell replacement therapy, and basic research [17].

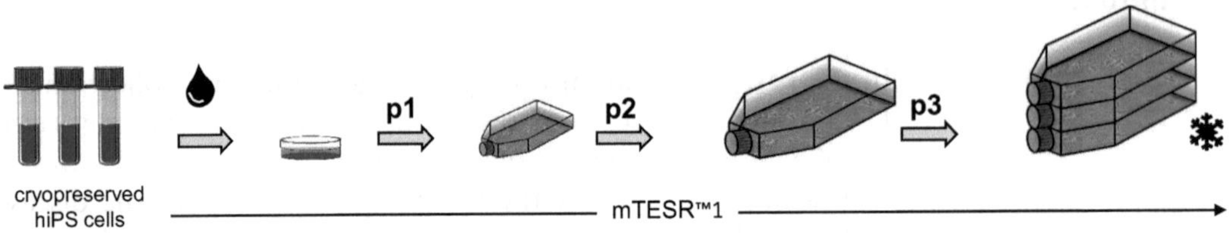

days	1-4	4-7	7-10	10-14
culture ware	3.5 cm dish	T-75 flask	T-175 flask	3 x T-175 flasks
hiPS cell harvest	1 x 10^6	1 x 10^7	3 x 10^7	9 x 10^7

Fig. 1 Process overview: large-scale expansion of hiPS cells for high-throughput adaption. Scheme of hiPS cell expansion. During large-scale cell expansion, the number of hiPS cells increased over 14 days from 1×10^6 to 9×10^7 cells, covering the scope of an HTS campaign. The expanded hiPS cells were frozen at this stage in aliquots of ~ 5×10^6 cells per vial for subsequent thawing and maintenance in T-175 flasks

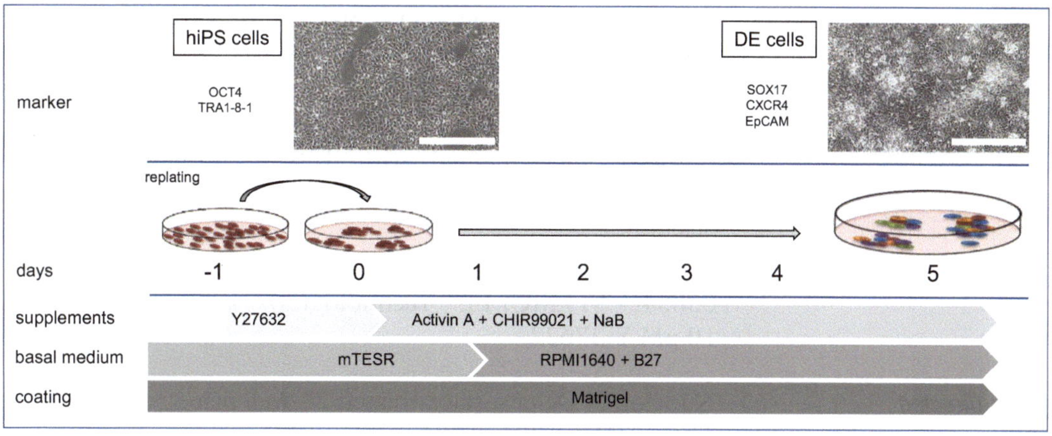

Fig. 2 Scheme of stepwise differentiation protocol toward hiPS-derived DE cells. Optimized differentiation of pluripotent stem cells to DE cells based upon and adapted from literature, for instance, Gotoh et al. [12]: scheme of the stepwise differentiation triggered through various proteins and small molecules. Previously published protocols had to be modified for the selected hiPS cell line: coating (Matrigel), seeding density (0.2×10^6 cells per cm^2) and concentrations of NaB (0.5 mM) were adjusted to get optimal conditions for DE differentiation. The hiPS cells were maintained on Matrigel coated culture dishes in mTeSR™1 medium until initiating differentiation. At day 0, the cells were replated on Matrigel coated plates and seeded in an appropriate cell density (0.2×10^6 cells per cm^2). The adapted differentiation protocol forces the hiPS cells toward endoderm differentiation through activation of WNT signaling with the protein activin A and inhibition of the glycogen synthase kinase 3β (GSK-3β). After 5 days of differentiation with daily medium change the cells were evaluated by RT-PCR, FACS and immunofluorescence for their expression of SOX17, EpCAM and CXCR4 known as classical DE markers. Scale bar = 400 μm; Y27632: ROCK inhibitor, CHIR99021: GSK3β inhibitor, NaB: Sodium Butyrate

2 Materials

2.1 Human iPS Cell Culture

1. Human iPS cell line SFC-086-03-01, generated and characterized within StemBANCC; *see* EBiSC website: https://cells.ebisc.org/STBCi052-A/.
2. 1× Dulbecco's Phosphate buffered saline (1× DPBS; catalog number (cat#) 14190-144; Thermo Fisher Scientific, Waltham, MA).
3. T175 Nunclon™ Delta surface-treated EasYFlasks (cat# 178883; Thermo Fisher Scientific).
4. Matrigel™ (cat# 354277; Corning Inc., Corning, NY).
5. UltraPure™ EDTA solution (Ethylenediaminetetraacetic acid; cat# 15575020; Thermo Fisher Scientific), 0.5 M stock solution: dilute to 5 mM in 1× DPBS. Store aliquots of 50 mL at 4 °C for up to 4 weeks.
6. Rho-associated, coiled-coil containing protein kinase (ROCK) inhibitor Y-27632 (cat# Y0503-5MG; Sigma-Aldrich; St. Louis, MO), 10 mM stock solution: Dissolve 5 mg of

Y-27632 Rock inhibitor in 1.563 mL of sterile water. Store aliquots of 500 μL at 4 °C for up to 4 weeks.

7. Human iPS cell maintenance medium: mTeSR™1 complete medium (cat# 05850; STEMCELL Technologies; Vancouver, Canada), supplemented with 100 U/mL penicillin and 100 μg/mL streptomycin (cat# 15140-122; Thermo Fisher Scientific).
8. Human iPS cell cryopreservation medium: mFreSR™ (cat# 05854; STEMCELL Technologies), supplemented with 10 μM Y-27632.

2.2 Directed Differentiation to DE Cells

1. Nunclon™ Delta surface-treated 24-well plates (cat# 142475; Thermo Fisher Scientific).
2. Accutase (cat# A6964; Sigma-Aldrich).
3. Activin A (cat# 338-AC-050; R&D Systems; Minneapolis, MN), 100 μg/mL stock solution: Dissolve 500 μg in 500 μL sterile 1× DPBS/0.1% BSA. Store aliquots of 50 μL at −20 ° C for up to 1 year. Once thawed, keep aliquots at 4 °C for up to 1 week (*see* **Note 7**).
4. CHIR99021 (cat# CT-99021; Axon Medchem; Groningen; Netherlands), 3 mM stock solution: Dissolve 5 mg of CHIR99021 in 3.581 mL of DMSO. Store aliquots of 50 μL at −20 °C for up to 1 year. Once thawed, keep aliquots at 4 °C for up to 1 week.
5. Natrium butyrate (NaB; cat# 303410-100G; Sigma Aldrich), 25 mM stock solution: Dissolve 100 mg of natrium butyrate in 36.333 mL of sterile water. Store at 4 °C for up to 2 months.
6. Seeding Medium: Human iPS cell maintenance medium supplemented with 10 μM of Y-27632 (cat# Y0503-5MG; Sigma-Aldrich).
7. Endoderm Inducing Medium (EIM): RPMI (cat# 11875-093; Thermo Fisher Scientific) (*see* **Note 6**), supplemented with 1 × B-27 (cat# 17504-044; Thermo Fisher Scientific) (*see* **Note 6**), 100 U/mL penicillin and 100 μg/mL streptomycin (cat# 15140-122; Thermo Fisher Scientific), 0.10 μg/mL Activin A (cat# 338-AC-050; R&D Systems), 1 μM CHIR99021 (cat# CT-99021; Axon Medchem), 10 μM of Y-27632 (cat# Y0503-5MG; Sigma-Aldrich), 0.5 mM natrium butyrate (cat# 303410-100G; Sigma-Aldrich) (*see* **Note 8**).
8. Definitive Endoderm Medium (DEM): RPMI1640 (cat# 11875-093; Thermo Fisher Scientific) (*see* **Note 6**), supplemented with 1 × B-27 (cat# 17504-044; Thermo Fisher Scientific) (Note 6), 100 U/mL penicillin and 100 μg/mL

streptomycin (cat# 15140-122; Thermo Fisher Scientific), 0.10 μg/mL activin A (cat# 338-AC-050; R&D Systems), 1 μM CHIR99021 (cat# CT-99021; Axon Medchem), 0.5 mM natrium butyrate (cat# 303410-100G; Sigma-Aldrich) (*see* **Note 8**).

2.3 Immuno-fluorescence Staining

1. Hoechst 33342 (cat# 33342; Thermo Fisher Scientific), 10 μg/mL stock solution: Dilute 1 μL of the Hoechst 33342 solution in 999 μL of 1× DPBS. Prepare stock solution prior to use.
2. Fixation solution: 37% (v/v) paraformaldehyde solution (cat# 252549-500 mL; Sigma-Aldrich), diluted in 1× DPBS to 4% (v/v).
3. Washing buffer: 1× DPBS.
4. Blocking solution: 100% bovine serum albumin (BSA, cat# A3059-100G; Sigma-Aldrich) diluted in 1× DPBS to 5% (v/w).
5. Permeabilization solution: 100% Triton™X-100 (cat# T8787-100 mL; Sigma-Aldrich) diluted in blocking solution to 0.3% (v/v).
6. Staining buffer: 100% bovine serum albumin (BSA, cat# A3059-100G; Sigma-Aldrich) diluted in 1× DPBS to 1% (v/w).

2.4 Flow Cytometry Analysis

1. Falcon™ FACS tubes (cat# FALC352058; Thermo Fisher Scientific).
2. Fixation solution: ice-cold 100% ethanol (cat# 24194-2.5 L; Honeywell; Morris Pleins, NJ).
3. Staining/Washing buffer: 100% bovine serum albumin (BSA, cat# A3059-100G; Sigma-Aldrich) diluted in 1× DPBS to 0.5% (v/w).
4. Permeabilization solution: 100% Triton™X-100 (cat# T8787-100 mL; Sigma-Aldrich) diluted in blocking solution to 0.3% (v/v).

2.5 One-Step RT-PCR

1. Realtime Ready (RTR) Cell Lysis Kit (cat# 05943523001; Roche; Basel, Switzerland).
2. RTR Master Mix: 1× RTR Virus Master Mixture (cat# 05992877001; Roche) and 1× RTR Catalog (cat# 05532957001; Roche; Table 1).Table 5 → Table 1Table 1 → Table 2Table 2 → Table 3Table 6 → Table 4Table 3 → Table 5Table 4 → Table 6Table 7 → Table 7
3. LightCycler® 480 Multiwell Plates 480 (cat# 04729749001; Roche).
4. LightCycler® 480 Sealing Foils (cat# 04729757001; Roche).

Table 1
Human RTR primer/probe assay sets human RTR catalog or designer assays are listed according to gene name, manufacturer, and assay ID

Gene name	Manufacturer	Assay ID
CXCR4	Roche	110817
SOX17	Roche	111185
EpCAM	Roche	103663
RN18S1	Roche	104092

3 Methods

During all experiments, maintenance and passaging, cells were cultured under humidified atmosphere of 5% CO_2 and 95% O_2 at 37 °C.

3.1 Human iPS Cell Maintenance

1. Prepare one Matrigel™-coated T175 cell culture flask. Dilute Matrigel™ in 1× DPBS according to the manufacturer's instructions. Add 12 mL Matrigel™ solution per T175 flask and incubate for 2 h at 37 °C (*see* **Note 1**).
2. Detach cells at 80% confluence by employing 6 mL per T175 flask of 0.5 mM EDTA for 6 min at 37 °C. Aspirate EDTA solution carefully and flush the loosened cells by tapping the flask. Add 15 mL of hiPS cell maintenance medium to the flask and transfer cell suspension to a conical 50 mL tube (*see* **Note 2**).
3. Transfer 5 mL of cell suspension into the previously prepared Matrigel™-coated T175 cell culture flask, resulting in a split ration of 1:3.
4. Gently rock the culture vessel to evenly distribute the cells and return to the incubator.
5. Exchange medium daily with fresh hiPS cell maintenance medium until next passaging step at 80% confluence.
6. For cryopreservation, detach cells as previously described in **step 2** and distribute to cryovials in 1 mL-aliquots with approximately 2×10^6 cells/mL in hiPS cell cryopreservation medium. Store cryovials for 24 h at −80 °C in 100% isopropyl alcohol freezing container. Subsequently transfer cryovials to −150 °C for long-term storage.

3.2 Directed Differentiation into DE Cells

1. On day −1, dilute Matrigel™ in 1× DPBS according to the manufacturer's instructions. Add 500 μL Matrigel™ solution per 24-well and incubate for 2 h at 37 °C.

2. Detach hiPS cells from one T175 maintenance culture flask at 80% confluence using 6 mL of Accutase solution for 4 min at 37 °C (*see* **Note 3**).
3. Flush the loosened cells by tapping the flask.
4. Add 14 mL of 1× DPBS and collect the single-cell suspension into a 50 mL conical tube.
5. Centrifuge the detached cells at 200 × *g* for 5 min.
6. Discard the supernatant and resuspend hiPS cells at 0.8×10^6 cells/mL in seeding medium, supplemented with 10 μM Y-27632.
7. Seed 500 μL/well of the hiPS single-cell suspension into the previously prepared Matrigel™-coated 24-well plate (*see* **Note 4**).
8. Gently rock the culture vessel to evenly distribute the cells and return to the incubator (*see* **Note 5**).
9. On day 0, replace medium with 500 μL/well of fresh EIM (*see* **Note 9**).
10. From days 1 to 5, replace medium with 500 μL/well of fresh DEM (*see* **Note 9**).

3.3 Characterization of Generated DE Cells

3.3.1 Immunofluorescence Staining

For verification of targeted differentiation toward DE, perform immunofluorescence staining (Fig. 3) on day 1 (hiPS cells) and day 5 (DE cells) as described below.

1. Aspirate medium and rinse cells once with 500 μL/well of washing buffer.
2. Aspirate washing buffer and cover cells 500 μL/well of fixation solution.
3. Incubate for 15 min (min) at room temperature (RT).
4. Aspirate fixation solution and wash cells 3× with 500 μL/well of washing buffer 5 min each.
5. Permeabilization step: cover cells with 500 μL/well of permeabilization solution for 15 min at RT.
   (a) Cells stained with membrane markers (CXCR4, EpCAM): skip permeabilization step use directly blocking solution for 60 min at RT.
6. Aspirate permeabilization solution and add 500 μL/well of blocking solution for 60 min at room temperature.
7. Aspirate blocking solution and add 250 μL/well of primary antibodies (ref. to Table 2), diluted in staining buffer, supplemented with 2 μg/mL Hochest33342.
8. Incubate over night at 4 °C.
9. Rinse cells three times with washing buffer for 5 min each.

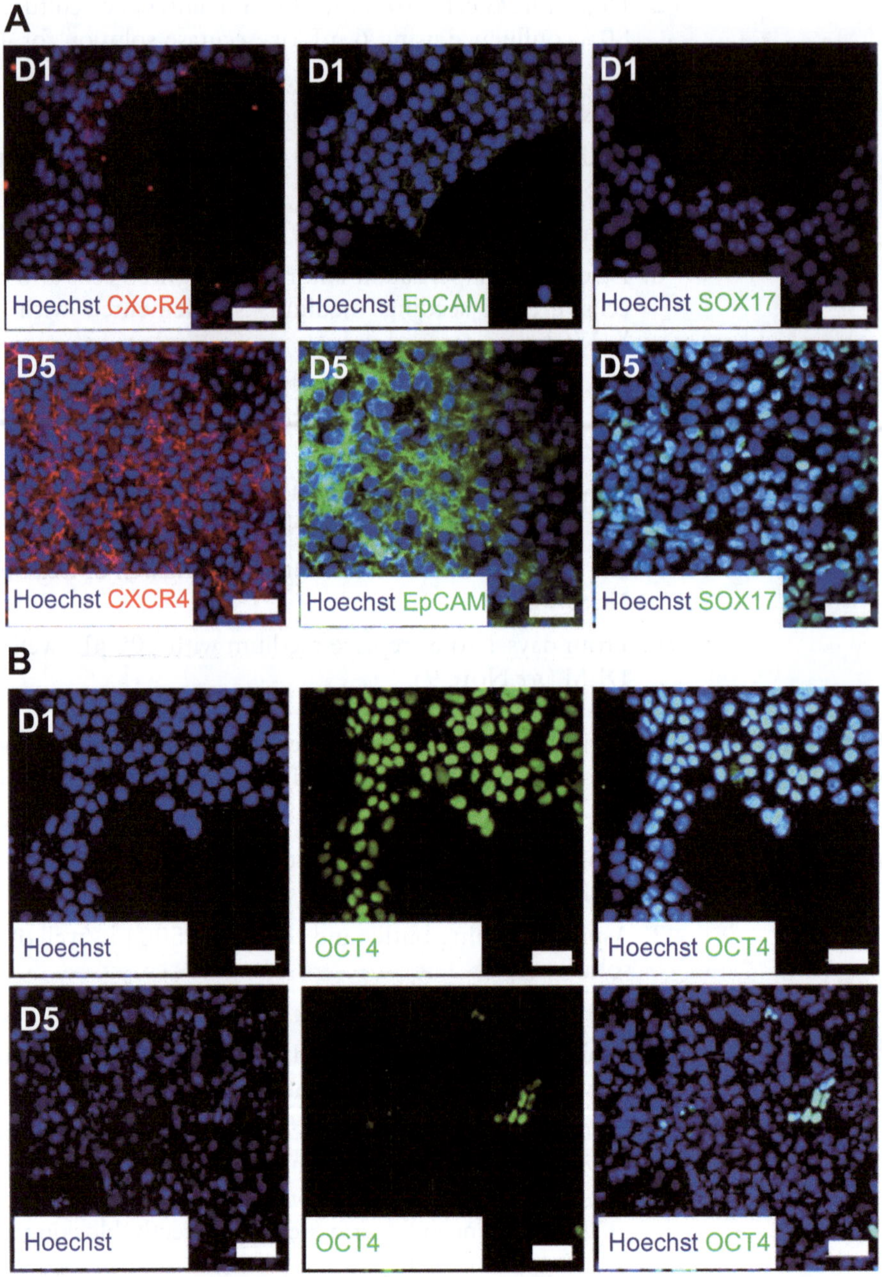

Fig. 3 Immunofluorescence characterization of hiPS cell-derived DE cells. (**a**) Representative immunofluorescence staining for classical DE marker expression in hiPS cells following 5 days of directed differentiation. Cells are stained at day 1 and 5 using antibodies for CXCR4 (red), EpCAM (green), and SOX17 (green). Nuclei are counterstained using Hoechst 33342 (blue). (**b**) Targeted differentiation leads to loss or pluripotency within 5 days. Cells are stained at day 1 and 5 using an antibody for OCT4 (green). Nuclei were counterstained using Hoechst 33342 (blue). Scale bars = 50 μm

Table 2
Primary antibodies are listed according to antigen, abbreviation (abbr.), dilution factor, manufacturer, and catalog number (cat#)

Antigen	Abbr.	Dilution	Manufacturer	Cat#
C-X-C chemokine receptor type 4	CXCR4	1:50	Stem Cell Technologies, Vancouver, Canada	60,089
Epithelial cell adhesion molecule	EpCAM	1:250	Abcam, Cambridge, UK	AB20160
SRY-box 17	SOX17	1:250	Abcam, Cambridge, UK	AB84990
Octamer-binding transcription factor 4	OCT4	1:200	Cell Signaling Technology	2840

Table 3
Secondary antibodies are listed according to target species, fluorochrome, dilution factor, manufacturer, and catalog number (cat#)

Target species	Fluorochrome	Dilution	Manufacturer	Cat#
Goat anti-mouse IgG	Alexa Fluor 488	1:500	Thermo Fisher Scientific	A-11029
Goat anti-rabbit IgG	Alexa Fluor 488	1:500	Thermo Fisher Scientific	A-11034
Goat anti-rabbit IgG2a	Alexa Fluor 488	1:500	Thermo Fisher Scientific	A-21131

10. Aspirate washing buffer and 500 μL/well of species-specific secondary Alexa Fluor™ antibodies (ref. to Table 3), diluted in staining buffer.
11. Incubate for 90 min at room temperature in the dark.
12. Wash cells three times with 500 μL/well of washing buffer for 5 min each.
13. Perform imaging with confocal microscopy (LSM 700, Zeiss) using a 488 nm and 647 nm laser.
14. Store PFA-fixed/ stained cells in 1× DPBS at 4 °C for up to 2 weeks.

3.3.2 Flow Cytometry Analysis

During directed differentiation into DE, perform FACS analysis (Fig. 4) starting on day 1 (hiPS cells) and on day 5 (DE cells), as described below.

1. Wash cells three times with 500 μL/well of 1× DPBS.
2. Aspirate supernatant and add 100 μL/well of detachment reagent. Let the cells detach for 6 min at 37 °C.
3. Flush the cells with 1 mL/well of 1× DPBS and collect the single-cell suspension into 2× 50 mL conical tubes.

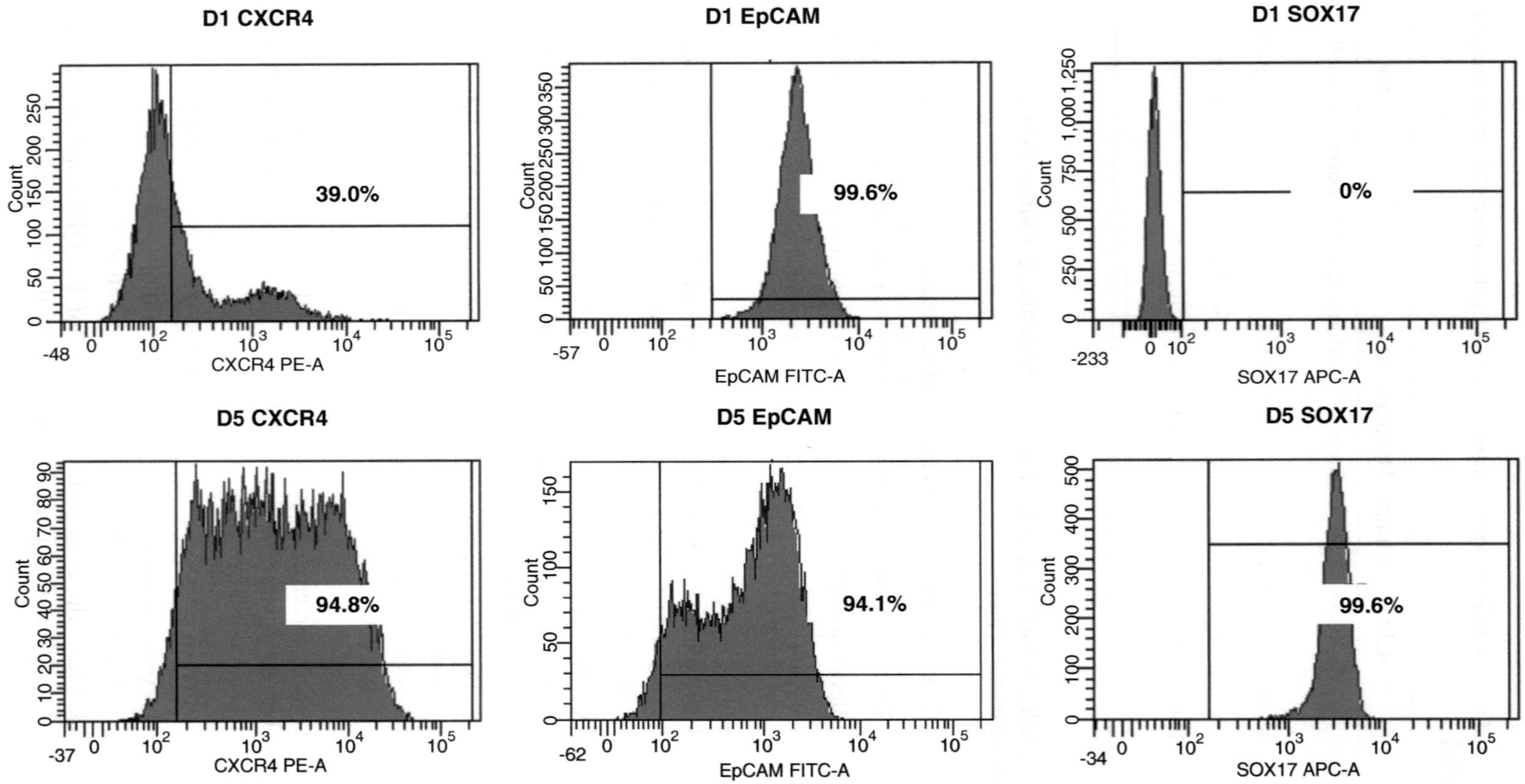

Fig. 4 Flow-cytometric characterization and quantification of hiPS cell-derived DE cells. Representative FACS histograms of CXCR4, EpCAM and SOX17 expression before (day 1; upper panel) and after successful differentiation (day 5; lower panel). Isotype controls specifically for the antibodies used here were used to set gates. Proportion of positives cells are indicated in each histogram

4. Centrifuge cells 5 min at 120 × *g*.

   Staining of membrane proteins (CXCR4, EpCAM)

   (a) Discard the supernatant of one 50 mL conical tube and resuspend cell pellet in 5 mL of staining buffer.
   (b) Aliquot cell suspension evenly across 5 FACS tubes (1 mL/tube) and centrifuge tubes for 5 min at 120 × *g*.
   (c) Discard supernatant and resuspend cells in 100 μL of staining buffer.
   (d) Label tubes according to Table 4.
   (e) Add the primary antibodies and the isotype controls to the tubes according to the appropriate dilution factor (ref. to Tables 3, 5, and 6).

Table 4
Tube labels for staining of membrane proteins

Tube1	Unstained control cells
Tube2	Mouse IgG2a, kappa PE
Tube3	Mouse IgG1, kappa FITC
Tube4	CXCR4
Tube5	EpCAM

Table 5
Flow cytometry antibodies

Antibody	Species	Conjugate	Cat#	Manufacturer	Final dilution
CXCR4	Mouse IgG2a, kappa	PE[a]	60089PE	Stemcell Technologies	1:20
SOX17	Goat, IgG1	APC[b]	IC1924A	R&D Systems	1:50
EpCAM	Mouse IgG1, kappa	FITC[c]	347197	BD	1:5

[a]PE = Phycoerythrin, [b]Allophycocyanin, [c]Fluorescein Isothiocyanate

Table 6
Isotype controls

Antibody	Isotype control	Conjugate	Cat#	Manufacturer	Final dilution
CXCR4	Mouse IgG2a, kappa	PE	60071PE	Stemcell Technologies	1:20
SOX17	Goat, IgG1	APC	IC108A	R&D Systems	1:50
EpCAM	Mouse IgG1, kappa	FITC	60070FI	Stemcell Technologies	1:5

(f) Vortex the FACS tubes and incubate for 30 min on ice in the dark.

(g) Add 3 mL/well of washing buffer (including unstained control cells) and centrifuge tubes for 5 min at 120 × *g*.

(h) Discard supernatant and repeat **step e**.

(i) Resuspend cell pellets in 300 μL of staining buffer. Keep samples on ice and in the dark until the flow-cytometric analysis starts.

(j) Perform flow-cytometric analysis of the samples. Use the unstained and isotype controls to set up the gating strategy.

Staining of intracellular marker proteins (SOX17)

(a) Discard the supernatant of one 50 mL conical tube and add dropwise 1 mL of ice-cold fixation solution while consistently vortexing the tube at 125 rpm (*see* **Note 10**).

(b) Incubate cells on ice for 15 min.

(c) Resuspend cell pellet in 3 mL of washing buffer and centrifuge falcon for 5 min at 800 120 × *g*.

(d) Add 3 mL of permeabilization buffer and incubate on ice for 15 min.

(e) Centrifuge falcon for 5 min at 120 × *g*.

(f) Resuspend cell pellet in 3 mL of washing buffer and divide the cell suspension evenly across 3 FACS tubes (1 mL/ tube) label tubes according to Table 7.

(g) Centrifuge tubes for 5 min at 120 × *g*.

(h) Add 100 μL of staining buffer and add the primary antibody and the isotype control to the tubes according to the appropriate dilution factor (Tables 3 and 5).

(i) Shake the FACS tubes and incubate for 30 min on ice in the dark.

(j) Add 3 mL of washing buffer to each tube (including unstained control cells) and centrifuge tubes for 5 min at 120 × *g*.

(k) Discard supernatant and repeat **step e**.

Table 7
Tube labels for staining of intracellular proteins

Tube1	Unstained control cells
Tube2	SOX17
Tube3	Mouse IgG1, APC

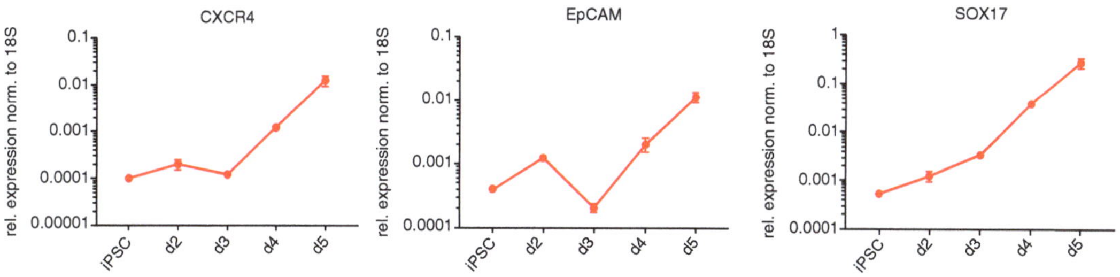

Fig. 5 Gene expression analysis of hiPS cell-derived DE cells over time. Relative expression levels at daily intervals during differentiation to DE confirm up-regulation of DE markers (CXCR4, EpCAM, SOX17) by day 5. RT-PCR data are given as relative expression [$2^{(-\Delta CP)}$] values normalized to the housekeeping gene 18S RNA. $n = 8$ replicates, error bars represent standard error of the mean

(l) Resuspend cell pellets in 300 μL of staining buffer. Keep samples on ice and in the dark until the flow-cytometric analysis starts.

(m) Perform flow-cytometric analysis of the samples. Use the unstained and isotype controls to set up the gating strategy.

3.3.3 One-Step RT-PCR

During directed differentiation into DE, perform RT-PCR gene expression analysis (Fig. 5) starting on day 0 (hiPS cells) and on all consecutive days, as described below. Store RT-PCR lysates at −20 °C and avoid repeated freeze-thaw-cycles (*see* **Note 11**).

1. Wash cells three times with 500 μL/well of 1× DPBS.
2. Aspirate supernatant and add 100 μL/well of 1× RTR Cell Lysis Buffer, supplemented with 2× RNase inhibitor.
3. Incubate cells for 5 min at RT.
4. Prepare RTR master mixture, composed of 1× RTR Virus Master and 1× RTR Assay according to the manufacturer's instructions.
5. Transfer 9 μL/384-well of the RTR master mixture into a 384-well LightCycler™ 480 plate.
6. Add 1 μL of the above generated RT-PCR lysates (**step 2**) into each well of the 384-well LightCycler™ 480 plate.
7. Seal the LightCycler™ 480 plate and centrifuge for 4 min at 1500 × *g*.
8. Measure gene expression using a LightCycler™ 480 instrument. Apply the following RT-PCR program: (1) Reverse Transcription: 50 °C, 8 min, 1 cycle; (2) Preincubation: 95 °C, 30 s, 1 cycle; (3) Amplification: 95 °C, 1 s, followed by 60 °C, 20 s, 45 cycles; (4) Cooling: 40 °C, 30 s, 1 cycle.

4 Notes

1. Matrigel™ is highly prone to clump while thawing. To avoid clumping, keep all consumables such as tubes and pipette tips cold to prevent Matrigel™ from clumping.
2. Thaw mTESR™1 supplement at room temperature (15–25 °C) or overnight at 2–8 °C in the dark as light sensitive proteins could be affected. Do not thaw in a 37 °C water bath.
3. hiPS cells should be seeded at approximately 80–90% confluence with good morphological appearance. Avoid using overgrown cultures as this may impair subsequent differentiation processes.
4. A single cells suspension of hiPS cells prior to DE differentiation is mandatory to adjust recommended seeding density, which is a key factor in determining the efficiency of differentiation toward DE. The optimal cell seeding density needs to be determined for each hiPS cell line individually, to guarantee a highly efficient differentiation process.
5. After plating of cells, hiPS cells should be dispersed evenly in the culture vessels to prevent uneven density and possible inhomogeneous differentiation processes.
6. A high batch to batch variability is known for the B-27 supplement. To ensure comparable and reproducible results, pretest several lots in terms of DE induction.
7. The quality of the activin A protein can vary between different batches and vendors. Try to avoid changing the supplier during DE differentiation as this could have a great influence on the yield of DE cells.
8. The optimal concentration of CHIR99021 may vary between 0.5 μM and 1.5 μM due to the sensitivity of each hiPS cell line. Therefore consider testing different concentrations within the given range, prior of establishing differentiation process.
9. Do not prewarm EIM or DEM in water bath at 37 °C, as Activin A might lose its activity, which could result in deficient cell growth and differentiation potential of cells.
10. When performing flow-cytometric analysis for intracellular markers, try to avoid cell clumps by gently adding dropwise ice-cold fixation solution and consistently vortexing the tube.
11. Store RT-PCR lysates at −20 °C for up to half a year. Avoid freezing–thawing cycles of lysates as the RNase inhibitor gets inactivated during the freezing process. Therefore, keep lysates on ice and prepare to RT-PCR as quick as possible.

Acknowledgments

The research leading to these results has received support from the Innovative Medicines Initiative Joint Undertaking under (grant no. 115439), resources of which are composed of financial contribution from the European Union's Seventh Framework Programme (FP7/2007-2013) and EFPIA companies. This publication reflects only the author's views and neither the IMI JU nor EFPIA nor the European Commission are liable for any use that may be made of the information contained therein. We sincerely thank Marcel Leist and Dirk Stenkamp for their excellent advice as well as Michael Traub for his technical assistance.

References

1. Takahashi K, Yamanaka S (2006) Induction of pluripotent stem cells from mouse embryonic and adult fibroblast cultures by defined factors. Cell 126:663–676
2. Traub S, Stahl H, Rosenbrock H et al (2017) Upscaling of hiPS cell-derived neurons for high-throughput screening. SLAS Discov 22:274–286
3. Heilker R, Traub S, Reinhardt P et al (2014) iPS cell derived neuronal cells for drug discovery. Trends Pharmacol Sci 35:510–519
4. Kim PT, Ong CJ (2012) Differentiation of definitive endoderm from mouse embryonic stem cells. Results Probl Cell Differ 55:303–319
5. Zorn AM, Wells JM (2009) Vertebrate endoderm development and organ formation. Annu Rev Cell Dev Biol 25:221–251
6. Wells JM, Melton DA (1999) Vertebrate endoderm development. Annu Rev Cell Dev Biol 15:393–410
7. Kraus MR, Grapin-Botton A (2012) Patterning and shaping the endoderm in vivo and in culture. Curr Opin Genet Dev 22:347–353
8. Kubo A, Shinozaki K, Shannon JM et al (2004) Development of definitive endoderm from embryonic stem cells in culture. Development 131:1651–1662
9. Wong AP, Chin S, Xia S et al (2015) Efficient generation of functional CFTR-expressing airway epithelial cells from human pluripotent stem cells. Nat Protoc 10:363–381
10. Huang SX, Green MD, De Carvalho AT et al (2015) The in vitro generation of lung and airway progenitor cells from human pluripotent stem cells. Nat Protoc 10:413–425
11. D'amour KA, Agulnick AD, Eliazer S et al (2005) Efficient differentiation of human embryonic stem cells to definitive endoderm. Nat Biotechnol 23:1534–1541
12. Gotoh S, Ito I, Nagasaki T et al (2014) Generation of alveolar epithelial spheroids via isolated progenitor cells from human pluripotent stem cells. Stem Cell Reports 3:394–403
13. Ghaedi M, Mendez JJ, Bove PF et al (2014) Alveolar epithelial differentiation of human induced pluripotent stem cells in a rotating bioreactor. Biomaterials 35:699–710
14. Wang P, Mcknight KD, Wong DJ et al (2012) A molecular signature for purified definitive endoderm guides differentiation and isolation of endoderm from mouse and human embryonic stem cells. Stem Cells Dev 21:2273–2287
15. Ng VY, Ang SN, Chan JX et al (2010) Characterization of epithelial cell adhesion molecule as a surface marker on undifferentiated human embryonic stem cells. Stem Cells 28:29–35
16. Grapin-Botton A (2008) Endoderm specification. In: StemBook, Cambridge MA
17. Green MD, Chen A, Nostro MC et al (2011) Generation of anterior foregut endoderm from human embryonic and induced pluripotent stem cells. Nat Biotechnol 29:267–272

Part II

High Throughput Metabolomics and Proteomic Cell-Based Assays

Chapter 10

LC-MS/MS-Based Metabolomics for Cell Cultures

Anna Artati, Cornelia Prehn, and Jerzy Adamski

Abstract

Metabolomics, a comprehensive analysis of metabolites in biological specimens (e.g., cells, body fluids, tissues, exhaled air, plants), offers promising tools in health, nutrition, biotechnology, and food sciences. Here we describe methods of LC-MS/MS-based analyses for cell metabolomics. Using methods employed in this section, over 1000 endogenous and exogenous metabolites can be detected, annotated, and quantified relatively by nontargeted analysis approach, whereas targeted metabolomics analysis enables us to quantify 188 endogenous metabolites.

Key words Metabolomics, Mass spectrometry, LC-MS/MS-based metabolomics, Cell culture metabolomics, Nontargeted metabolomics, Targeted metabolomics

1 Introduction

Metabolomics is one of omics technologies, which deals with comprehensive evaluation of metabolites in biological samples. It provides a snapshot of the interrelation of genetic structures, expression of genes, protein functions, and environmental influences, as metabolites are the endpoints of biological processes. This technology offers promising tools to study the mechanisms of health and diseases, nutrition effects, biotechnological processes, or to perform crop and food quality analyses. Common approaches in metabolomics study are targeted and nontargeted metabolite analysis. Targeted metabolomics allows for quantification of a preselected set of known metabolites, whereas nontargeted metabolomics provides a global view of metabolites of different chemical classes, which are simultaneously present in organisms.

Not only does the nontargeted metabolomics approach allow for relative quantification of metabolites, which are already annotated in databases, but it is also able to deliver information about unidentified metabolites, which are not yet annotated. These unidentified metabolites can be important in the context of a phenotype investigation and can be further identified with the

Carl-Fredrik Mandenius and James A. Ross (eds.), *Cell-Based Assays Using iPSCs for Drug Development and Testing*, Methods in Molecular Biology, vol. 1994, https://doi.org/10.1007/978-1-4939-9477-9_10,
© Springer Science+Business Media, LLC, part of Springer Nature 2019

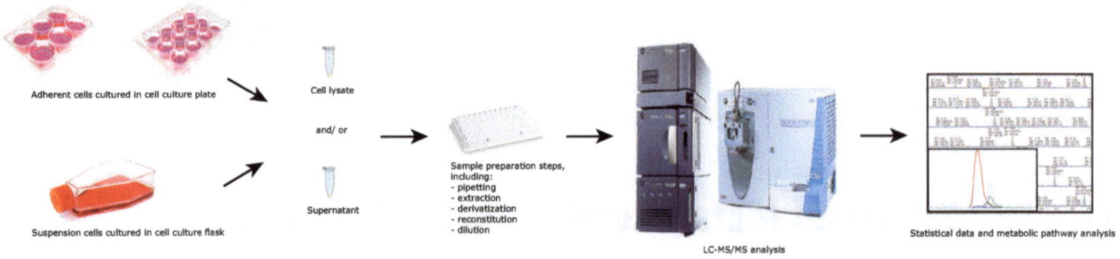

Fig. 1 Workflow of LC-MS/MS-based nontargeted and targeted analysis of cell metabolomics

help of a set of characteristic parameters (i.e., retention times and mass spectra). Targeted and nontargeted approaches have been shown to have some overlap in metabolites, and revealed a very good correlation of quantification [1].

Metabolomics has already been validated and applied to a wide range of biological matrices including body fluids (plasma, serum, urine, cerebrospinal fluid, saliva) [2, 3], tissues, stool [4], lung lavage [5], exhaled air [6], cells, and even for plants. In this report, we focus only on metabolomics for cell cultured based studies.

Cell metabolomics consists of four sequential steps: (a) cell harvesting, metabolism quenching and sample collection, (b) sample preparation, (c) metabolic detection and quantification or profiling of the metabolites, and (d) bioinformatics data analysis (Fig. 1). It is very important that metabolites in the sample remain the same as they are at the time of sampling. Therefore, an optimum quenching and harvesting method have to be ensured. Cell harvesting is aimed to prevent further metabolism by inactivating intracellular enzymes to avoid metabolite degradation and alteration of the sample composition. Trypsinization was a common method used to detach adherently growing cells. However, some studies reported a possibility of cell membrane damage [7] and metabolite leakage when cells were harvested by trypsinization. Therefore, we harvest adherent cells by scraping the cells using extraction solvents.

Since the polarity of intracellular metabolites varies widely, optimization of the extraction and analytical methodology is extremely important to recover and to detect as many metabolites as possible. Moreover, sample preparation should be highly reproducible, robust, and fast to allow for high-throughput studies [8].

Analytical techniques used for metabolomics studies are mostly based on nuclear magnetic resonance (NMR) or mass spectrometry (MS) systems, including liquid chromatography (LC)-MS, gas chromatography (GC)-MS, ultrahigh performance liquid chromatography (UHPLC)-MS, flow injection assay (FIA)-MS, and capillary electrophoresis (CE)-MS [9, 10]. Nontargeted metabolomics often requires a high degree of parallel analyses (e.g., simultaneous validated analyses on LC- and GC-MS) to cover as many metabolites as possible and to avoid a bias on specific chemical classes. It also needs special algorithms for metabolite identification with specific databases [11–13].

Here, we describe LC-tandem MS (LC-MS/MS)-based analytical methods for targeted and nontargeted cell metabolomics.

2 Materials (*See* Note 8)

2.1 Harvesting of Cells and Collection of Supernatant or Cells

1. Eighty percent (v/v) methanol extraction solvent containing recovery standard compounds (*see* **Note 1**) for nontargeted metabolomics analysis, and 80% (v/v) methanol extraction solvent without recovery standard compounds for targeted metabolomics analysis. Store at 4 °C.
2. Phosphate buffered saline (PBS).
3. Safe lock Eppendorf tube 1.5 mL or 2.0 mL.
4. Rubber-tipped cell scraper.
5. Screw-cap micro tube 2.0 mL, PP, reference number 72.694.005 (Sarstedt, Nümbrecht, Germany) for nontargeted metabolomics analysis, and screw-cap micro tube 0.5 mL, PP, reference number 72.730.005 (Sarstedt, Nümbrecht, Germany) for targeted metabolomics analysis.

2.2 Sample Preparation for LC-MS/MS-Based Metabolomics Analysis

1. Ninety-six-well plate, 2 mL, Nunc, Thermo Fisher Scientific, Schwerte, Germany.
2. Human reference plasma, Seralab, West Sussex, UK.
3. Precellys24 glass beads 0.5 mm, Peqlab, Erlangen, Germany.
4. Precellys24 homogenizer, Peqlab, Erlangen, Germany.
5. Ninety-six-well microplate, Sarstedt, Nümbrecht, Germany.

2.3 LC-MS/MS Analysis for Nontargeted Metabolomics

1. Reconstitution solvent: the solvents for both ionization, that is, positive and negative ionization modus, contained labeled standard compounds that allows us to monitor instrument performance and as retention index markers. The labeled compounds are dissolved in watery mobile phases of the respective ionization modus (*see* **Note 2**). Store at 4 °C.
2. Q Exactive MS, Thermo Fisher Scientific, Dreieich, Germany.
3. Acquity UPLC system, Waters, Eschborn, Germany.
4. BEH C18 separation column 2.1 × 100 mm, 1.7 μm particle, Waters, Eschborn, Germany.
5. UPLC mobile phase for acidic system: solvent A1: 0.1% formic acid in water, and solvent B1: 0.1% formic acid in methanol.
6. UPLC mobile phase for basic system: solvent A2: 6.5 mM ammonium bicarbonate pH 8.0, and solvent B2: 6.5 mM ammonium bicarbonate in 95% methanol.

2.4 LC-MS/MS Analysis for Targeted Metabolomics

1. Absolute *IDQ*™ p180 Kit, BIOCRATES Life Sciences AG, Innsbruck, Austria.
2. Precellys24 glass beads 0.5 mm, Peqlab, Erlangen, Germany.
3. Precellys24 homogenizer, Peqlab, Erlangen, Germany.
4. Five percent phenylisothiocyanate.
5. Methanol containing 5 mM ammonium acetate.
6. API 4000 System triple quadrupol MS, Sciex Deutschland GmbH, Darmstadt, Germany.
7. One Thousand Two Hundred (1200) Series HPLC, Agilent Technologies Deutschland GmbH, Böblingen, Germany.
8. HTC PAL auto sampler, CTC Analytics, Zwingen, Switzerland.
9. Zorbax Eclipse XDB-C18 separation column 3 × 100 mm, 3.5 μm particle, Agilent Technologies Deutschland GmbH, Böblingen, Germany.

3 Methods

Although it depends on the cell line and it should be evaluated prior to the study, 1–5 × 10^6 cells per mL cell suspension is a good start for the analysis. To reach these cell numbers it might be necessary to pool cells from some wells. Furthermore, metabolomics data normalization must be considered, for example, by conducting Hoechst assay [14] with which the DNA content and thus the cell number of the sample can be indirectly determined. This step should be done prior to the analytics.

3.1 Harvesting and Collection of Adherently Growing Cells for Metabolomics Analysis

Adherent cells for metabolomics analysis should be grown in cell culture plates (e.g., 6- or 12-well) and not in flasks.

3.1.1 Collection of Supernatant for Nontargeted Metabolomics Analysis (See **Note 3**)

1. Collect 1 mL of supernatant into 1.5 mL or 2.0 mL safe lock Eppendorf tube.
2. Centrifuge the supernatant at 1000 × *g* and room temperature for 10 min.
3. Carefully transfer 0.5 mL of the cell-free supernatant into a fresh 1.5 mL or 2.0 mL safe lock Eppendorf tube and immediately place the tube on dry ice.
4. Store samples at −80 °C immediately after sample collection process is finished until it is ready for analysis.

3.1.2 Harvesting and Collection of Cells for Nontargeted Metabolomics Analysis (See ***Note 4****)*

1. Remove cell supernatant from the well.
2. Wash the cells two times with PBS (37 °C). Use at least 1 mL more PBS than culture medium to ensure proper washing of the well (e.g., if the growth medium volume for the 6-well is 2 mL, then use 3 mL PBS for washing). Remove the PBS after the second washing step as completely as possible.
3. Add 400 μL of precooled (dry ice) 80% (v/v) methanol extraction solvent containing recovery standard compounds to each well and ensure that all cells are covered immediately.
4. Detach the cells using a rubber-tipped cell scraper and transfer the cell suspension into precooled (dry ice) 2.0 mL screw-cap micro tubes. The samples should be handled on dry ice during the whole procedure.
5. Add another 100 μL of precooled (dry ice) 80% (v/v) methanol extraction solvent containing recovery standard compounds to wash the well and transfer the wash volume to the micro tube as well.
6. As 1 mL of cell suspension per tube is required for the analysis, suspensions of two 6-wells must be pooled to achieve a sample of 1 mL cell suspension.
7. Store samples at −80 °C immediately after sample collection process is finished until it is ready for analysis.

3.1.3 Collection of Supernatant for Targeted Metabolomics Analysis (See ***Note 3****)*

1. Collect 1 mL of supernatant into 1.5 mL or 2.0 mL safe lock Eppendorf tube.
2. Centrifuge the supernatant at 1000 × *g* and room temperature for 10 min.
3. Carefully transfer 0.5 mL of the cell-free supernatant into a fresh 1.5 mL or 2.0 mL safe lock Eppendorf tube and immediately place the tube on dry ice.
4. Store samples at −80 °C immediately after sample collection process is finished until it is ready for analysis.

3.1.4 Harvesting and Collection of Cells for Targeted Metabolomics Analysis (See ***Note 5****)*

1. Remove cell supernatant from the well.
2. Wash the cells two times with PBS (37 °C). Use at least 1 mL more PBS than culture medium to ensure proper washing of the well (e.g., if the growth medium volume for the 12-well is 1 mL, then use 2 mL PBS for washing). Remove the PBS after the second washing step as completely as possible.
3. Add 200 μL of the precooled (dry ice) 80% (v/v) methanol extraction solvent and ensure that all cells are covered immediately.
4. Detach the cells using a rubber-tipped cell scraper and transfer the cell suspension into precooled (dry ice) 0.5 mL screw-cap

micro tubes. The samples should be handled on dry ice during the whole procedure.

5. Add another 100 μL of precooled (dry ice) 80% (v/v) methanol extraction solvent to wash the well and transfer the wash volume to the micro tube as well.
6. As 0.3 mL of cell suspension per tube is required for the analysis, suspension of one 12-well is sufficient for one sample.
7. Store samples at −80 °C immediately after sample collection process is finished until it is ready for analysis.

3.2 Harvesting and Collection of Suspension Growing Cells for Metabolomics Analysis

Suspension cells for metabolomics analysis should be grown in cell culture flasks.

3.2.1 Collection of Supernatant for Nontargeted Metabolomics Analysis (See Note 3)

1. Gently mix the cell suspension.
2. Transfer 1 mL of the cell suspension into a 1.5 mL or 2.0 mL safe lock Eppendorf tube.
3. Centrifuge the cell suspension with centrifugal force depend on the cell types and lines at room temperature so that the cell pellet is firmly separate from the supernatant.
4. Carefully transfer 0.5 mL of the cell-free supernatant into a fresh 1.5 mL or 2.0 mL safe lock Eppendorf tube and immediately place the tube on dry ice.
5. Store samples at −80 °C immediately after sample collection process is finished until it is ready for analysis.

3.2.2 Harvesting and Collection of Cells for Nontargeted Metabolomics Analysis

1. Gently mix the cell suspension.
2. Remove the desired volume of cell suspension from the flask and count the cells in an aliquot using an appropriate cell counting method.
3. Transfer the respective volume containing the desired cell number to a 2.0 mL screw-cap micro tube.
4. Centrifuge the cells with centrifugal force depend on the cell types and lines at room temperature so that the cell pellet is firmly separate from the supernatant.
5. Remove the supernatant as completely as possible and gently resuspend the cell pellet in 2 mL warm PBS (37 °C).
6. Perform two washing steps with PBS, remove the PBS of the second washing step as completely as possible, and immediately place the tube with the dry cell pellet on dry ice.
7. Store samples at −80 °C immediately after sample collection process is finished until it is ready for analysis.

3.2.3 Collection of Supernatant for Targeted Metabolomics Analysis (See ***Note 3****)*

1. Gently mix the cell suspension.
2. Transfer 1 mL of the cell suspension into a 1.5 mL or 2.0 mL safe lock Eppendorf tube.
3. Centrifuge the cell suspension with centrifugal force depend on the cell types and lines at room temperature so that the cell pellet is firmly separate from the supernatant.
4. Carefully transfer 0.5 mL of the cell-free supernatant into a fresh 1.5 mL or 2.0 mL safe lock Eppendorf tube and immediately place the tube on dry ice.
5. Store samples at −80 °C immediately after sample collection process is finished until it is ready for analysis.

3.2.4 Harvesting and Collection of Cells for Targeted Metabolomics Analysis

1. Gently mix the cell suspension.
2. Remove the desired volume of cell suspension from the flask and count the cells in an aliquot using an appropriate cell counting method.
3. Transfer the respective volume containing the desired cell number to a 0.5 mL screw-cap micro tube.
4. Centrifuge the cells with centrifugal force depend on the cell types and lines at room temperature so that the cell pellet is firmly separate from the supernatant.
5. Remove the supernatant as completely as possible and gently resuspend the cell pellet in 0.5 mL warm PBS (37 °C).
6. Perform two washing steps with PBS, remove the PBS of the second washing step as completely as possible, and immediately store the dry cell pellet on dry ice.
7. Store samples at −80 °C immediately after sample collection process is finished until it is ready for analysis.

3.3 Sample Preparation for LC-MS/MS-Based Metabolomics Analysis

3.3.1 Extraction of Supernatant for Nontargeted Metabolomics Analysis

1. Transfer 100 μL of the supernatant into a 2 mL 96-well plate. In addition to samples of the study, place a 100 μL human reference plasma sample and 100 μL aliquots of pool of small amount of all samples of the study in 1 and 6 wells, respectively, of the 96-well plate. These samples served as technical replicates throughout the data set to assess process variability. Water, serves as process blank, should be placed into 6 wells (100 μL each) of the plate (*see* **Note 7**).
2. Add 475 μL 80% (v/v) methanol extraction solvent containing recovery standard compounds into each sample to extract the metabolites and to precipitate the protein. Shake the 2 mL 96-well plate which contained the sample extracts at 675 rpm for 2 minutes.
3. Centrifuge the samples for 5 min at 2500 rpm. After centrifugation, split the supernatant into four aliquots of 100 μL each onto two 96-well microplates (Sarstedt).

4. Dry the supernatant under nitrogen stream.
5. Store the samples at −80 °C until the LC-MS/MS measurement is performed.

3.3.2 Extraction of Adherently Growing Cells for Nontargeted Metabolomics Analysis

1. Add approximately 140 mg glass beads into 1 mL of cell lysate suspension in the 2.0 mL screw-cap micro tube.
2. Homogenize the cell suspension in the micro tube using Precellys24 homogenizer by grinding them with the glass beads, at 5500 rpm (4 °C) in 2 cycles, 25 s each with a 5 s break (*see* **Note 6**).
3. Centrifuge the cell homogenate for 5 min at 11,000 × *g* at 4 °C.
4. Pool small portion of the lysate cell extract of all samples of the study into a tube. This sample serves as quality control sample to assess process variability. Aliquot 100 μL of the pooled sample into six wells in a 96-well microplate. In addition to the pooled sample, place extract of a human reference plasma sample and water blank samples into 1 and 6 wells, respectively, of the microplate (*see* **Note 7**).
5. Transfer 100 μL of lysate cell extract of each sample of the study into the rest of the wells in a 96-well microplate.
6. Dry the lysate cell extract under nitrogen stream.
7. Store the samples at −80 °C until the LC-MS/MS measurement is performed.

3.3.3 Extraction of Suspension Growing Cells for Nontargeted Metabolomics Analysis

1. Add precooled (dry ice) 80% (v/v) methanol extraction solvent containing recovery standard compounds into cell lysate in the 2.0 mL screw-cap micro tube, to reach 1–5 × 10^6 cells per mL cell suspension.
2. Add approximately 140 mg glass beads per mL extraction solvent added into the cell lysate in the micro tube.
3. Homogenize the cell suspension in the micro tube using Precellys24 homogenizer by grinding them with the glass beads, at 5500 rpm (4 °C) in 2 cycles, 25 s each with a 5 s break (*see* **Note 6**).
4. Centrifuge the cell homogenate for 5 min at 11,000 × *g* at 4 °C.
5. Pool small portion of the lysate cell extract of all samples of the study into a tube. This sample serves as quality control sample to assess process variability. Aliquot 100 μL of the pooled sample into six wells in a 96-well microplate. In addition to the pooled sample, place extract of a human reference plasma sample and water blank samples into 1 and 6 wells (100 μL each), respectively, in the microplate (*see* **Note 7**).

6. Transfer 100 μL of lysate cell extract of each sample of the study into the rest of the wells in the 96-well microplate.
7. Dry the lysate cell extract under nitrogen stream.
8. Store the samples at −80 °C until the LC-MS/MS measurement is performed.

3.4 LC-MS/MS Analysis

3.4.1 LC-MS/MS Analysis for Nontargeted Metabolomics

1. Reconstitute the dried extract samples in the 96-well microplate either with positive reconstitution solvent for LC-MS/MS measurement in positive ionization mode or with negative reconstitution solvent for LC-MS/MS measurement in negative ionization mode (*see* **Note 7**).
2. The LC-MS/MS analysis is performed on a Q Exactive MS coupled to an Acquity UPLC system. Two separate BEH C18 (2.1 × 100 mm, 1.7 μm particle) separation columns are used for acidic (i.e., solvent A1 and solvent B1) and for basic (i.e., solvent A2 and solvent B2) mobile phase conditions, optimized for positive and negative electrospray ionization (ESI), respectively. After injection of the sample extracts, the columns are developed in a gradient of 99.5% A to 98% B in 11 min run time at 350 μL/min flow rate. The eluent flow is directly connected to the ESI source of the Q Exactive MS. Full scan mass spectra (80–1000 *m/z*) and data dependent MS/MS scans with dynamic exclusion are recorded in turns.
3. Metabolites are identified by Metabolon, Inc. from the LC-MS/MS data by automated multiparametric comparison with a proprietary library, containing retention times, *m/z* ratios, and related adduct/fragment spectra that enable us to detect, annotate, and relatively quantify over 1000 endogenous and exogenous metabolites. In addition, the library includes unknown chemicals whose identities are yet to be determined.

3.4.2 LC-MS/MS Analysis for Targeted Metabolomics

The targeted metabolomics analysis described in this section is based on LC-ESI-MS/MS and FIA-ESI-MS/MS measurements using Absolute *IDQ*™ p180 Kit. Measurements are performed according to the manufacturer's manual AS-P180, as follow, briefly [15]:

1. Pipette internal standards for the LC-MS/MS procedure onto the filter inserts of the 96-well plate of the p180-kit.
2. This plate is prespiked with internal standards for the FIA-MS/MS procedure by the manufacturer.
3. Add approximately 70 mg glass beads into 0.3 mL of cell lysate suspension in the 0.5 mL screw-cap micro tube.
4. Homogenize the cell suspension in the micro tube using Precellys24 homogenizer by grinding them with the glass beads, at 5500 rpm (4 °C) in 2 cycles, 25 s each with a 5 s break (*see* **Note 6**).

5. Centrifuge the cell homogenate for 5 min at 11,000 × *g* at 4 °C.
6. Place 10 μL of lysate cell extract or supernatant onto the filter in the cavities of the 96-well filter plate of the p180-kit.
7. Dry the samples in the 96-well plate under nitrogen stream for 30 min.
8. Derivatize amino acids and biogenic amines in the samples with an excess of 5% phenylisothiocyanate for 20 min and dry under nitrogen stream for 40 min.
9. Extract metabolites and internal standards with 300 μL methanol containing 5 mM ammonium acetate by incubating them for 30 min at RT with continuous shaking.
10. Metabolites are eluted by centrifugation for 5 min at 500 × *g* at RT.
11. Dilute one part of the eluate with p180-kit running solvent (1/5 (v/v)) for FIA-MS/MS analysis, the other part with water (50/50, (v/v)) for LC-MS/MS analysis.
12. LC-MS/MS analysis is performed on an API 4000 System triple quadrupole MS coupled to a 1200 Series HPLC and a HTC PAL auto sampler and a HTC PAL auto sampler. The LC run is performed using a Zorbax Eclipse XDB-C18 separation column (3 × 100 mm, 3.5 μm particle). Mass detection and compound identification are performed by multiple reactions monitoring (MRM). For the LC part, compound identification and quantification were based on scheduled multiple reaction monitoring measurements (sMRM). The LC-MS/MS procedure enables the quantification of 42 metabolites, covering 21 amino acids (AA) and 21 biogenic amines (BA). The FIA-MS/MS procedure is used for the quantification of 146 additional metabolites, covering 40 acylcarnitines (AC) including free carnitine, 38 phosphatidylcholines with acyl/acyl side chains (PCaa), 38 phosphatidylcholines with acyl/alkyl side chains (PCae), 14 lysophosphatidylcholines (lysoPC), 15 sphingolipids (SM), and the sum of hexoses.

4 Notes

1. Standard compounds are added into 80% (v/v) methanol extraction solvent applied in nontargeted metabolomics analysis to enable us monitoring the extraction efficiency. We find that the extraction solvent is stable for about 2 weeks at 4 °C.
2. We find that the reconstitution solvent is stable for about 2 weeks at 4 °C.

3. This step should be omitted if the study does not require analysis of supernatant. For control purposes, blank medium, which has been incubated too but without cells, is required.
4. The volumes described here is for a single well of a 6-well plate. If 12-well plates are used, the volumes should be downscaled.
5. The volumes described here is for a single well of a 12-well plate. If 6-well plates are used, the volumes should be upscaled.
6. In the case of Hoechst assay application for data normalization, collect 20 μL cell homogenate in this step, to determine DNA content and thus to determine cell number of the sample.
7. To ensure the quality of analysis and the reliability of the metabolite annotation, quality control assessment is crucial in nontargeted metabolomics analysis. The assessment should include monitoring the quality of chromatograms, reproducibility of retention times, accuracy of the masses, and fragmentation pattern of the labeled standard compounds spiked into each sample. In the method described in this session, the labeled standard compounds are dissolved in reconstitution solvent.
8. Due to the inherent sensitivity of MS equipment, only the highest quality plastic materials and reagents/solvents (at least LC-MS grade) are used in the preparation and execution of LC-MS/MS-based metabolomics experiments.

Acknowledgments

The research leading to these results has received support from the Innovative Medicines Initiative Joint Undertaking under grant agreement no. 115439, resources of which are composed of financial contribution from the European Union's Seventh Framework Programme (FP7/2007-2013) and EFPIA companies' in-kind contribution.

References

1. Suhre K, Meisinger C, Doring A, Altmaier E, Belcredi P, Gieger C et al (2010) Metabolic footprint of diabetes: a multiplatform metabolomics study in an epidemiological setting. PLoS One 5(11):e13953
2. Fiehn O (2002) Metabolomics--the link between genotypes and phenotypes. Plant Mol Biol 48(1–2):155–171
3. Wishart DS, Knox C, Guo AC, Eisner R, Young N, Gautam B et al (2009) HMDB: a knowledgebase for the human metabolome. Nucleic Acids Res 37(Database issue): D603–D610
4. Fahy E, Subramaniam S, Murphy RC, Nishijima M, Raetz CR, Shimizu T et al (2009) Update of the LIPID MAPS comprehensive classification system for lipids. J Lipid Res 50(Suppl):S9–S14
5. Giavalisco P, Kohl K, Hummel J, Seiwert B, Willmitzer L (2009) 13C isotope-labeled metabolomes allowing for improved compound annotation and relative quantification in liquid chromatography-mass spectrometry-

based metabolomic research. Anal Chem 81 (15):6546–6551
6. Waters NJ, Garrod S, Farrant RD, Haselden JN, Connor SC, Connelly J et al (2000) High-resolution magic angle spinning (1)H NMR spectroscopy of intact liver and kidney: optimization of sample preparation procedures and biochemical stability of tissue during spectral acquisition. Anal Biochem 282(1):16–23
7. Batista U, Garvas M, Nemec M, Schara M, Veranic P, Koklic T (2010) Effects of different detachment procedures on viability, nitroxide reduction kinetics and plasma membrane heterogeneity of V-79 cells. Cell Biol Int 34 (6):663–668
8. Dettmer K, Nurnberger N, Kaspar H, Gruber MA, Almstetter MF, Oefner PJ (2011) Metabolite extraction from adherently growing mammalian cells for metabolomics studies: optimization of harvesting and extraction protocols. Anal Bioanal Chem 399(3):1127–1139
9. Griffiths WJ, Karu K, Hornshaw M, Woffendin G, Wang Y (2007) Metabolomics and metabolite profiling: past heroes and future developments. Eur J Mass Spectrom (Chichester) 13(1):45–50
10. Zhao X, Fritsche J, Wang J, Chen J, Rittig K, Schmitt-Kopplin P et al (2010) Metabonomic fingerprints of fasting plasma and spot urine reveal human pre-diabetic metabolic traits. Metabolomics 6(3):362–374
11. Lawton KA, Berger A, Mitchell M, Milgram KE, Evans AM, Guo L et al (2008) Analysis of the adult human plasma metabolome. Pharmacogenomics 9(4):383–397
12. Ohta T, Masutomi N, Tsutsui N, Sakairi T, Mitchell M, Milburn MV et al (2009) Untargeted metabolomic profiling as an evaluative tool of fenofibrate-induced toxicology in Fischer 344 male rats. Toxicol Pathol 37 (4):521–535
13. Sreekumar A, Poisson LM, Rajendiran TM, Khan AP, Cao Q, Yu J et al (2009) Metabolomic profiles delineate potential role for sarcosine in prostate cancer progression. Nature 457 (7231):910–914
14. Muschet C, Moller G, Prehn C, de Angelis MH, Adamski J, Tokarz J (2016) Removing the bottlenecks of cell culture metabolomics: fast normalization procedure, correlation of metabolites to cell number, and impact of the cell harvesting method. Metabolomics 12 (10):151
15. Zukunft S, Sorgenfrei M, Prehn C, Moller G, Adamski J (2013) Targeted metabolomics of dried blood spot extracts. Chromatographia 76(19–20):1295–1305

Chapter 11

Preparation of iPSCs for Targeted Proteomic Analysis

Anna Baud, Wendy E. Heywood, Daniel Little, Paul Gissen, and Kevin Mills

Abstract

Induced pluripotent stem cells have great potential as a human model system in regenerative medicine, disease modeling, and drug screening. However, extensive analysis of iPSC are required before their therapeutic applications. With recent developments in mass spectrometry and proteomics, this technique can become a great alternative to traditional genomic approaches for iPSC analysis. Here, we describe preparation of iPSC for targeted proteomic analysis, and measurement of pluripotency markers allowing for classification into either pluripotent or nonpluripotent cells.

Key words Induced pluripotent stem cells, Pluripotency, Proteomics, Mass spectrometry

1 Introduction

Induced pluripotent stem cells (iPS cells) are somatic cells which acquire properties similar to the embryonic stem cells [1]. There are multiple ways to generate iPSCs: retroviral transfection of transcription factors OCT4, SOX-2, c-MYC, and KLF-4 [2], plasmids [3], nonintegrating viruses (Sendai virus) [4, 5], drug-inducible systems [6], transposition [7], protein transduction [8], or mature microRNAs [9]. Despite the different techniques available for generating iPSCs, the reprogramming process yields low number of iPS cells, and there is a need for an assay able to distinguish between pluripotent and nonpluripotent cells. Several assays for iPSC quality control are available: visualization of colony-morphology using light microscopy or extracellular marker staining, embryoid body or teratoma formation, as well as assays based on gene expression, that is, PluriTest [10] or ScoreCard [11]. However, these tests have major drawbacks, such as long performance time and significant cost.

With recent advances in targeted mass spectrometry, this technique is now enabling fast, reliable, and absolute quantitation of proteins of interest, and supports not only basic but also clinical

Carl-Fredrik Mandenius and James A. Ross (eds.), *Cell-Based Assays Using iPSCs for Drug Development and Testing*, Methods in Molecular Biology, vol. 1994, https://doi.org/10.1007/978-1-4939-9477-9_11,
© Springer Science+Business Media, LLC, part of Springer Nature 2019

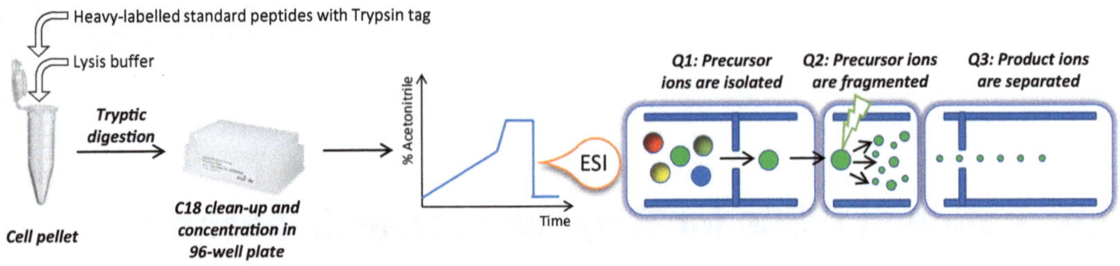

Fig. 1 Typical LC-MRM pluripotency assay workflow. Adapted from [14]

research [12, 13]. Taking advantage of that, and in order to increase the simplicity and efficiency of iPSCs characterization, we have developed a peptide based multiple reaction monitoring-mass spectrometry assay (MRM-LC-MS/MS) [14]. In this assay, iPS cell pellets are digested and resulting peptides are analyzed by UPLC-MS/MS (Fig. 1). At the first stage of method development, the list of potential pluripotency biomarkers was determined from the literature and current pluripotency tests (staining markers and self-renewal markers included in ScoreCard assay). Then, after sample preparation and analytical method refinement, the expression of five proteins, OCT4, SOX-2, LIN28, PODXL, and CD44, was demonstrated as being able to discriminate between pluripotent and nonpluripotent cells. The pluripotency signature measured by the mass spectrometry was confirmed by the generation of three-germ layer embryoid bodies and the successful terminal differentiation of these clones into neuronal cells [15, 16].

To summarize, a simple and high-throughput measurement of pluripotency signature can be achieved using as little as 200,000 cells by means of targeted mass spectrometry. Moreover, this assay is significantly simpler and faster than conventional genomic tests. Our test can be implemented in any laboratory, which has access and expertise in using triple quadrupole mass spectrometers and this method can be easily amended for measuring other proteins of interest in iPSC pellets, based on literature search or mass spectrometry discovery experiments.

2 Materials

Prepare all solutions using ultrapure water (prepared by purifying deionized water, to attain a sensitivity of 18 MΩ-cm at 25 °C) and analytical and mass spectrometry grade reagents. Prepare and store all reagents at room temperature (unless indicated otherwise).

2.1 Standard Peptides

1. The list of quantotypic peptides selected for pluripotency test is given in Table 1 (*see* **Note 1**).
2. 50:50 (v:v) acetonitrile (CH_3CN)–H_2O.

Table 1
List of peptides and transitions selected for the pluripotency test

Protein	Peptide	Transition	Cone voltage	Collision energy
LIN28	AAEEAPEEAPEDAAR	778.39 > 1084.28	43	20
LIN28	AAEEAPEEAPEDAAR	778.39 > 729.24	43	20
LIN28	MGFGFLSMTAR	609.41 > 882.40	58	18
LIN28	MGFGFLSMTAR	609.41 > 678.29	58	16
LIN28 SIL	MGFGFLSMT[A(13C3;15 N)] RAGVA	611.44 > 886.61	17	22
LIN28 SIL	MGFGFLSMT[A(13C3;15 N)] RAGVA	611.44 > 686.45	17	20
CD44	LVINSGNGAVEDR	672.47 > 904.52	39	26
CD44	LVINSGNGAVEDR	672.47 > 566.43	39	20
CD44	YGFIEGHVVIPR	462.82 > 612.52	15	14
CD44	YGFIEGHVVIPR	462.82 > 510.39	15	14
OCT4	AETLVQAR	444.34 > 687.62	34	10
OCT4	AETLVQAR	444.34 > 586.32	34	14
OCT4	WVEEADNNENLQEICK	645.33 > 847.18	14	15
OCT4	WVEEADNNENLQEICK	645.33 > 677.33	16	15
OCT4 SIL	AETL[V(13C5; 15 N)]QAR	447.39 > 693.57	17	22
OCT4 SIL	AETL[V(13C5; 15 N)]QAR	447.39 > 479.36	17	22
PODXL	LASVPGSQTVVVK	643.05 > 914.66	43	20
PODXL	LASVPGSQTVVVK	643.05 > 457.87	43	20
PODXL	ATFNPAQDK	496.28 > 819.56	43	16
PODXL	ATFNPAQDK	496.28 > 558.36	43	18
SOX-2	SEASSSPPVVTSSSHSR	568.06 > 743.49	2	10
SOX-2	SEASSSPPVVTSSSHSR	568.06 > 577.38	2	16
SOX-2	DMISMYLPGAEVPEPAAPSR	711.1 > 824.43	34	22
SOX-2	DMISMYLPGAEVPEPAAPSR	711.1 > 598.33	34	22
SOX-2 SIL	SEASSSPP[V(13C5;15 N)] VTSSSHSR	569.94 > 623.82	2	16
SOX-2 SIL	SEASSSPP[V(13C5;15 N)] VTSSSHSR	569.94 > 580.31	2	16

2.2 Standard Isotope Labeled (SIL) Analog of Standard Peptide

1. The list of SIL peptides selected for pluripotency test is given in Table 1 (*see* **Note 2**).

2.3 iPS Cell Harvest

1. 0.5 M EDTA pH 8.0, diluted to 0.5 mM in PBS.
2. PBS.

2.4 Digestion

1. Digestion buffer: 100 mM Tris–HCl, 6 M urea, 2 M thiourea, 2% ASB-14.
2. DL-dithiothreitol (DTT): 30 mg/mL in 100 mM Tris, pH 7.8 (*see* **Note 3**).
3. Iodoacetamide (IAA): 36 mg/mL in 100 mM Tris, pH 7.8. Protect from light (*see* **Note 4**).
4. Sequencing grade modified trypsin: 0.1 μg/μL in 50 mM ammonium bicarbonate buffer pH.
5. Store at −20 before ready to use.

2.5 Solid Phase Exchange (SPE)

1. Commercially available SPE C18 100 mg/1 mL column.
2. 0.2% trifluoroacetic acid (TFA).
3. 0.1% TFA.
4. 70% CH_3CN in 0.1% TFA.

2.6 Chromatography and Mass Spectrometry

1. Solvent A: 0.1% formic acid in dd H_2O.
2. Solvent B: 0.1% formic acid in CH_3CN.
3. Waters CORTECS UPLC C18 Column, 90 Å, 1.6 μm, 2.1 mm × 50 mm attached to a C18 VanGuard precolumn.
4. Acquity UPLC system (Waters) coupled to a XevoTM TQ-S triple quadrupole mass spectrometer (Waters) equipped with electrospray source.

3 Methods

Carry out all procedures at room temperature, unless stated otherwise. Always wear gloves to reduce keratin contamination.

3.1 Preparation of Standard Peptides and SIL Analogs of Standard Peptides

1. Resuspend synthetic peptides to 1 mg/mL stock concentration according to manufactures instructions. By default if instructions are not available, resuspend peptides in 50:50 (v:v) acetonitrile (CH_3CN)–H_2O.
2. Prepare the 1:10 dilutions of the peptide from the stock concentration and pool 1000 pmoles of each peptide into a low binding microcentrifuge tube.

3. Prepare several pools of 1000 pmols of SIL analogs of standard peptides.
4. Dry down in a speed-vac concentrator the final pool and store at −20 °C. Prepare several pools for future use.
5. Resuspend an aliquot of pooled 1000 pmol peptides in digestion buffer to obtain concentrations of 10 pmol/μL.

3.2 iPS Cell Harvest

1. Warm 0.5 mM EDTA to 37 °C.
2. Aspirate media from almost confluent well of iPS cells and add PBS to wash cells then immediately aspirate.
3. Add warmed 0.5 mM EDTA to cover surface of dish and incubate at 37 °C for 5 min.
4. Aspirate EDTA carefully to leave cells attached to dish, pipette PBS around surface of dish approximately three times to detach cells.
5. Collect detached cells and centrifuge at 300 × *g*, carefully remove PBS from cell pellet.

3.3 Digestion

Typical assay workflow is given in Fig. 1.

1. Dissolve harvested cell pellet (~two million cells/pellet) in 100 μL of digestion buffer (*see* **Note 5**). Vortex well and sonicate for 3 min.
2. Add 1 μL of 10 pmol/μL SIL analog of standard peptides with trypsin tag. Incubate for 1 h.
3. Add 7.5 μL of 30 mg/mL DTT. Incubate for 1 h.
4. Add 15 μL of 36 mg/mL IAA. Incubate for 30 min in dark (*see* **Note 4**).
5. Add 750 μL of dd H_2O.
6. Add 58 μL of 0.1 μg/μL trypsin.
7. Complete with H_2O to a final volume of 1000 μL.
8. Incubate in water bath at 37 °C for 12–16 h.

3.4 Purification of iPS Cell Digests Using SPE

1. Take 500 μL of digest and mix it thoroughly with 500 μL 0.2% TFA.
2. Prime the SPE column with 2 mL of 70% CH_3CN in 0.1% TFA.
3. Prime the SPE column with 2 mL of 0.1% TFA.
4. Apply the sample on the column and collect the flow-through.
5. Apply the flow-through on the column.
6. Wash the column with 1 mL of 0.1% TFA.

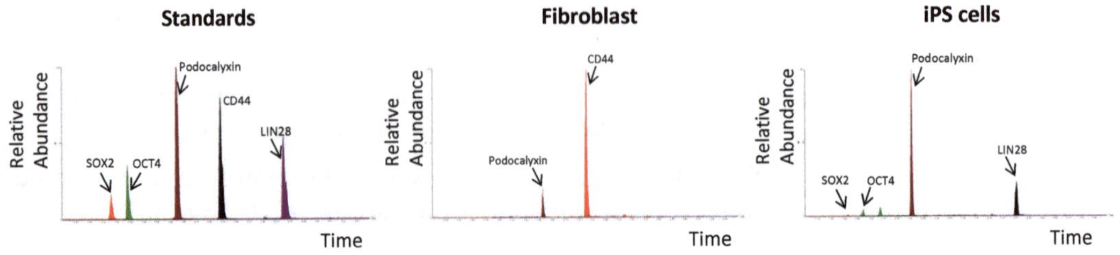

Fig. 2 Typical chromatographic results of the LC-MRM pluripotency assay. Adapted from [14]

7. Elute peptides with 500 μL of 70% CH_3CN in 0.1% TFA.
8. Freeze-dry the peptides.
9. Resuspend peptides in 50 μL of Solvent A.

3.5 LC-MRM Assay

Typical chromatographic results are given in Fig. 2.

1. Transfer purified peptide digests into glass insert vials.
2. Set up analytical UPLC conditions: the flow rate 0.8 mL/min, and a 7-min linear gradient starting at 100% solvent, ramping to 40% solvent B in 3.5 min, then to 99.9% Solvent B over 0.25 min (held for 1.5 min) and finally decreased to 100% solvent A in 0.25 min and reconditioned for 1.5 min.
3. Set up mass spectrometry conditions: positive ion mode, dynamic multiple reaction monitoring with a dwell time set up to automatic, and minimum 8 data points per peak.

   Transition list for peptides selected for pluripotency test are given in Table 1.
4. Set up sequence of analyzed samples, including standard curve, quality controls (QC), and blanks.
5. Inject 10 μL of sample per analysis.

3.6 Standard Curve

Representative standard curves for the pluripotency LC-MRM assay are given in Fig. 3.

1. Digest and purify one pooled 1000 pmol aliquot of standard peptides spiked with SIL. After purification, freeze-dry and resuspend peptide standards in Solvent A.
2. Dilute digested standards in Solvent A to the following concentrations: 0, 0.001, 0.005, 0.01, 0.05, 0.1, 0.2, 0.5, and 1 pmol/μL.
3. Transfer diluted standards into glass insert vials.
4. Inject 10 μL of standard per analysis.
5. Analyze standard peptides using the same LC-MRM method as for samples.

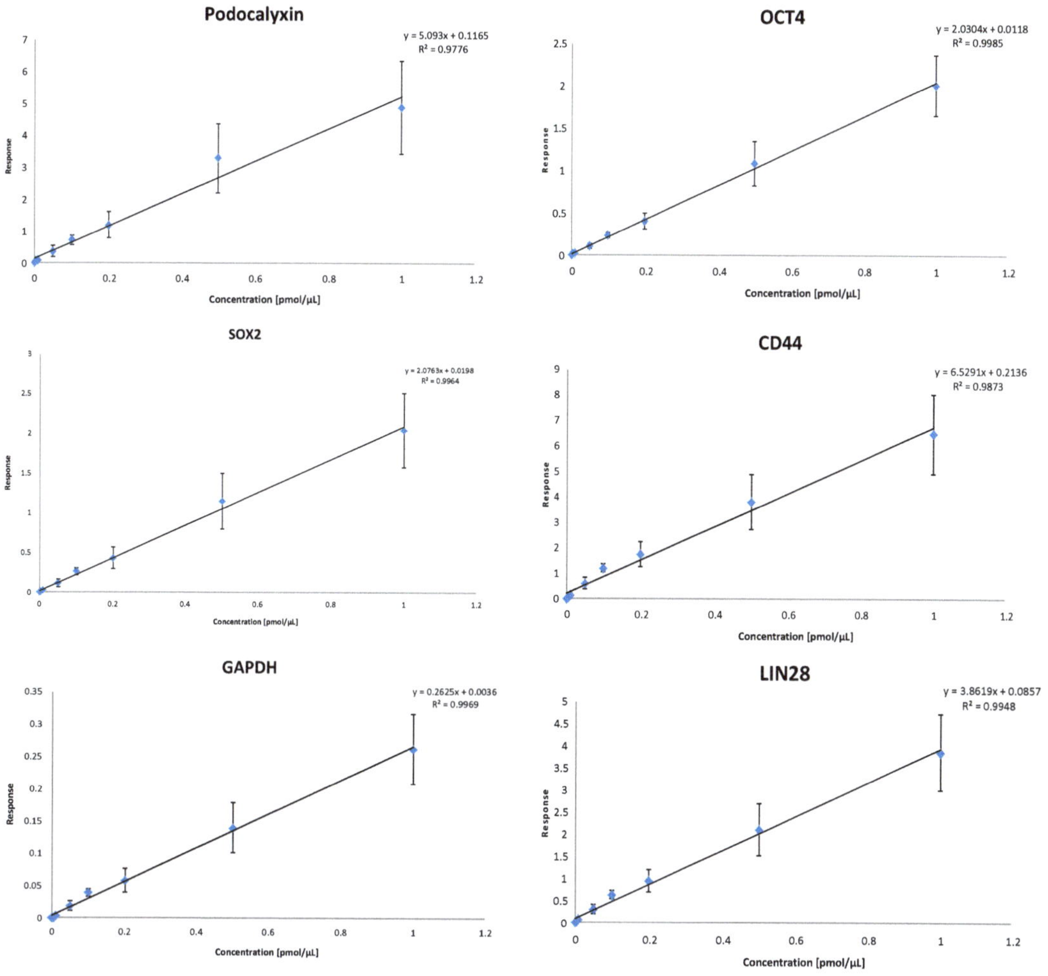

Fig. 3 Representative standard curves for the pluripotency LC-MRM assay

3.7 Quality Control

1. Pool 10 µL of each sample and spike in your standard peptides, then transfer into glass insert vial.
2. Before analyzing samples, inject 3 QC samples to ensure good sensitivity of the instrument.
3. Inject QC samples in intervals between samples, that is, between every ten samples.

3.8 Data Analysis

1. Use Waters MassLynx and TargetLynx V4.1 software to analyze the raw data.
2. Set up a method to ratio integrated peaks to SIL peptide analogs.
3. Verify the automatic integration to ensure accuracy.
4. Interpolate response values from the standard curve.

4 Notes

1. Marker peptide should be unique to the protein of interest (proteotypic) and representing the quantitative abundance of the protein (quantotypic). Standard marker peptides can be custom synthetized by various commercial companies.
2. In order to monitor for digestion efficiency, a trypsin tag should be added to the standard peptide sequence. The trypsin tag of 3–5 amino acids can be determined from the sequence of the protein, found in the Uniprot database. SIL peptides with trypsin tag can be custom-synthesized by various commercial companies.
3. DTT should be prepared fresh on the day and kept on ice until use.
4. Iodoacetamide is sensitive to light. Protect the stock solution of iodoacetamide and samples containing iodoacetamide from light by wrapping tubes in aluminum foil and keeping them in dark cabinet. Try to manipulate quickly in order to limit exposition to light. The stock solution should be prepared fresh on the day.
5. If two million cells are not available, the protocol can be adjusted to the lesser amount of cells, by decreasing the volumes of added reagents during digestion. The minimum of cells required is 200,000.

Acknowledgments

The authors declare no competing financial interests. The research leading to these results has received support from the Innovative Medicines Initiative Joint Undertaking under grant agreement no. 115439, resources of which are composed of financial contribution from the European Union's Seventh Framework Programme (FP7/2007-2013) and EFPIA companies' in-kind contribution. All research at Great Ormond Street Hospital NHS Foundation Trust and UCL Great Ormond Street Institute of Child Health is made possible by the NIHR Great Ormond Street Hospital Biomedical Research Centre. The views expressed are those of the authors and not necessarily those of the NHS, the NIHR the Department of Health, IMI JU, or EFPIA or the European Commission and are not liable for any use that may be made of the information contained therein. The authors would like to gratefully acknowledge the support of Leonard Wolfson and Peto foundations.

References

1. Takahashi K, Yamanaka S (2006) Induction of pluripotent stem cells from mouse embryonic and adult fibroblast cultures by defined factors. Cell 126:663–676
2. Takahashi K, Tanabe K, Ohnuki M, et al (2007) Induction of pluripotent stem cells from adult human fibroblasts by defined factors. Cell 131:861–872
3. Yu J, Hu K, Smuga-otto K, Tian S, Stewart R, Igor I, Thomson JA (2009) Human Induced Pluripotent Stem Cell Free of Vector Transgene Sequences. Science 324(5928):797–801
4. Fusaki N, Ban H, Nishiyama A, Saeki K, Hasegawa M (2009) Efficient induction of transgene-free human pluripotent stem cells using a vector based on Sendai virus, an RNA virus that does not integrate into the host genome. Proc Jpn Acad Ser B Phys Biol Sci 85(8):348–362
5. Nishimura K, Sano M, Ohtaka M, Furuta B, Umemura Y, Nakajima Y, Ikehara Y, Kobayashi T, Segawa H, Takayasu S, Sato H, Motomura K, Uchida E, Kanayasu-Toyoda T, Asashima M, Nakauchi H, Yamaguchi T, Nakanishia M (2011) Development of defective and persistent Sendai virus vector: A unique gene delivery/expression system ideal for cell reprogramming. J Biol Chem 286(6):4760–4771
6. Hockemeyer D, Soldner F, Cook EG, Gao Q, Mitalipova M, Jaenisch R (2008) A Drug-Inducible System for Direct Reprogramming of Human Somatic Cells to Pluripotency. Cell Stem Cell 3:346–353
7. Samavarchi-Tehrani P, Golipour A, David L, Sung HK, Beyer TA, Datti A, Woltjen K, Nagy A, Wrana JL (2010) Functional genomics reveals a BMP-Driven mesenchymal-to-Epithelial transition in the initiation of somatic cell reprogramming. Cell Stem Cell 7 (1):64–77
8. Kim D, Kim C, Moon J, Chung Y, Chang M, Han B, Ko S, Yang E, Cha KY, Lanza R, Kim K (2009) Generation of Human Induced Pluripotent Stem Cells by Direct Delivery of Reprogramming Proteins. Stem Cell 4(6):472–476
9. Miyoshi N, Ishii H, Nagano H, Haraguchi N, Dewi DL, Kano Y, Nishikawa S, Tanemura M, Mimori K, Tanaka F, Saito T, Nishimura J, Takemasa I, Mizushima T, Ikeda M, Yamamoto H, Sekimoto M, Doki Y (2011) Reprogramming of Mouse and Human Cells to Pluripotency Using Mature MicroRNAs. Stem Cell 8(6):633–638
10. Müller F-J, Schuldt BM, Williams R, Mason D, Altun G, Papapetrou EP, Danner S, Goldmann JE, Herbst A, Schmidt NO, Aldenhoff JB, Laurent LC, Loring JF (2011) A bioinformatic assay for pluripotency in human cells. Nat Methods 8(4):315–317
11. Tsankov AM, Akopian V, Pop R, Chetty S, Gifford CA, Daheron L, Tsankova NM, Meissner A (2015) A qPCR ScoreCard quantifies the differentiation potential of human pluripotent stem cells. Nat Biotechnol 33(11):1–15
12. Picotti P, Aebersold R (2012) Selected reaction monitoring–based proteomics: workflows, potential, pitfalls and future directions. Nat Methods 9(6):555–566
13. Ebhardt HA, Root A, Sander C, Aebersold R (2015) Applications of targeted proteomics in systems biology and translational medicine. Proteomics 15(18):3193–3208
14. Baud A, Wessely F, Mazzacuva F, McCormick J, Camuzeaux S, Heywood WE, Little D, Vowles J, Tuefferd M, Mosaku O, Lako M, Armstrong L, Webber C, Cader MZ, Peeters P, Gissen P, Cowley SA, Mills K (2017) A multiplex high-throughput targeted proteomic assay to identify induced pluripotent stem cells. Anal Chem 89:2440–2448
15. Fernandes HJR, Hartfield EM, Christian HC, Emmanoulidou E, Zheng Y, Booth H, Bogetofte H, Lang C, Ryan BJ, Sardi SP, Badger J, Vowles J, Evetts S, Tofaris GK, Vekrellis K, Talbot K, Hu MT, James W, Cowley SA, Wade-Martins R (2016) ER Stress and Autophagic Perturbations Lead to Elevated Extracellular α-Synuclein in GBA-N370S Parkinson's iPSC-Derived Dopamine Neurons. Stem Cell Reports 6:342–356
16. Haenseler W, Sansom SN, Buchrieser J, Newey SE, Moore CS, Nicholls FJ, Chintawar S, Schnell C, Antel JP, Allen ND, Cader MZ, Wade-Martins R, James WS, Cowley SA (2017) A Highly Efficient Human Pluripotent Stem Cell Microglia Model Displays a Neuronal-Co-culture-Specific Expression Profile and Inflammatory Response. Stem Cell Reports 8(6):1727–1742

Chapter 12

Measurement of Bile Acids as a Marker of the Functionality of iPSC-Derived Hepatocytes

Francesca Mazzacuva, Maëlle Lorvellec, Agostino Cilibrizzi, Kevin Mills, Peter Clayton, and Paul Gissen

Abstract

During the process of differentiation from induced pluripotent stem cells (iPSCs) to hepatocyte-like cells it is crucial to monitor the functionality of cells, in order to avoid an overestimation of the level of maturation. To this end, we propose bile acid profiling as a novel approach which is useful in determining the maturation of hepatocyte-like cells. The main advantages of the method are the simplicity and rapidity of test (i.e., single-step sample preparation followed by 3.5-min analysis), as well as the possibility to localize possible enzyme deficiencies by quantifying the accumulation of specific intermediates involved in the synthetic pathways.

Key words Bile acids, Liposomes, Mass spectrometry, Liquid chromatography, Cholesterol

1 Introduction

iPSC-derived hepatocytes are useful in vitro models of liver function and toxicity, as well as an alternative source of human hepatocytes [1, 2]. Although the production of hepatocytes from iPSCs has been extensively reported [3–6] and iPSC-derived hepatocytes seem to share many features with primary human hepatocytes, it as well known that they tend to be generally less mature [7]. Fetal hepatocytes may express proteins (e.g., CYP3A4) that are commonly used as adult markers [8, 9]; therefore, it is essential to characterize cellular function in detail, in order to avoid an overestimation of cell maturation [8].

In this context, we propose the bile acid (BA) profiling as a novel approach to monitor the stage of cellular differentiation. Indeed, BA synthesis and secretion is a specific function of hepatocytes, as 17 distinct enzymes are required to produce primary bile acids and at least 40 intermediates have been described for this biochemical pathway [10]. The main advantages of our method

Carl-Fredrik Mandenius and James A. Ross (eds.), *Cell-Based Assays Using iPSCs for Drug Development and Testing*, Methods in Molecular Biology, vol. 1994, https://doi.org/10.1007/978-1-4939-9477-9_12,
© Springer Science+Business Media, LLC, part of Springer Nature 2019

are the rapidity and accuracy of responses which are achieved through a single-step sample preparation followed by a 3.5 min analysis with liquid chromatography–tandem mass spectrometry (LC-MS/MS). Moreover, as the enzymes involved in BA synthesis are located in different cellular organelles, the accumulation of specific BA intermediate can potentially give useful information about any enzyme deficiency [10, 11] (e.g., C27 hydroxylation occurs mainly in the mitochondria, whereas further ring structure modification is performed in the cytoplasm; side chain modification and conjugation mainly are carried out in peroxisomes [12]). In some circumstances, when BAs are already present in the media, because used as nutrients for iPSC-derived hepatocytes, it is essential to incubate the cells with isotopically labeled cholesterol prior to the measurement of the production of isotopically labeled BAs.

2 Materials

1. Bile acids and their deuterium labeled internal standards were purchased from CDN Isotopes.
2. 5-Cholesten-3β-ol-2,3,4-$^{13}C_3$ ($^{13}C_3$-cholesterol) was purchased from Sigma-Aldrich.
3. Phospholipon® 90 (P90H) was donated by Phospholipid GmbH.
4. HPLC grade methanol was purchased from Fisher Scientific.
5. Water was purified using a Milli-Q system (Millipore).

3 Methods

3.1 Direct Quantitation of Bile Acids from Media

1. Extract bile acids by adding 4 volumes of methanol (containing 20 nM of labeled internal standards) to 50 μL of media (15 min; −20 °C) (*see* **Note 1**). The negative control consisted of cell-free media and the positive control was media from primary human hepatocytes.
2. Remove the supernatant from the cell pellet after centrifugation (5 min; 9000 × *g*).
3. Inject 20 μL into the LC-MS/MS system and perform the quantitation by using calibration curves in methanol as previously reported [10] (Fig. 1).

3.2 Quantitation of Isotopically Labeled Bile Acids After the Administration of $^{13}C_3$ Cholesterol to Cells

A direct and precise quantitation of the amount of BAs secreted from the cells is not always possible as some of the media, specifically used for hepatocyte cultures, contain a variable amount of BAs. To overcome quantitation problems and perform an accurate measure of BAs effectively secreted from the cells, we have fed the

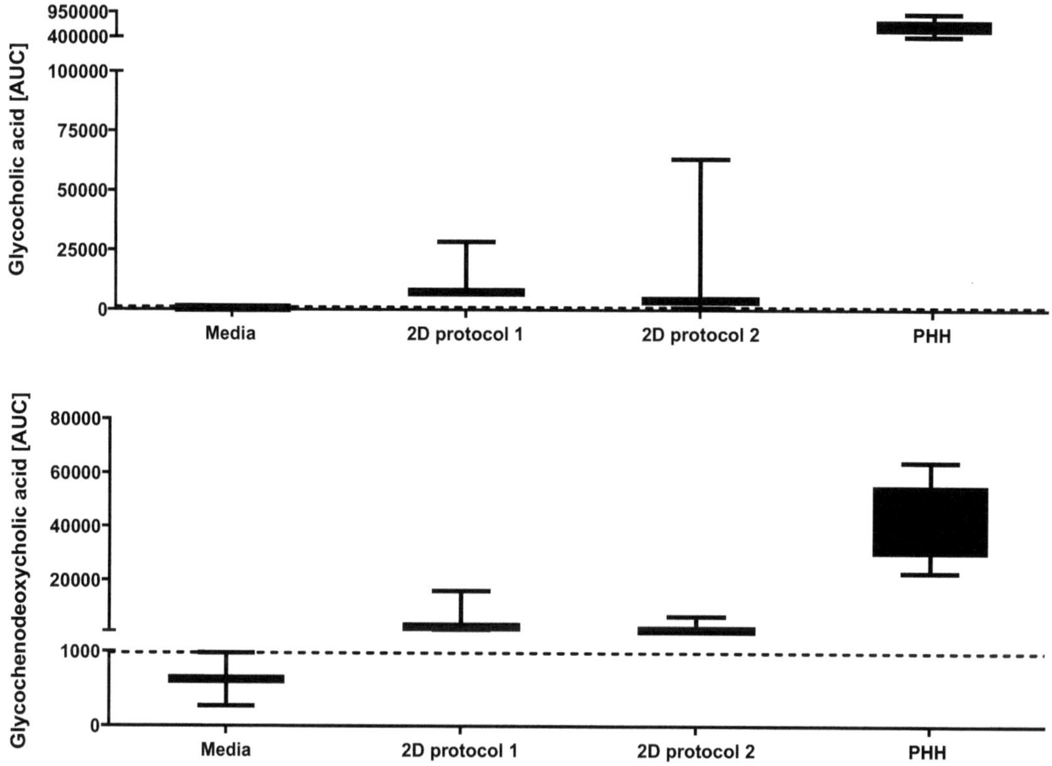

Fig. 1 Representative graphs displaying the direct quantitation of glycocholic acid and glycochenodeoxycholic acid (*y* axes) secreted from two batches of iPSC-derived hepatocytes (obtained following two different 2D protocols adapted from the method originally published by Sullivan and coauthors [6] (*x* axes); media is used as a negative control and primary human hepatocytes (PHH) are used as a positive control

cells with $^{13}C_3$-cholesterol and measured the amount of $^{13}C_3$-labeled BAs in the media. As cholesterol is not soluble in aqueous media, $^{13}C_3$-cholesterol liposomes were prepared, according to the protocol below, and administered to the cells (*see* **Note 2**).

3.3 Preparation of $^{13}C_3$-Cholesterol Liposomes

1. Resuspend 12 mg of P90H and 2.5 mg of $^{13}C_3$ cholesterol in 2 mL of sterile physiological water in a glass bottle. Add the antimicrobial Primocin.
2. Sonicate at room temperature for 10 min (power 14) (it is recommended to conduct **steps 1** and **2** under a fume hood, in order to avoid cell contamination).
3. Let the suspension cool down at room temperature before using it to incubate the cells (*see* **Note 3**).
4. Dilute the liposomes to 2 mM in the media before filtering with a 0.4 μm filter and further dilute to 1 mM.
5. Incubate the cells for 24 h with the liposomal suspension.

6. Harvest the media and extract the $^{13}C_3$-BAs as described at **steps 1** and **2** of Subheading 3.1.
7. Inject 20 μL into the LC-MS/MS system and perform the quantitation by using calibration curves prepared in methanol as previously reported [10] (*see* **Notes 4** and **5**).

4 Notes

1. The isotopically labeled BAs used as internal standards are: $^{2}H_4$-cholic acid, $^{2}H_4$-glycocholic acid, $^{2}H_5$-taurocholic acid, $^{2}H_4$-chenodeoxycholic acid, $^{2}H_4$-glycochenodeoxycholic acid, $^{2}H_4$-taurochenodeoxycholic acid.
2. Alternatively, isotopically labeled unconjugated BAs can be used in the place of $^{13}C_3$-cholesterol. For instance, by feeding primary human hepatocytes (positive control) with $^{2}H_6$-deoxycholic acid or $^{2}H_9$-chenodeoxycholic acid, we could measure the production of their glycine conjugates. However, these compounds resulted in high cell toxicity, by acting as powerful detergents and thus leading to cell lysis. In addition, the conjugation with glycine gives only limited information about cellular functionality.
3. Use the liposomal suspension within a few days to avoid possible deterioration. If not used within the same day, store the suspension at −4 °C in the fridge.
4. Considering $^{13}C_3$-cholesterol is used for the synthesis of BAs, it is required to add 3 mass units to all the precursor ions of the MRM panel.
5. For an accurate quantitation of $^{13}C_3$-BAs, it is important to consider the interference produced by the unlabeled BAs which are contained into the media. The calculated isotopic contribution of unlabeled BAs that can interfere with the MS signal of $^{13}C_3$-BAs is reported in Table 1 (highlighted in yellow). Therefore, it is essential to firstly quantitate the unlabeled BAs, in order to calculate their isotopic interference on $^{13}C_3$-BAs and subtract it (Fig. 2).

Acknowledgments

This work was supported by StemBANCC (support from the Innovative Medicines Initiative joint undertaking under grant 115439-2, whose resources are composed of financial contribution from the European Union [FP7/2007-2013] and EFPIA companies' in-kind contribution).

Table 1
Theoretic isotopic pattern of representative BAs

	(m/z)	%
Cholic acid	407	100
	408	26.8
	409	4.4
	410	0.5
Chenodeoxycholic acid	391	100
	392	26.7
	393	4.2
	394	0.5
Glycocholic acid	464	100
	465	29.4
	466	5.4
	467	0.7
Glycochenodeoxycholic acid	448	100
	449	29.4
	450	5.2
	451	0.6
Taurocholic acid	514	100
	515	30.3
	516	10.2
	517	2
Taurochenodeoxycholic acid	498	100
	499	30.2
	500	10
	501	1.9

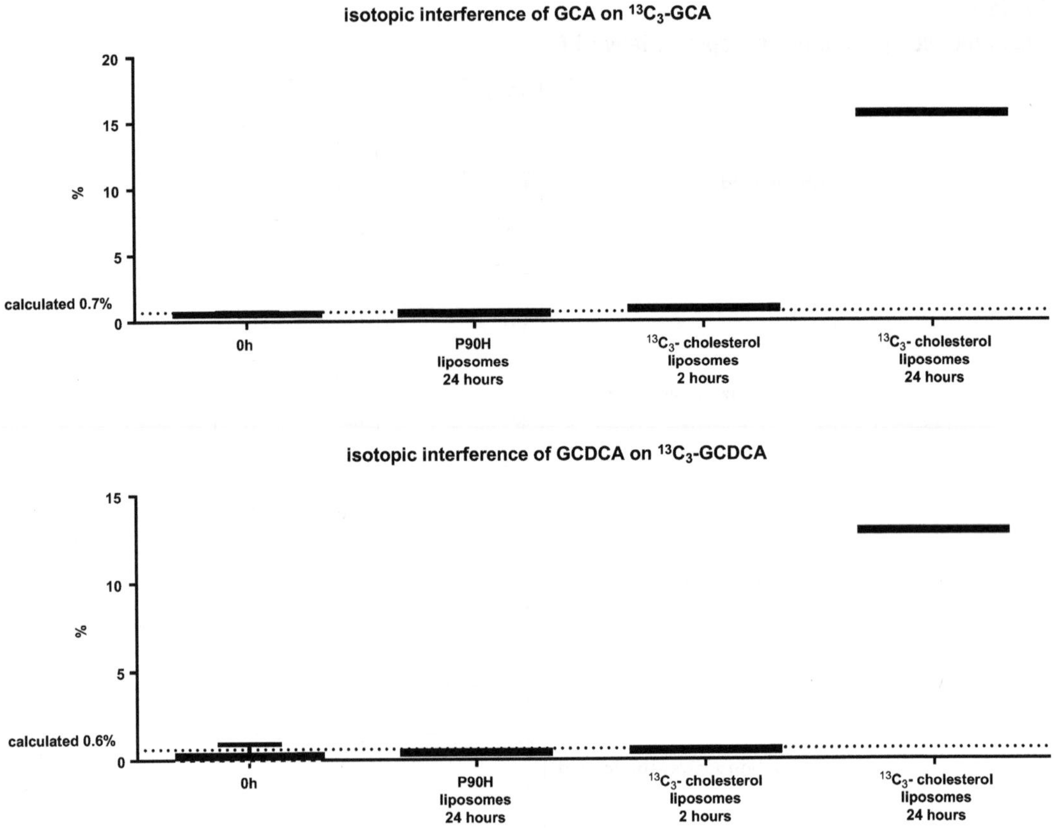

Fig. 2 Representative graphs displaying the isotopic interference produced by the amount of glycocholic acid (GCA) or glycochenodeoxycholic acid (GCDCA), contained into the media, on the quantitation of $^{13}C_3$-glycocholic acid ($^{13}C_3$-GCA) or $^{13}C_3$-glycochenodeoxycholic acid ($^{13}C_3$-GCDCA). Cells have been incubated with liposomes made only with P90G (negative control) and liposomes made with P90G and $^{13}C_3$-cholesterol (for 2 and 24 h in order to select the appropriate incubation time to measure the production of BAs).The contribution of GCA in the medium to the measured $^{13}C_3$-GCA remains constant and the true $^{13}C_3$-GCA produced by the cells from $^{13}C_3$-cholesterol is clearly measurable above this background after 24 h incubation

References

1. Baxter MA, Rowe C, Alder J et al (2010) Generating hepatic cell lineages from pluripotent stem cells for drug toxicity screening. Stem Cell Res 5:4–22
2. Zhang Z, Liu J, Liu Y et al (2013) Generation, characterization and potential therapeutic applications of mature and functional hepatocytes from stem cells. J Cell Physiol 228:298–305
3. J C, Zhao Y, Liu Y et al (2007) Directed differentiation of human embryonic stem cells into functional hepatic cells. Hepatology 45:1229–1239
4. Agarwal S, Holton KL, Lanza R (2008) Efficient differentiation of functional hepatocytes from human embryonic stem cells. Stem Cells 26:1117–1127
5. Basma H, Soto-Gutiérrez A, Yannam GR et al (2009) Differentiation and transplantation of human embryonic stem cell-derived hepatocytes. Gastroenterology 136:990–999
6. Sullivan GJ, Hay DC, Park IH et al (2010) Generation of functional human hepatic endoderm from human induced pluripotent stem cells. Hepatology 51:329–335

7. Si-Tayeb K, Noto FK, Nagaoka M et al (2010) Highly efficient generation of human hepatocyte-like cells from induced pluripotent stem cells. Hepatology 51:297–305
8. Baxter M, Withey S, Harrison S et al (2015) Phenotypic and functional analyses show stem cell-derived hepatocyte-like cells better mimic fetal rather than adult hepatocytes. J Hepatol 62:581–589
9. Rowe C, Gerrard DT, Jenkins R et al (2013) Proteome-wide analyses of human hepatocytes during differentiation and dedifferentiation. Hepatology 58:799–809
10. Mazzacuva F, Mills P, Mills K et al (2016) Identification of novel bile acids as biomarkers for the early diagnosis of Niemann-Pick C disease. FEBS Lett 590:1651–1562
11. Vilarinho S, Sari S, Mazzacuva F et al (2016) *ACOX2* deficiency: A disorder of bile acid synthesis with transaminase elevation, liver fibrosis, ataxia, and cognitive impairment. PNAS 113:11289–11293
12. Bove KE, Heubi JE, Balistreri WF et al (2004) Bile acid synthetic defects and liver disease: a comprehensive review. Pediatr Dev Pathol 7:315–334

Chapter 13

Mass Spectrometry Measurement of Albumin–Alpha Fetoprotein Ratio as an Indicator of iPSC-Derived Hepatocyte Differentiation

Francesca Mazzacuva, Maëlle Lorvellec, Agostino Cilibrizzi, Kevin Mills, Wendy E. Heywood, Peter Clayton, and Paul Gissen

Abstract

During the process of differentiation from induced pluripotent stem cells (iPSCs) to hepatocytes it is crucial to monitor the levels of cellular maturation. We present a new method to evaluate the stage of differentiation based on the monitoring of the ratio between two plasma proteins typically secreted by hepatocytes, that is, albumin and alpha-fetoprotein. This ratio is particularly useful for the direct comparison of cells grown in different conditions, avoiding typical processes of standardization for the cell number (i.e., variation of cell quantity due to the use of different seeding densities and different growth vessels/supports or difficulties in establishing the effective cell viability due to the use of bioreactors or other 3D devices). Our analysis is performed via liquid chromatography–tandem mass spectrometry which allows a precise, selective, and reproducible quantitation of low-abundance proteins.

Key words Albumin, Alpha fetoprotein, Liquid chromatography, Mass spectrometry, Plasma proteins, Bioreactor

1 Introduction

To monitor the process of differentiation from iPSC to hepatocytes, we propose the quantitation of two plasma proteins, albumin and alpha fetoprotein (AFP), using liquid chromatography coupled to tandem mass spectrometry (LC-MS/MS). Both proteins are secreted from hepatocytes. However, albumin production (negligible in fetal cells) has been shown to increase during the maturation of hepatocytes, while AFP is a marker of immature liver cells, being secreted only from hepatoblasts and fetal hepatocytes [1, 2].

With the aim to perform a direct comparison between cells grown in different conditions, we demonstrate that the albumin–alpha fetoprotein ratio is a new effective indicator of the stage of hepatocyte maturation. Specifically, the use of this ratio allows to

Carl-Fredrik Mandenius and James A. Ross (eds.), *Cell-Based Assays Using iPSCs for Drug Development and Testing*, Methods in Molecular Biology, vol. 1994, https://doi.org/10.1007/978-1-4939-9477-9_13,
© Springer Science+Business Media, LLC, part of Springer Nature 2019

overcome typical standardization issues, due to the presence of a different number of cells on the plate. This aspect is crucial in the case of bioreactors or other devices, where the cell count is not always possible or easily performed.

Several methods have been reported in literature for the analysis of plasma proteins from cultured cells, for instance two-dimensional electrophoresis, ELISA, RIA and mass spectrometry analysis [3–6]. Among them, the separation via liquid chromatography followed by detection and quantitation via mass spectrometry presents numerous advantages. Firstly, it offers high selectivity and specificity (i.e., absence of cross-interference between proteins having a similar sequence) through multiple reaction monitoring (MRM) analysis [7]. Additionally, the limited presence of contaminants and matrix effects (due to the separation via liquid chromatography), as well as the high signal-to-noise ratio (which allows for an accurate, precise, and reproducible quantitation also of low-abundance proteins), are also key advantages.

2 Materials

1. Tryptic peptides and their respective isotopically labeled internal standards (GenScript, NJ, USA).
2. Sequencing grade modified trypsin (Promega).
3. HPLC grade acetonitrile (Fisher Scientific).
4. Water was purified using a Milli-Q system (Millipore).
5. The UPLC-MS/MS instrument consisted of a Waters ACQUITY UPLC coupled to a Xevo TQ-S triple quadrupole mass spectrometer with an electrospray ionization source. The mass spectrometer was operated in positive ion mode and data were acquired using MassLynx V4.1 software. Chromatographic separations were conducted through a Waters ACQUITY UPLC™ BEH C18 column (1.7 μm, 2.1 × 50 mm) kept at 40 °C. The mobile phases, freshly prepared before the use, were 0.01% formic acid (A) and 0.01% formic acid in acetonitrile (B). The flow rate was 700 $\mu L{\cdot}min^{-1}$ and separations were conducted under the following chromatographic conditions: 97% solvent A for 0.2 min, decreased to 60% over 6.8 min before being further decreased to 0 in 0.01 min; solvent A was then maintained for 1.50 min at 0 before being quickly increased to 97% over 0.01 min. Column equilibration time was 1.5 min, with a total run time of 10 min. The injection volume was 20 μL. Mass spectrometric conditions were as follows: capillary voltage 2.7 kV, desolvation temperature 600 °C, cone gas flow 150 L/h, desolvation gas flow 1200 L/h, collision gas flow

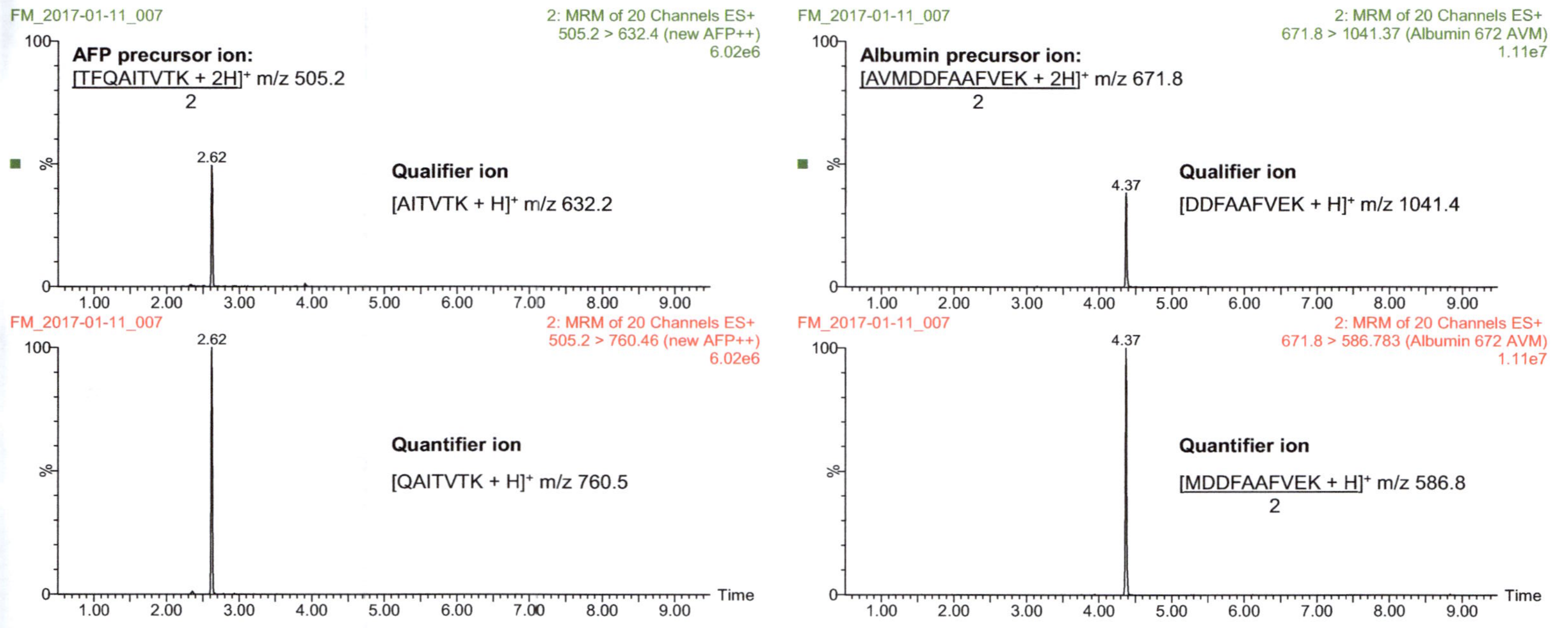

Fig. 1 Representative chromatograms for AFP and albumin obtained from iPSC-derived hepatocytes

Table 1
Protein, peptide sequence, retention time, precursor ion, cone voltage, product ion, and collision energy of albumin, AFP, and their isotopically labeled internal standards

Protein	Peptide sequence	Retention time (min)	Precursor ion (*m/z*)	Cone voltage (V)	Product ion (*m/z*)	Collision energy (V)
Albumin	AVMDDFAAFVEK	4.37	671.8	30	586.8 (quantifier)	16
					1041.4 (qualifier)	20
IS-albumin	EQLK. AVMDDFAAF[V (13C5,15 N)]EK		574.8		589.9 (quantifier)	16
					1047.6 (qualifier)	20
AFP	TFQAITVTK	2.62	505.2	50	760.5 (quantifier)	14
					632.4 (qualifier)	14
IS-AFP	NFG.TRTFQA[I (13C6,15 N)] TVTK		508.3		767.3 (quantifier)	14
					639.3 (qualifier)	14

0.15 mL/min and nebulizer gas flow 7 bar. Dwell time was set at 3 ms for each analyte. Stable isotope-labeled peptides were employed as internal standards. Quantitation was performed by using one unique peptide per protein and two transitions per peptide (Fig. 1). Multiple reaction monitoring (MRM) parameters are presented in Table 1.

3 Methods

3.1 Sample Preparation and LC-MS/MS Analysis

1. Precipitate the proteins by adding 4 volumes of methanol to 50 μL of media (15 min; −20 πC). Negative controls: cell media; positive controls: primary human hepatocytes (*see* **Notes 1–3**).
2. Remove the supernatant after centrifugation (5 min; 20817 rcf).
3. Dissolve the pellets at room temperature in 20 μL of denaturing buffer (100 mM Tris base, 6 M urea, 2 M thiourea and 2% ASB-14, adjusted to pH 7.8 with HCl) containing 1 μM of labeled internal standards (*see* **Notes 4** and **5**).
4. Add 3 μL of 0.2 M 1,4-dithioerythritol freshly solubilized in the denaturing buffer (30 min; rt) (*see* **Note 6**).
5. Add 6 μL of 0.2 M iodoacetamide freshly solubilized in the denaturing buffer (15 min; rt) (*see* **Notes** 7 and **8**).
6. Add 150 μL of water to quench the reaction.
7. Add 10 μL of 0.1 mg/mL trypsin (over night; 37 πC) (*see* **Note 9**).

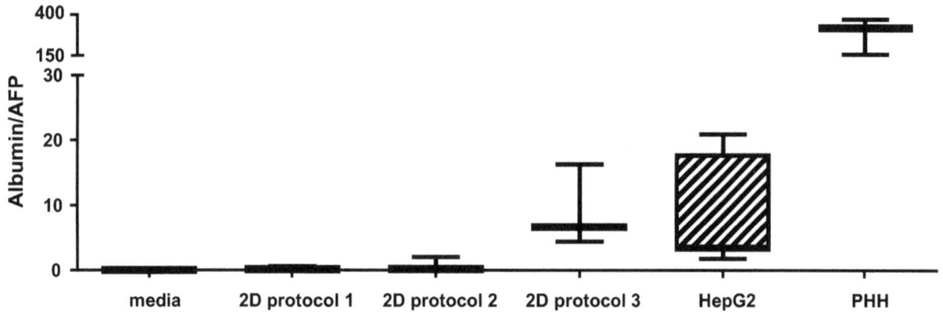

Fig. 2 Representative graph displaying albumin–AFP ratio (*y* axes) vs. cell type (*x* axes): media, used as a negative control; comparison between three batches of iPSC-derived hepatocytes (obtained following three different 2D protocols adapted from the method originally published by Sullivan and coauthors [1]); HepG2 cells; primary human hepatocytes (PHH), used as a positive control

8. Spin (5 min; 13,000 rpm) and transfer the supernatant into injection vials.
9. Inject 20 μL into the LC-MS/MS system (*see* **Note 10**).
10. Quantify the two proteins by using calibration curves prepared into the same matrix (Fig. 2).

4 Notes

1. This procedure describes the analysis of albumin and AFP from the supernatant of cells grown in culture. In the case of cells grown in a 3D system (e.g., decellularized liver scaffold), we found it was essential to add further purification/concentration steps, in order to remove collagen and other contaminants released from the tissues. These steps are particularly crucial in the presence of big volumes and low protein abundance (i.e., bioreactors).
   (a) Centrifuge 15 mL of supernatant by using 100 KDa cutoff 15 mL Amicon® Ultra centrifugal filters (25 °C; 20 min; 4000 × *g*). Wash the concentrated solution with water (25 °C; 2 min; 4000 × *g*). This step helps reducing the amount of collagen and other big molecular weight proteins.
   (b) Centrifuge the filtrates by using 50 KDa cutoff 15 mL Amicon® Ultra centrifugal filters (15 °C; 20 min; 4000 × *g*). This step helps concentrate albumin and AFP (MW between 66 and 69 KDa) by removing lower MW contaminants (e.g., salts from the buffer).

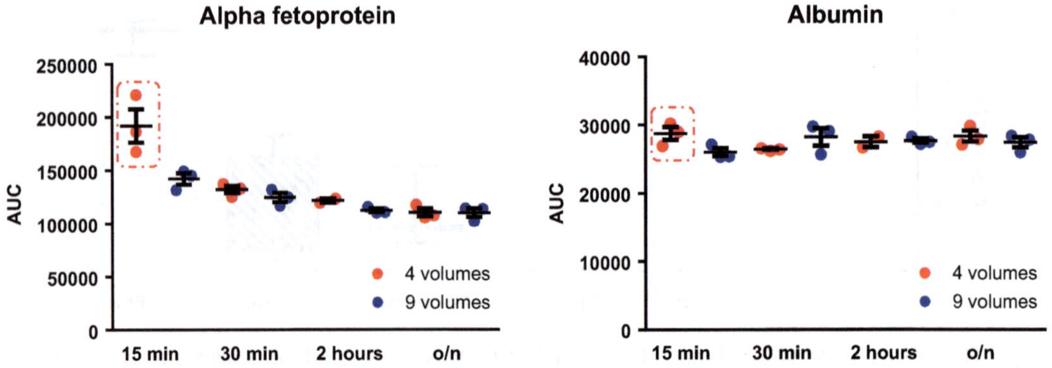

Fig. 3 Optimization of the precipitation of AFP and albumin by using 4 or 9 volumes of methanol for 15 min, 30 min, 2 h, or o/n

(c) Precipitate all the remaining concentrated volume (about 200–250 μL) by adding 4 volumes of methanol, as described in **step 1** of Subheading 3.1.

2. For the precipitation of albumin and AFP, comparable results can be obtained by using cold acetone in the place of methanol in the same conditions.

3. Four or nine volumes of methanol have been used at different times to improve the precipitation of albumin and AFP. Optimal results were obtained by using 4 volumes of methanol for 15 min. By increasing volumes or time, we observed a decrease in the amount of precipitated albumin and AFP, probably due to their higher solubilization in the solvent (Fig. 3).

4. Freshly dilute the internal standards (IS) into the buffer from a 0.2 e^{-4} M stock solution in MilliQ water.

   IS contains a tag that is recognized and cleaved by trypsin. This tag is useful to exclude problems with trypsin in case the unlabeled peptides are not detected.

5. Vortex and quickly spin before leaving the mixture on a shaker at rt for at least 10 min to favor the solubilization.

6. Addition of 1,4-dithioerythritol is essential to reduce disulfide bridges, preventing the formation of intramolecular and intermolecular bonds. Reaction is performed on a shaker.

7. Iodoacetamide is used to alkylate SH residues, preventing the reformation of SS bonds, and thus improving the digestion step with trypsin. Iodoacetamide is unstable and light-sensitive; therefore, it is important to prepare a fresh solution immediately before use and perform the carbamidomethylation in the dark (e.g., using aluminum foil). The reaction is carried out on a shaker.

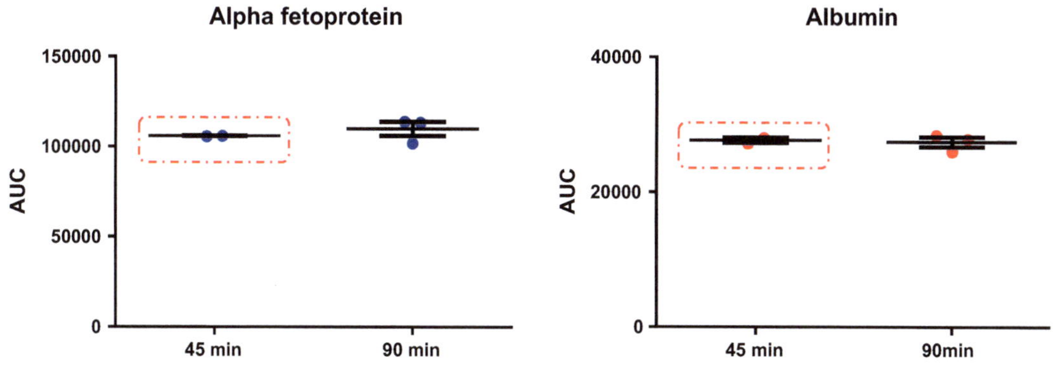

Fig. 4 Optimization of the reaction times with 1,4-dithioerythritol and iodoacetamide

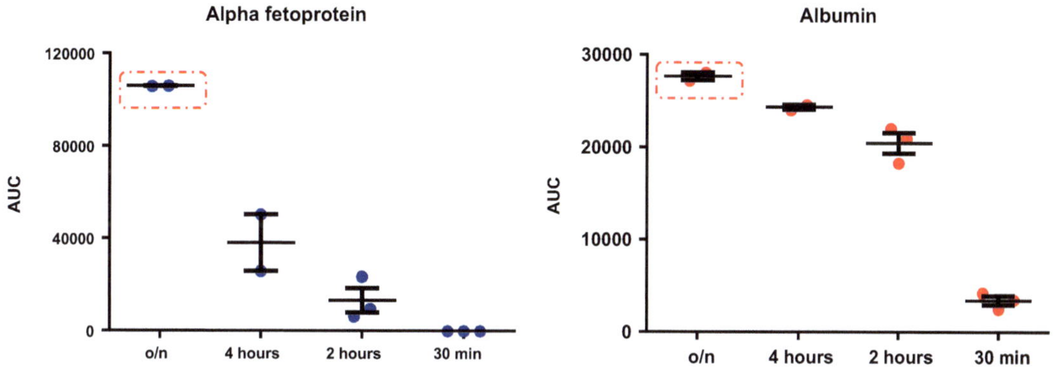

Fig. 5 Optimization of the tryptic digestion of AFP and albumin

8. 1,4-Dithioerythritol and iodoacetamide were reacted for 45 and 90 min in total (30 + 15 min vs 60 + 30 min) and results were comparable (Fig. 4).
9. Tryptic digestion was performed for 30 min, 2 h, 4 h, and o/n in order to choose the best condition for the digestion of albumin and AFP (Fig. 5).
10. Tryptic peptides, with sequences unique to human albumin (AVMDDFAAFVEK) and human AFP (TFQAITVTK) and their respective isotopically labeled internal standards (EQLK.AVMDDFAAF[V(13C5,15 N)]EK; NFG.TRTFQA[I (13C6,15 N)]TVTK), were chosen according to the criteria reported by Liebler and Zimmerman [8]. Bovine (present in the media) and mouse peptides (present from the tissue scaffolds in the case of cells grown in a 3D system) were excluded by homology comparison of the amino acid sequences with human sequences.

Acknowledgments

This work was supported by StemBANCC (support from the Innovative Medicines Initiative joint undertaking under grant 115439-2, whose resources are composed of financial contribution from the European Union [FP7/2007-2013] and EFPIA companies' in-kind contribution).

References

1. Sullivan GJ, Hay DC, Park IH et al (2010) Generation of functional human hepatic endoderm from human induced pluripotent stem cells. Hepatology 51:329–335
2. Guigoz Y, Werffeli P, Favre D et al (1987) Aggregate cultures of foetal rat liver cells: development and maintenance of liver gene expression. Biol Cell 60:163–171
3. Kurosawa H, Yuminamochi E, Yasuda R, Amano Y (2003) Morphology and albumin secretion of adult rat hepatocytes cultured on a hydrophobic porous expanded polytetrafluoroethylene membrane. J Biosci Bioeng 95:59–64
4. Buyl K, De Kock J, Bolleyn J et al (2015) Measurement of albumin secretion as functionality test in primary hepatocyte cultures. In: Vinken M, Rogiers V (eds) Protocols in in vitro hepatocyte research. Methods in molecular biology (methods and protocols), vol 1250. Humana Press, New York
5. Yu LR, Zeng R, Shao XX et al (2000) Identification of differentially expressed proteins between human hepatoma and normal liver cell lines by two-dimensional electrophoresis and liquid chromatography-ion trap mass spectrometry. Electrophoresis 21:3058–3068
6. Pressey JG, Pressey CS, Robinson G et al (2011) 2D-difference gel electrophoretic proteomic analysis of a cell culture model of alveolar rhabdomyosarcoma. J Proteome Res 10:624–636
7. Shi T, Su D, Liu T et al (2012) Advancing the sensitivity of selected reaction monitoring-based targeted quantitative proteomics. Proteomics 12:1074–1092
8. Liebler DC, Zimmerman LJ (2013) Targeted quantitation of proteins by mass spectrometry. Biochemistry 52:3797–3806

Part III

High Content Imaging

Chapter 14

Seeding Induced Pluripotent Stem Cell-Derived Neurons onto 384-Well Plates

Daniel Little, Christin Luft, Oliver Pezzini-Picart, Olukunbi Mosaku, Robin Ketteler, Michael J. Devine, and Paul Gissen

Abstract

Induced pluripotent stem cell (iPSC) derived neurons are an excellent in vitro model of neurological diseases that are often used in early stage drug discovery projects. Thus far, the use of iPSC-derived cells in small molecule drug screening has been limited, and one of the reasons for this has been the challenge of miniaturization of iPSC culture and differentiation in low volume microwell plate formats. Here we describe a method of seeding iPSC-derived neurons into 384-well plates towards the end of the differentiation procedure. This method covers coating the plates with substrates to aid attachment, dissociation of the cells into a single cell suspension, and seeding onto 384-well plates to give an even distribution of neurons. This method facilitates the use of iPSC-derived neurons for high-content imaging, whole-well assays, and small-molecule drug screening.

Key words Induced pluripotent stem cells, Neuronal cells, 384-well plates, Microplates, Cell culture, Drug discovery

1 Introduction

Over the past decade, tremendous progress has been made in our understanding of neurodegenerative diseases at the cellular level. In large part, this has been facilitated by the availability of induced pluripotent stem cell (iPSC) derived neuronal cell cultures, which—for the first time—overcame problems associated with accessibility to ex situ studies of patient brain material that has hampered progress in the past. iPSC-derived neurons have attracted attention as an exciting new model of neurological diseases. Since these cells are of human origin and can be generated using cells from a patient with a particular disease they have been proposed to be more relevant than animal cells or immortalized cell lines. For drug discovery applications, cells need to be seeded into microplates to enable the screening of large numbers or compounds in a short

Carl-Fredrik Mandenius and James A. Ross (eds.), *Cell-Based Assays Using iPSCs for Drug Development and Testing*, Methods in Molecular Biology, vol. 1994, https://doi.org/10.1007/978-1-4939-9477-9_14,
© Springer Science+Business Media, LLC, part of Springer Nature 2019

period of time. The use of microplates also reduces the amount of expensive reagents needed to perform experiments. This is particularly important when using iPSC-derived neurons because the differentiation of these cells is time-consuming and costly, therefore efficient use of this precious resource is important. Some recent studies have used iPSC-derived neurons in 384-well plates [1–4] and multiple protocols exist which include seeding of iPSC-derived neurons [5–8]; however, there are no detailed protocols for seeding iPSC-derived neurons onto 384-well plates.

The described procedure occurs in three steps: coating, seeding, and assaying. To coat the plates this protocol uses poly-L-ornithine, a synthetic positively charged amino acid chain widely used to enhance cell attachment to plasticware and glassware alongside laminin and fibronectin, which are extracellular matrix proteins that are regularly used to maintain attachment of neuronal cell cultures. Here we use a two-step coating method whereby a tissue culture-treated plastic microplate is first coated for 24 h with poly-L-ornithine. This is then removed and the plate is coated for a further 24 h with a combination of laminin and fibronectin. This coating method ensures that cells are evenly distributed throughout the well and do not become detached during subsequent media changes or downstream applications. The coating solution is then removed immediately before iPSC-derived neurons are seeded onto the coated plate. For this, the cell suspension is passed through a 40 μm nylon strainer to remove any cell aggregates that would negatively impact the even distribution of cells. Cell seeding is performed using a liquid dispenser, which results in equivalent seeding efficiency as manual pipetting (Fig. 1). Once seeded, the cells can undergo multiple wash steps involved in downstream applications, such as immunofluorescence staining, without detaching from the plate (Fig. 1).

2 Materials

2.1 Coating Solutions

1. Poly-L-ornithine, molecular weight 30,000–70,000 0.01% sterile-filtered solution, stock concentration 100 μg/ml, store at 4 °C, working concentration 15 μg/ml, diluted in PBS on the day of use.
2. Laminin and fibronectin, stock concentration of each 1 mg/ml, store at −20 °C, combine and dilute in PBS to give working concentration of 5 μg/ml of each on the day of use.

2.2 Cell Dissociation

1. Accutase solution, sterile-filtered, used as supplied.
2. Cell lifter.
3. Falcon 40 μm cell strainer.

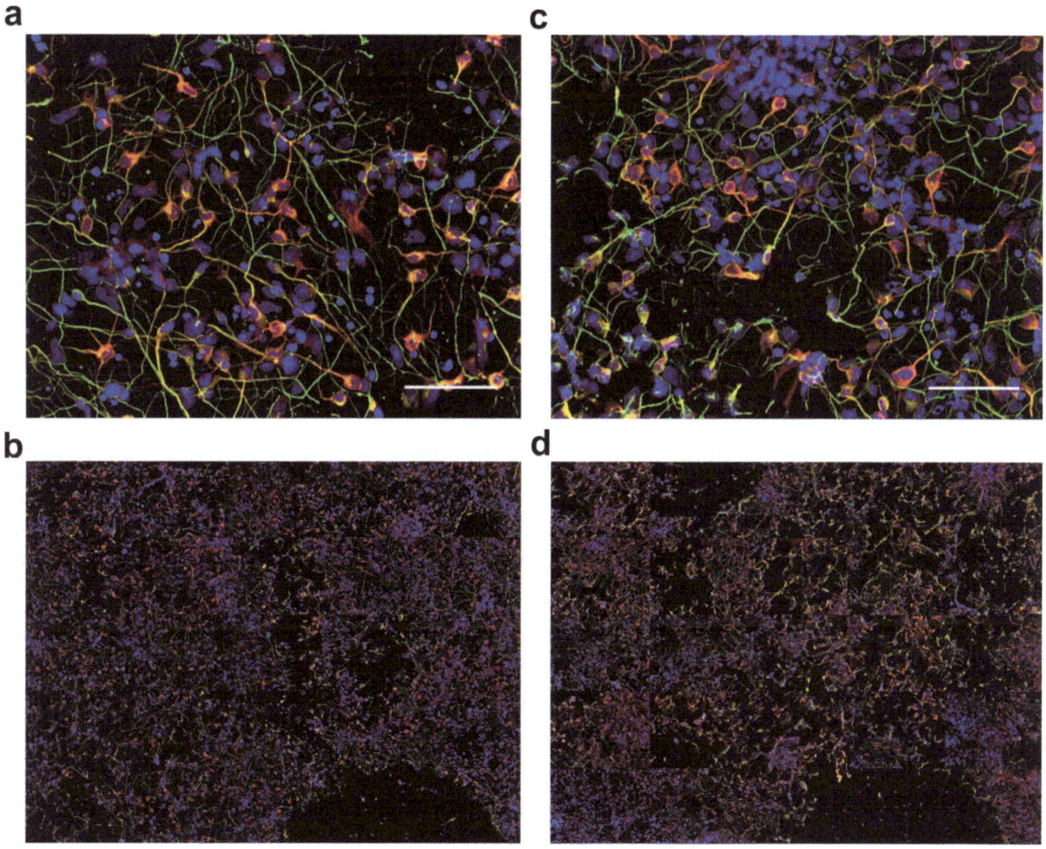

Fig. 1 iPSC-derived neurons seeded on a 384 well plate. iPSC-derived neurons were seeded according to this protocol on Corning Falcon 384-well black/clear tissue culture treated flat bottom plates then fixed with 4% PFA and stained for neuronal specific tubulin β3 (green), neuronal specific microtubule associated protein 2 (red) and with Hoechst 33342 (blue) to highlight nuclei. Cells were either seeded manually (**a** and **b**) or seeded using a liquid dispenser (**c** and **d**). Lower panels (**b** and **d**) each show 25 images taken from one well stitched together. Scale bar represents 20 μm

2.3 Neuronal Medium (See Note 1)

Neurobasal medium, 50% DMEM/F12 medium, 1 × B27 supplement, 1 × N2 supplement, 400 μM L-glutamine, 2-mercaptoethanol, BDNF (20 ng/ml), GDNF (20 ng/ml).

3 Methods

3.1 Coating Plates with Poly-L-Ornithine, Laminin, and Fibronectin

1. Pipette 15 μg/ml poly-L-ornithine in PBS into a tissue culture-treated 384-well plate, 30 μl per well. Place plates into incubator at 37 °C and incubate for 24 h.
2. Aspirate poly-L-ornithine, wash wells with sterile H_2O and leave to air dry completely.
3. Add 5 μg/ml laminin and fibronectin solution to each well, incubate and 4° for 24 h (*see* **Note 2**).

3.2 Dissociation of iPSC-Derived Neurons

1. Use neurons that have been differentiated in a 6-well plate (*see* **Note 3**) for approximately 45 days from start of neuronal induction using preferred differentiation protocol (*see* **Note 4**).
2. Completely aspirate media from well, add 1 ml PBS, then aspirate and add 1 ml Accutase then incubate for 10 min at 37 °C.
3. Add 1 ml neuronal medium to each well, use a cell lifter to gently scrape along the surface of the place to detach the cells while tilting the place to collect the cells at the bottom edge of the well (*see* **Note 5**).
4. Collect the cell suspension into a tube and centrifuge at 200 × *g* for 4 min.
5. Remove supernatant and add 2–3 ml fresh N2B27 to the centrifuged cells.
6. Pipet gently up and down using a p1000 pipette to break up the cell pellet, do this 5–10 times until you can see an even distribution of cells in the solution. Avoid introducing air bubbles into the cell suspension.
7. Pass the cell suspension through a 40 μm nylon cell strainer to remove any remaining cell aggregates.

3.3 Seeding Dissociated Neurons

1. Count cells using preferred counting method.
2. Dilute cells in the appropriate amount of neuronal medium to make a cell solution of 400,000 cells per ml. Add Y-27632 Rho kinase inhibitor to cell solution (10 μM final concentration) to improve cell viability.
3. Transfer cell solution to a reagent reservoir.
4. Remove laminin and fibronectin solution from coated plate.
5. Add 30 μl of cell solution to each well of the coated plate using a 12-channel pipette (*see* **Note 6**). Each well will contain approximately 12,000 cells (*see* **Note 7**).

3.4 Observation of Plated Neurons

1. The following day change media by aspirating the media fully and adding 30 μl of fresh neuronal medium to each well.
2. Change media every 3–4 days (twice a week).
3. Inspect cells regularly by light microscopy, cellular processes should start to appear after 2–3 days.
4. Check for cell density and presence of aggregates, cells should be evenly spaced and not too dense.
5. For optimal images cells should be imaged or fixed at 7 to 10 days after plating (*see* **Note 8**).

4 Notes

1. The neuronal medium used here is based on the method used in Chambers et al. [5] Nat. Biotechnol. Use whichever neuronal media the cells are growing in at the time of seeding.
2. Plates coated with laminin and fibronectin can be stored at 4 °C for up to 1 week.
3. Neurons can be differentiated in 6-, 12-, or 24-well plates, for best results keep neurons at high density and passage at least every 2 weeks before final seeding in 384-well plates.
4. Seeding can be performed on neurons aged between 35 and 65 days. The day on which it is best to plate the cells may differ depending on the differentiation protocol, if there is too much proliferation after plating try plating at a later date, if there is a lot of death after plating try plating at an earlier date.
5. This can also be achieved without a cell scraper by gently pipetting media onto the surface of the well to dissociate cells, pipetting around the well approximately three times to dissociate all of the cells.
6. It is also possible to use a liquid dispenser to transfer the cell solution to the 384-well plate.
7. It is important to dispense cells to the bottom of the well to ensure no air bubbles are trapped between the bottom of the well and the cell suspension, position pipette tip at the bottom corner of the well and dispense cells slowly.
8. Cells can be left longer if need be, depending on confluency, the presence of proliferative cells and the assay to be used. Cells may become too confluent or may aggregate if left longer than 10 days. To reduce aggregation of cells, 5μg/ml laminin can be added to neuronal media at approximately 7 days after plating.

References

1. Burkhardt MF, Martinez FJ, Wright S, Ramos C, Volfson D, Mason M, Garnes J, Dang V, Lievers J, Shoukat-Mumtaz U, Martinez R, Gai H, Blake R, Vaisberg E, Grskovic M, Johnson C, Irion S, Bright J, Cooper B, Nguyen L, Griswold-Prenner I, Javaherian A (2013) A cellular model for sporadic ALS using patient-derived induced pluripotent stem cells. Mol Cell Neurosci 56:355–364. https://doi.org/10.1016/j.mcn.2013.07.007
2. Ryan KR, Sirenko O, Parham F, Hsieh JH, Cromwell EF, Tice RR, Behl M (2016) Neurite outgrowth in human induced pluripotent stem cell-derived neurons as a high-throughput screen for developmental neurotoxicity or neurotoxicity. Neurotoxicology 53:271–281. https://doi.org/10.1016/j.neuro.2016.02.003
3. Xu X, Lei Y, Luo J, Wang J, Zhang S, Yang X, Sun M, Nuwaysir E, Fan G, Zhao J, Lei L, Zhong Z (2013) Prevention of β-amyloid induced toxicity in human iPS cell-derived neurons by inhibition of Cyclin-dependent kinases and associated cell cycle events. Stem Cell Res 10:213–227. https://doi.org/10.1016/j.scr.2012.11.005
4. Little D, Luft C, Mosaku O, Lorvelle M, Yao Z, Paillusson S, Kriston-Vizi J, Gandhi S, Abramov AY, Ketteler R, Devine MJ, Gissen P (2018) A

single cell high content assay detects mitochondrial dysfunction in iPSC-derived neurons with mutations in SNCA. Sci Rep. https://doi.org/10.1038/s41598-018-27058-0 SREP-18-02363

5. Chambers SM, Fasano CA, Papapetrou EP, Tomishima M, Sadelain M, Studer L (2009) Highly efficient neural conversion of human ES and iPS cells by dual inhibition of SMAD signaling. Nature Biotechnol 27:275–280. https://doi.org/10.1038/nbt.1529
6. Kriks S, Shim J, Piao J, Ganat Y, Wakeman D, Xie Z, Carrillo-Reid L, Auyeung G, Antonacci C, Buch A, Yang L, Beal MF, Surmeier DJ, Kordower J, Tabar V, Studer L (2011) Dopamine neurons derived from human ES cells efficiently engraft in animal models of Parkinson's disease. Nature 480(7378):547–551. https://doi.org/10.1038/nature10648
7. Liu Y, Liu H, Sauvey C, Yao L, Zarnowska ED, Zhang S (2013) Directed differentiation of forebrain GABA interneurons from human pluripotent stem cells. Nat Protoc 8:1670–1679. https://doi.org/10.1038/nprot.2013.106
8. Shi Y, Kirwan P, Livesey F (2012) Directed differentiation of human pluripotent stem cells to cerebral cortex neurons and neural networks. Nature Protoc 7:1836–1846. https://doi.org/10.1038/nprot.2012.116

Chapter 15

High-Content Autophagy Analysis in iPSC-Derived Neurons Using Immunofluorescence

Daniel Little, Christin Luft, Olukunbi Mosaku, Robin Ketteler, Michael J. Devine, and Paul Gissen

Abstract

Autophagy is the process by which cellular proteins and organelles are degraded and recycled and is essential to the survival of cells. Defective autophagic degradation has been linked to many neurodegenerative diseases and in particular lysosomal storage diseases. Here we describe a high-content assay to detect defects in the autophagy pathway in induced pluripotent stem cell-derived neurons. This assay utilizes immunofluorescence to stain autophagosomes and uses automated image analysis to measure changes in autophagosome levels in response to modulators of autophagy.

Key words Autophagy, Induced pluripotent stem cells, Neurons, High-content screening, Image analysis

1 Introduction

Autophagy is a critical cellular process that results in the degradation and recycling of proteins [1]. In macroautophagy, referred to hereafter as simply autophagy, the process starts with the packaging of proteins within a double-membraned structure called an autophagosome. The autophagosome is transported to and fuses with a lysosome, resulting in the degradation of the protein cargo by lysosomal enzymes. Many neurodegenerative diseases are associated with defective autophagy, including Parkinson's disease, Alzheimer's disease, amyotrophic lateral sclerosis, and multiple lysosomal storage diseases [2–9]. Here we use induced pluripotent stem cell (iPSC)-derived neuronal cells, which provide an excellent model of neurodegenerative disorders, as they are human cells that can be generated from patients with a specific disease [10]. iPSC-derived neurons are thus an attractive model for identifying or testing potential therapeutics for neurodegenerative diseases [11],

Carl-Fredrik Mandenius and James A. Ross (eds.), *Cell-Based Assays Using iPSCs for Drug Development and Testing*, Methods in Molecular Biology, vol. 1994, https://doi.org/10.1007/978-1-4939-9477-9_15,
© Springer Science+Business Media, LLC, part of Springer Nature 2019

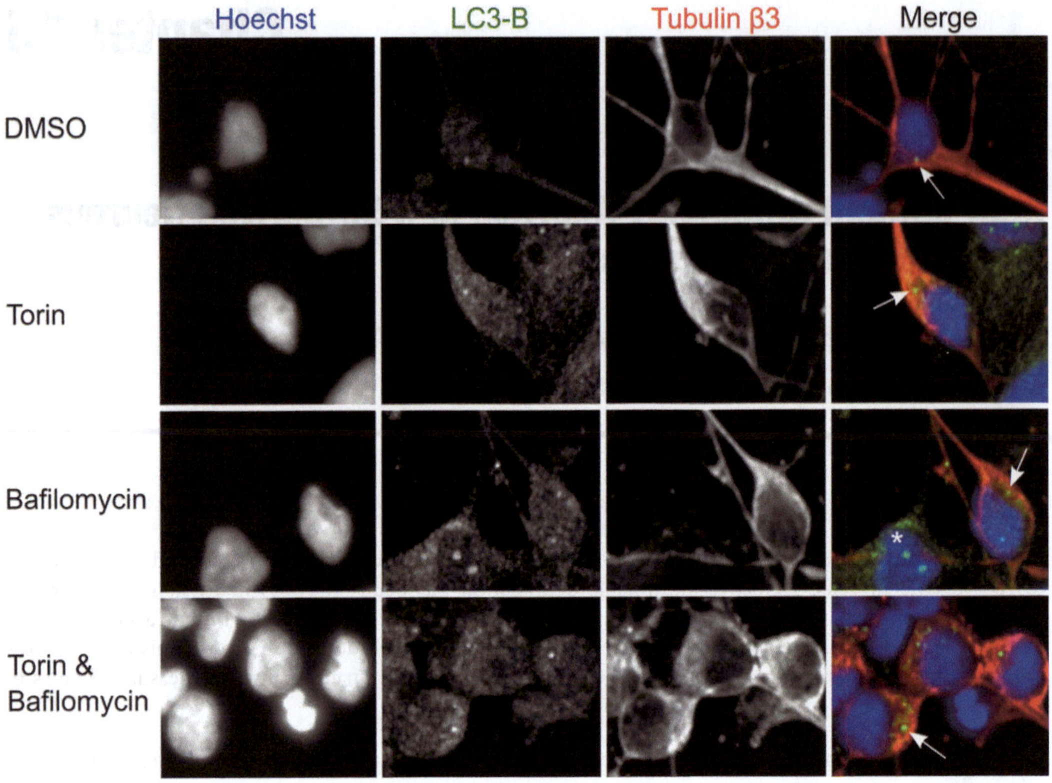

Fig. 1 Immunofluorescence for markers of autophagosomes and neurons. iPSC-derived neurons stained for autophagosome marker LC3-B (green), neuronal specific tubulin B3 (red) and DNA stain Hoechst 33342 (blue), following exposure to either DMSO, Torin, Bafilomycin, or Torin and Bafilomycin. White arrows highlight LC3 puncta, while the white asterisk highlights a nonneuronal cell which is excluded from analysis

and this assay could be used to test modulators of autophagy as a potential therapeutic target for neurodegenerative diseases [12].

This assay utilizes high content image analysis principles to identify changes in LC3B-labelled autophagosomes within iPSC-derived neurons. Firstly, cells are exposed to control compounds: Torin 1 to induce autophagy [13], Bafilomycin A1 to inhibit lysosomal function [14], and a combination of the two. Cells are then fixed and stained for Tubulin β3 to identify neuronal cells (and exclude nonneuronal cells from analysis), and LC3B, to identify autophagosomes (Fig. 1). The stained cells are then imaged, and the images are analyzed (Fig. 2).

LC3B is lipidated from cytoplasmic LC3B-I to LC3B-II when recruited from the cytoplasm to the membrane of the forming autophagosome [15]. LC3B-II then decorates the autophagosome membrane as it is transported to and fuses with a lysosome. By segmenting the bright LC3B-II puncta from the fainter cytoplasmic LC3B-I staining, autophagosomes can be identified and measured. Torin is used to induce the formation of new

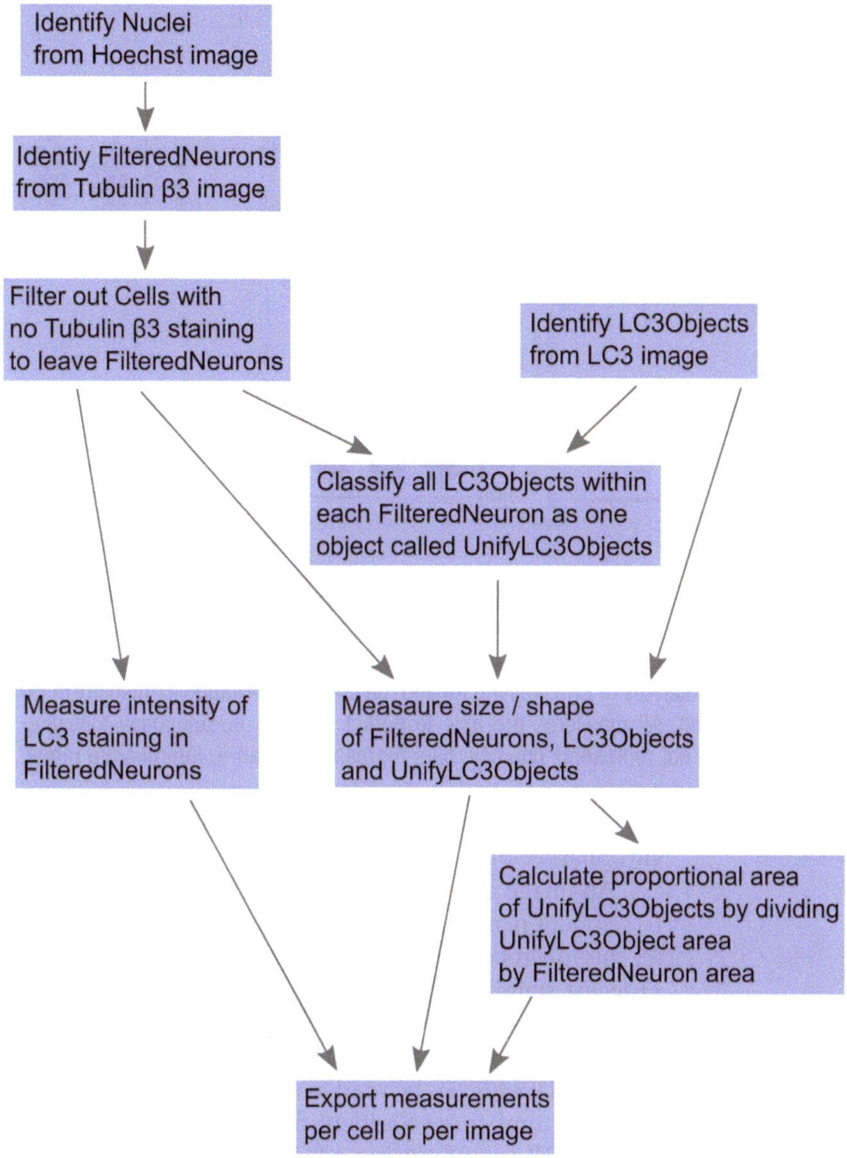

Fig. 2 Flowchart showing outline of image analysis pipeline

autophagosomes, resulting in an increase in LC3B puncta. Bafilomycin A1 is used to inhibit lysosomal function by increasing lysosomal pH, causing a reduction in autophagosome turnover and thus an increase in LC3B puncta. By combining the two control compounds a greater increase in the amount of autophagosomes is detected (Fig. 3). In addition to demonstrating the efficacy of the image analysis, these compounds can also help identify where in this pathway a defect may be occurring in a disease model. For example, if Torin does not increase the number of punctae then formation of autophagosomes is disrupted. If Bafilomycin A1 does not increase

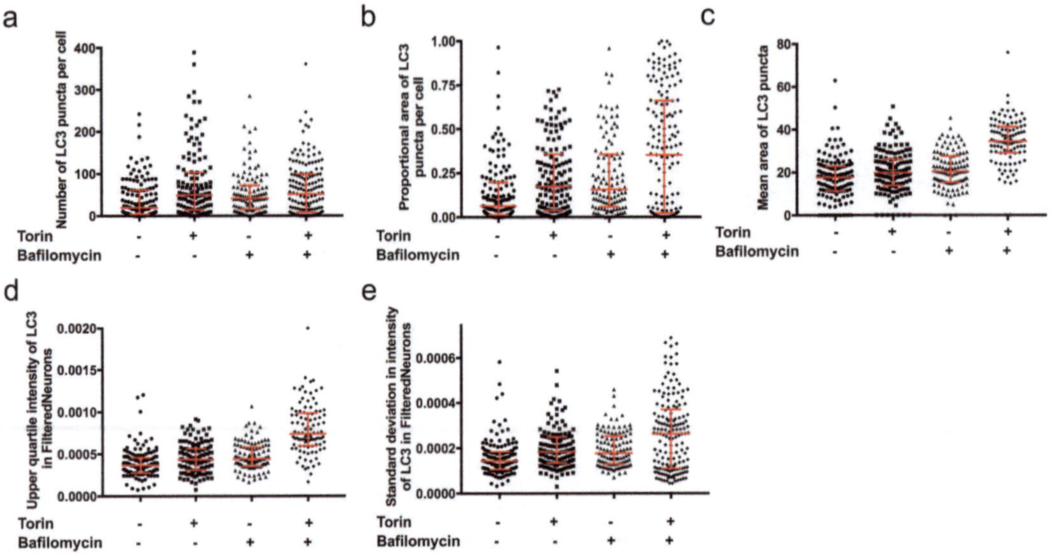

Fig. 3 Effect of control compounds on autophagy by different measures Three measures based upon segmentation of LC3 puncta were taken; the number of puncta per cell (**a**), the proportional area of LC3 puncta (**b**) and the mean area of LC3 puncta (**c**). Two measures independent of LC3 segmentation were also taken; the upper quartile intensity of LC3 staining (**d**) and the standard deviation in intensity of LC3 staining (**e**) within each FilteredNeuron. All measures are able to detect changes in autophagy following exposure to control compounds. Each dot represents one cell, bars represent median $\pm$ interquartile range

the number of punctae there is either a problem with formation or autophagosomes, or lysosome function is already inhibited.

An image analysis pipeline using CellProfiler software [16, 17] to segment neuronal cells and measure autophagosomes is outlined in this protocol (Fig. 3). Essentially, nuclei are identified from Hoechst staining, then the associated cytoplasm is identified from neuronal specific tubulin β3 staining, any nuclei without tubulin β3 staining are filtered from further analysis (Fig. 1). The proportion of neuronal cells can be used as a measure of cell viability, alongside the total number of cells identified. Autophagosomes are identified from LC3B staining using a fixed threshold and a variety of measures are taken for each cell. To measure changes in autophagy the total number of LC3 puncta per cell is measured (Fig. 3a), as well as the total area of LC3 puncta normalized to cell area (Fig. 3b) and the mean area of each LC3 puncta (Fig. 3c). Accurate LC3 puncta identification is dependent on setting the correct threshold for segmentation. Two measurements of intensity, upper quartile intensity (Fig. 3d) and standard deviation in intensity (Fig. 3e) of LC3 staining within FilteredNeurons can be measured to avoid the necessity of segmenting LC3 puncta while still revealing changes upon exposure to control compounds.

2 Materials

2.1 Neuronal Cell Culture Medium

1. Dependent on neuronal differentiation protocol used to generate iPSC-derived neurons.

2.2 Control Compounds

1. Torin 1 in DMSO, used at 100 nM final concentration.
2. Bafilomycin A1 in DMSO, used at 20 nM final concentration.
3. DMSO.

2.3 Fixation and Immunofluorescence

1. Methanol 100%, chilled.
2. 0.1% Triton in PBS.
3. Goat serum.
4. Anti-LC3B antibody raised in rabbit (Sigma, L7543).
5. Anti-tubulin β3 antibody raised in mouse (Biolegend, 801,201).
6. Goat anti-rabbit Alexa Fluor 488 antibody (ThermoFisher A-11034).
7. Goat anti-mouse Alexa Fluor 546 antibody (ThermoFisher A-11031).
8. Hoechst 33342, 1 μg/ml final concentration.

2.4 Image Analysis

1. CellProfiler software available from www.cellprofiler.org.

3 Methods

3.1 Seeding Cells

1. Seed iPSC-derived neurons into plates suitable for live fluorescent microscopy at a density of 120,000 cells/cm^2 by preferred protocol (*see* **Note 1**).
2. At least 8 wells are needed to use the control compounds plus DMSO in duplicate.
3. Incubate for 7 days, changing media every 2 days.

3.2 Addition of Control Compounds

1. Prepare enough neuronal cell culture medium for all wells and divide into four tubes. To one tube add Torin (100 nM final concentration), to the second tube add Bafilomycin A1 (20 nM final concentration), to the third tube add both Torin and Bafilomycin A1 (100 nM and 20 nM final concentration respectively) and to the final tube add an equivalent amount of DMSO.
2. Aspirate media from cells and add control solutions to two wells each.
3. Incubate for 24 h (*see* **Note 2**).

3.3 Fixation

1. Aspirate media from all wells and add PBS to each well.
2. Aspirate PBS immediately and add ice-cold 100% methanol to each well.
3. Incubate at −20 °C for 8 min.
4. Aspirate methanol and add PBS.

3.4 Immuno-fluorescence

1. Remove PBS and add blocking solution of 0.1% Triton in PBS with 5% goat serum.
2. Incubate for 1 h at room temperature.
3. Prepare primary antibody solution; to PBS with 5% goat serum add anti-LC3 antibody and anti-tubulin β3 antibody at a dilution of 1:1000 for both.
4. Remove blocking solution and add primary antibody solution.
5. Incubate for 2 h at room temperature or overnight at 4 °C.
6. Aspirate primary antibody solution and add PBS, incubate for 10 min, then remove and repeat a further two times.
7. Prepare secondary antibody solution of goat anti rabbit Alexa Fluor 488 antibody and goat anti mouse Alexa Fluor 546, both diluted 1:500 in PBS.
8. Remove PBS and add secondary antibody solution, incubate at room temperature for 1 h.
9. Aspirate secondary antibody solution and add PBS with Hoechst-33,342 (1 μg/ml final concentration) and incubate for 10 min.
10. Remove Hoechst solution and add PBS, incubate for 10 min, then remove and repeat.

3.5 Fluorescence Microscopy

1. Configure microscope to image LC3 (Alexa Fluor 488, Ex/Em), tubulin β3 (Ex/Em), and Hoechst 33342 (Ex350/Em461).
2. Acquire images from multiple sites per well at 40× magnification (*see* **Note 3**).

3.6 Image Analysis Using CellProfiler Software

1. Settings such as diameter of objects and threshold correction factor will vary depending on image acquisition parameters, below are suggested settings.
2. Configure input modules:
   (a) On the Images module drag images you wish to analyze into the File List box.
   (b) Metadata can be extracted from file or folder names to extract information describing the images. Regular expression code can be entered to tell the software how

the image names relate to information describing the images.

(c) In the NamesAndTypes module assign names to the different channels using either file name or metadata.

(d) Groups can be used to split a list of images into subsets.

3. Configure analysis modules:

(a) IdentifyPrimaryObjects—Identifies nuclei from Hoechst image; these objects are named Nuclei. Use Otsu thresholding with three-class thresholding, minimize weighted variance, assign pixels in middle intensity class to foreground, automatic smoothing for thresholding with a threshold correction factor of 3. Distinguish clumped objects by shape and draw dividing lines between clumped objects by intensity. Lower and upper bounds on threshold can be adjusted to avoid incorrect segmentation.

(b) IdentifySecondaryObjects—Identifies neuronal cell soma from tubulin β3 image by propagation from Nuclei identified in module 1; these objects are named Cells. Use two-class thresholding and minimize weighted variance with automatic thresholding and threshold correction factor of 3.5. Lower and upper bounds on threshold can be adjusted to avoid incorrect segmentation of nonneuronal cells.

(c) MeasureObjectSizeShape—Measures the size of Nuclei and Cells identified in modules 1 and 2.

(d) CalculateMath—Divides the area of each Nuclei object by area of each related Cell to give CellNucleiRatio.

(e) FilterObjects—Removes Cells with a CellNucleiRatio lower than 1.1 to remove any Cells without tubulin β3 staining, that is, nonneuronal cells; the remaining objects are named FilteredNeurons.

(f) IdentifyPrimaryObjects—Identify LC3 puncta from LC3 image, these objects are named LC3Objects. First analyze a selection of control images using three-class Otsu thresholding, minimize weighted variance, assign pixels in middle intensity class to background, with automatic smoothing and a threshold correction factor of 5. Then determine the median threshold used in test images (excluding outliers) and apply a manual threshold to analyze images.

(g) ReassignObjectNumbers—Classifies all LC3Objects within a FilteredNeuron as one object, called UnifyLC3Objects; this enables the measurement of the total area of LC3Objects in each cell.

(h) MeasureObjectIntensity—Measures the intensity of LC3 image in FilteredNeurons, and UnifyLC3Objects.

(i) MeasureObjectSizeShape—Measures the size and shape of FilteredNeuron, LC3Objects and UnifyLC3Objects.

(j) CalculateMath—Divides area (in AreaShape category) of UnifyLC3, by area (in AreaShape category) of FilteredNeurons to give the proportion of the area of each cell occupied by mitochondria.

(k) RelateObjects—Relates Nuclei with associated FilteredCells and calculates per cell means for Nuclei size/shape measurements from module 3.

(l) RelateObjects—Relates LC3Objects with associated FilteredCells and calculates per cell means for LC3Object intensity and size/shape measurements from module 8 and 9.

(m) RelateObjects—Relates UnifyLC3Objects with associated FilteredCells and calculates per cell means for UnifyLC3Object intensity and size/shape measurements from module 8 and 9.

(n) ExportToSpreadsheet—exports data from previous modules to spreadsheets for each object and per image. Add image metadata columns to object data file and calculate per-image mean values for object measurements. Select measurements to export in particular within the FilteredNeurons category; LC3Objects count (in the Children subcategory), the proportional area of LC3Objects (in the Math subcategory), LC3Objects area (in the LC3Objects subcategory) and StdIntensity and UpperQuartileIntensity for LC3 (in the Intensity subcategory). Nuclei characteristics such as area, compactness and eccentricity can be measured to assess cytotoxicity. Within the Image category make sure to select appropriate metadata and final threshold of Nuclei, Cells, and LC3Objects. Also select count for Nuclei and FilteredNeurons to calculate the number of cells identified per well and the proportion of neuronal cells in the culture.

4. Determine correct segmentation parameters for each object identification module using test mode. OverlayOutlines and SaveImages modules can be used to check accuracy of segmentation by saving an outline of the segmentation on top of the relevant image. Check settings using a subset of control images, then run analysis on all images with manual thresholding for mitochondria identification.

4 Notes

1. It is recommended to seed iPSC-derived neurons at approximately 45 days from start of neuronal induction.
2. Control compounds can be incubated for a shorter period of time such as 8 or 16 h.
3. Z-stacks can be acquired to ensure the full extent of each cell is captured but this is not essential for the accuracy of the assay. If stacks are acquired perform the image analysis on projections of the stacks.

Acknowledgments

The research leading to these results has received support from the Innovative Medicines Initiative Joint Undertaking under (grant no. 115439), resources of which are composed of financial contribution from the European Union's Seventh Framework Programme (FP7/2007-2013) and EFPIA companies. This publication reflects only the author's views, and neither the IMI JU nor EFPIA nor the European Commission is liable for any use that may be made of the information contained therein.

References

1. Eskelinen EL, Saftig P (2009) Autophagy: a lysosomal degradation pathway with a central role in health and disease. Biochim Biophys Acta 1793(4):664–673. https://doi.org/10.1016/j.bbamcr.2008.07.014
2. Boland B, Kumar A, Lee S, Platt FM, Wegiel J, Yu WH, Nixon RA (2008) Autophagy induction and autophagosome clearance in neurons: relationship to autophagic pathology in Alzheimer's disease. J Neurosci 28(27):6926–6937. https://doi.org/10.1523/JNEUROSCI.0800-08.2008
3. Chung KM, Hernandez N, Sproul AA, Yu WH (2018) Alzheimer's disease and the autophagic-lysosomal system. Neurosci Lett. https://doi.org/10.1016/j.neulet.2018.05.017
4. Metaxakis A, Ploumi C, Tavernarakis N (2018) Autophagy in age-associated neurodegeneration. Cell 7(5). https://doi.org/10.3390/cells7050037
5. Nguyen DKH, Thombre R, Wang J (2018) Autophagy as a common pathway in amyotrophic lateral sclerosis. Neurosci Lett. https://doi.org/10.1016/j.neulet.2018.04.006
6. Nixon RA (2013) The role of autophagy in neurodegenerative disease. Nat Med 19:983–997. https://doi.org/10.1038/nm.3232
7. Settembre C, Fraldi A, Rubinsztein DC, Ballabio A (2008) Lysosomal storage diseases as disorders of autophagy. Autophagy 4:113–114
8. Winslow AR, Rubinsztein DC (2011) The Parkinson disease protein alpha-synuclein inhibits autophagy. Autophagy 7:429–431
9. Yang Q, Mao Z (2010) Parkinson disease: a role for autophagy? Neuroscientist 16:335–341. https://doi.org/10.1177/1073858409357118
10. Grskovic M, Javaherian A, Strulovici B, Daley G (2011) Induced pluripotent stem cells—opportunities for disease modelling and drug discovery. Nat Rev Drug Discov:1–15. https://doi.org/10.1038/nrd3577
11. Khurana V, Tardiff DF, Chung CY, Lindquist S (2015) Toward stem cell-based phenotypic screens for neurodegenerative diseases. Nat Rev Neurol 11:339–350. https://doi.org/10.1038/nrneurol.2015.79

12. Rubinsztein DC, Codogno P, Levine B (2012) Autophagy modulation as a potential therapeutic target for diverse diseases. Nat Rev Drug Discov 11:709–730. https://doi.org/10.1038/nrd3802
13. Thoreen CC, Kang SA, Chang JW, Liu Q, Zhang J, Gao Y, Reichling LJ, Sim T, Sabatini DM, Gray NS (2009) An ATP-competitive mammalian target of rapamycin inhibitor reveals rapamycin-resistant functions of mTORC1. J Biol Chem 284:8023–8032. https://doi.org/10.1074/jbc.M900301200
14. Mauvezin C, Neufeld TP (2015) Bafilomycin A1 disrupts autophagic flux by inhibiting both V-ATPase-dependent acidification and Ca-P60A/SERCA-dependent autophagosome-lysosome fusion. Autophagy 11:1437–1438. https://doi.org/10.1080/15548627.2015.1066957
15. Kabeya Y, Mizushima N, Ueno T, Yamamoto A, Kirisako T, Noda T, Kominami E, Ohsumi Y, Yoshimori T (2000) LC3, a mammalian homologue of yeast Apg8p, is localized in autophagosome membranes after processing. EMBO J 19:5720–5728. https://doi.org/10.1093/emboj/19.21.5720
16. Carpenter AE, Jones TR, Lamprecht MR, Clarke C, Kang IH, Friman O, Guertin DA, Chang JH, Lindquist RA, Moffat J, Golland P, Sabatini DM (2006) CellProfiler: image analysis software for identifying and quantifying cell phenotypes. Genome Biol 7:R100. https://doi.org/10.1186/gb-2006-7-10-r100
17. Kamentsky L, Jones TR, Fraser A, Bray MA, Logan DJ, Madden KL, Ljosa V, Rueden C, Eliceiri KW, Carpenter AE (2011) Improved structure, function and compatibility for CellProfiler: modular high-throughput image analysis software. Bioinformatics 27:1179–1180. https://doi.org/10.1093/bioinformatics/btr095

Chapter 16

High-Content Analysis of Mitochondrial Function in iPSC-Derived Neurons

Daniel Little, Christin Luft, Olukunbi Mosaku, Robin Ketteler, Michael J. Devine, and Paul Gissen

Abstract

Mitochondrial dysfunction is linked to many neurological diseases; therefore, the ability to measure mitochondrial function is of great use for researching disease and testing potential therapeutics. Here we describe a high-content assay to simultaneously measure mitochondrial membrane potential, morphology and cell viability in iPSC-derived neurons. Neurons are seeded into plates suitable for fluorescent microscopy, stained with the mitochondrial membrane potential-dependent dye TMRM, cytoplasmic dye Calcein AM, and nuclear stain Hoechst 33342. Images are acquired in live cells and analyzed using automated image analysis software.

Key words Mitochondria, Induced pluripotent stem cells, Neurons, High-content screening, Image analysis

1 Introduction

Mitochondrial dysfunction has been implicated in many neurodegenerative disorders such as Alzheimer's disease, Parkinson's disease, Huntington's disease, and motor neuron disease [1–4]. Here we present a method of analyzing mitochondrial function in induced pluripotent stem cell (iPSC) derived neurons in a manner that is compatible with high-content screening methods and could be used to test potential therapeutics for neurodegenerative diseases. Being derived from patients, neurons differentiated from iPSCs are felt to provide accurate disease models that can be used to identify or test potential therapeutics for neurodegenerative diseases [5, 6].

This method utilizes the dye tetramethylrhodamine (TMRM) to measure mitochondrial membrane potential and morphology and utilizes the dyes Calcein AM and Hoechst 33342 to aid cell segmentation and measure cell viability (Fig. 1). TMRM has been

Carl-Fredrik Mandenius and James A. Ross (eds.), *Cell-Based Assays Using iPSCs for Drug Development and Testing*, Methods in Molecular Biology, vol. 1994, https://doi.org/10.1007/978-1-4939-9477-9_16,
© Springer Science+Business Media, LLC, part of Springer Nature 2019

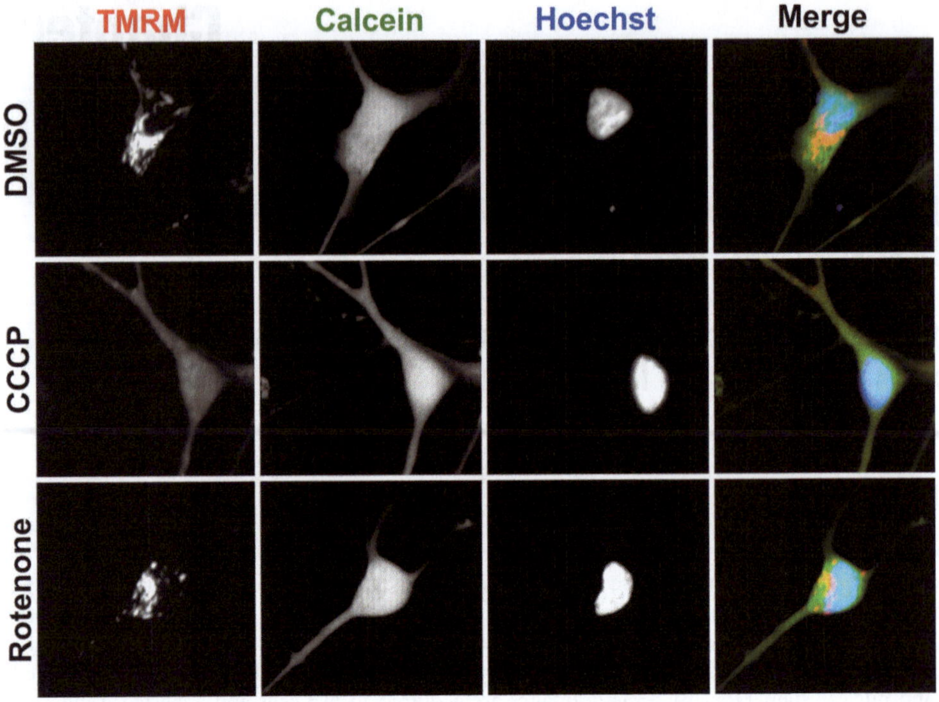

Fig. 1 Example of stains and effect of control compounds. iPSC-derived dopaminergic neurons stained with TMRM (red in merge), Calcein (green in merge), and Hoechst (blue in merge) exposed to DMSO, CCCP, or rotenone

widely used to measure mitochondrial membrane potential and has helped to implicate mitochondrial dysfunction as a mechanism of various neurodegenerative diseases [7–11]. The presence of the mitochondrial membrane potential ($\Delta\psi m$) is essential for TMRM to be taken up by mitochondria; when $\Delta\psi m$ is reduced, fluorescence intensity decreases when used in non-quench mode as it is here. Reduced $\Delta\psi m$ equates to reduced capacity for ATP generation because ATP generation is dependent on the electrochemical proton gradient across the mitochondrial inner membrane [12–14]. Mitochondrial function is directly related to morphology, because mitochondria must move, divide and fuse in response to cellular demands [2, 15, 16], which is particularly important in neuronal cells [17].

An image analysis pipeline using CellProfiler software [18, 19] to segment cells and measure features is outlined in this protocol (Fig. 2). Essentially, nuclei are identified from Hoechst staining, then the associated cytoplasm is identified from Calcein staining. Any nuclei without cytoplasmic staining are classed as apoptotic and filtered from further analysis (Fig. 3). The proportion of apoptotic cells can be used as a measure of cell viability, alongside the number of filtered cells identified (Fig. 4). Mitochondria are

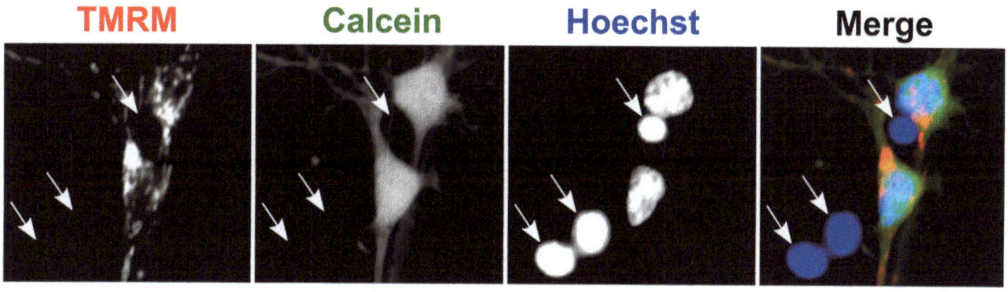

Fig. 2 Apoptotic cells used in cell viability assessment. In this assay apoptotic cells are classed as cells with Hoechst staining but no Calcein staining (white arrows); these are filtered from mitochondrial function analysis. The proportion of these apoptotic cells, compared to the total number of nuclei, can be used as a measure of cell viability

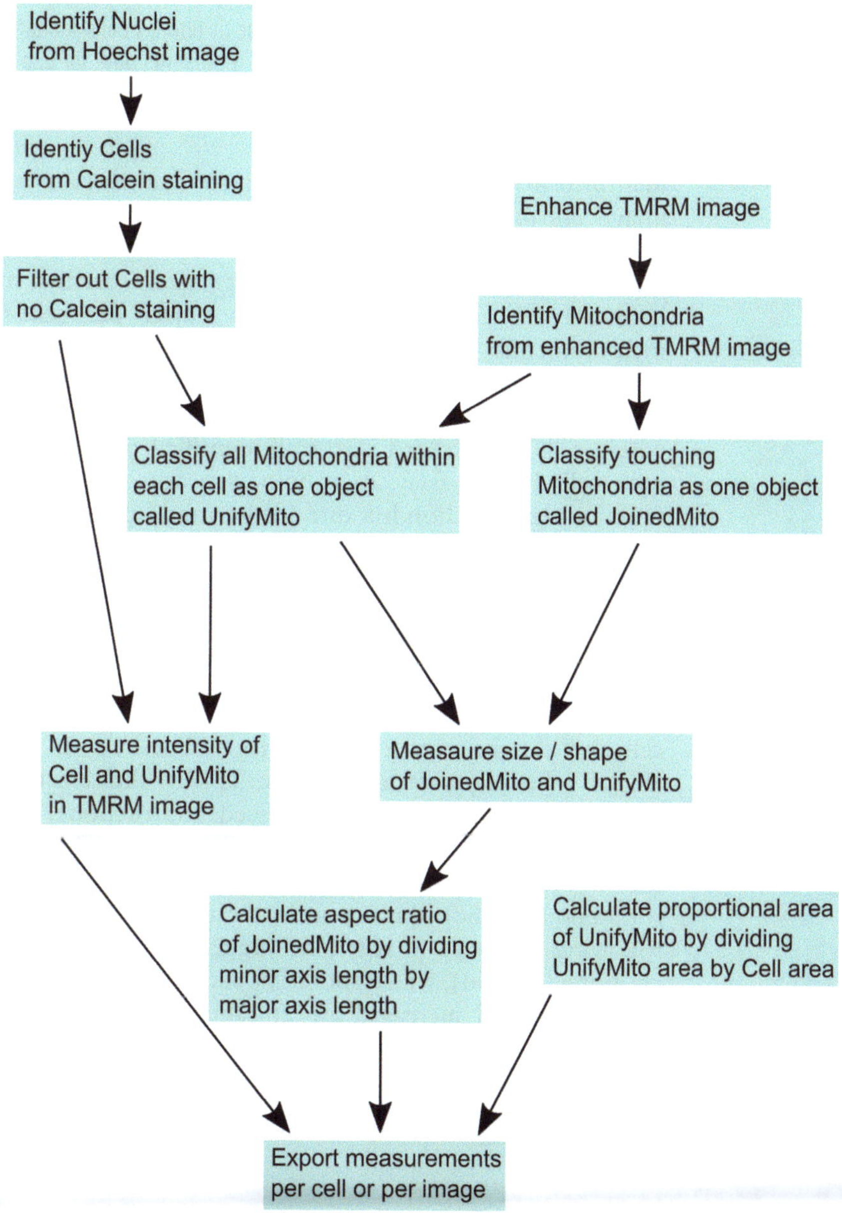

Fig. 3 Flowchart showing outline of image analysis pipeline

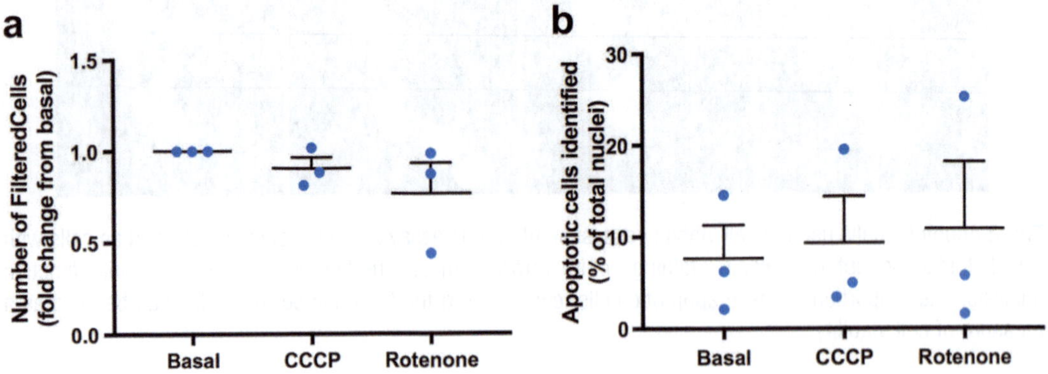

Fig. 4 Cell viability assessment. Cell viability is assessed by the number of filtered cells identified for each condition (**a**) and the proportion of apoptotic cells identified (**b**). Each dot represents data for one experiment; error bars represent mean + SEM

identified from TMRM staining using a fixed threshold following image enhancement using a white tophat filter to reduce local background, used previously [20]. Touching mitochondria are classed as one object for morphology measurements, and mitochondria are also grouped as one object per cell for some measurements.

To measure mitochondrial membrane potential the mean intensity of the TMRM fluorescence can be measured within whole cells or averaged across all identified mitochondria within each cell. In this assay, morphology measurements only include functioning mitochondria due to the mode of action of TMRM described above. CCCP is not used as a control for morphology measurements as very few mitochondria are identified. Mitochondrial morphology measurements used include mean area, aspect ratio, and major axis length. The total area of functioning mitochondria is also measured, normalized to the total area of the cell to give the proportional area of mitochondria. Data is collected per cell and can be averaged per image if desired, as shown here (Fig. 5). For $\Delta\psi m$ measurements, TMRM fluorescence intensity should be reduced in cells exposed to rotenone, and almost completely absent in CCCP wells (Fig. 5a, b). The proportional area of mitochondria should also be reduced in cells exposed to rotenone (Fig. 5c). Reduction in the mean area of mitochondria following exposure to rotenone suggests fragmentation of mitochondria (Fig. 5d), which should be associated with reduced major axis length and increased aspect ratio of mitochondria as they become rounder (Fig. 5e, f).

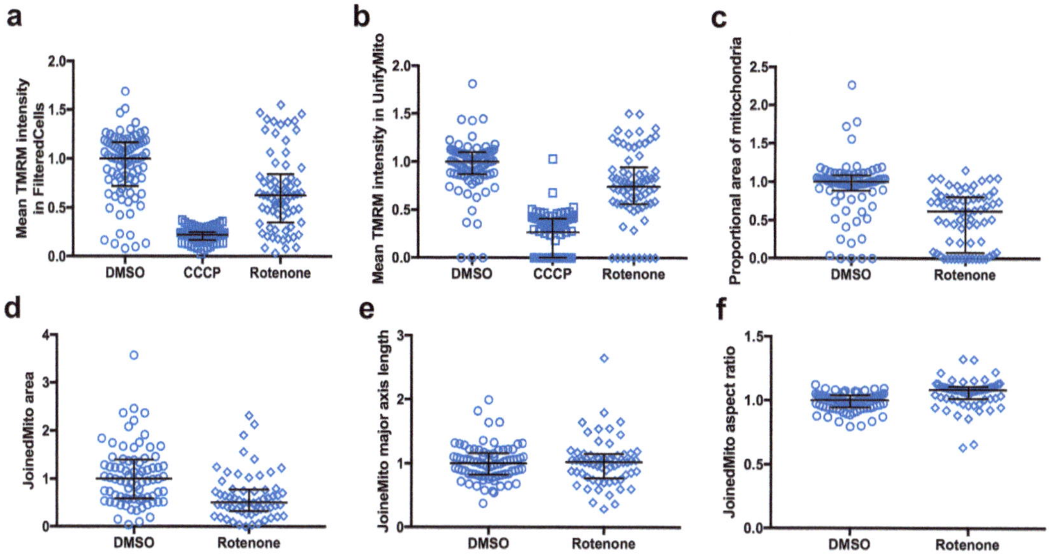

Fig. 5 Expected results of control compounds by different measures. TMRM intensity measurements used are mean TMRM intensity per Cell (**a**) and mean intensity in UnifyMito (**b**). Mitochondrial morphology measurements used are proportional area of mitochondria (**c**), calculated by dividing area of UnifyMito by area of Cell, JoinedMito area (**d**), JoinedMito major axis length (**e**), and JoinedMito major axis length (**f**). Measurements averaged per cell then per image for cells exposed to DMSO, CCCP, or rotenone; data represents fold change from DMSO. Each dot represents mean data for one image; error bars represent median ± interquartile range, $n = 3$ independent experiments

2 Materials

2.1 Cell Culture Medium for Fluorescence Microscopy

1. Neurobasal medium without phenol red.

2.2 Dyes for Fluorescence Microscopy

1. TMRM, 25 nM final concentration.
2. Calcein AM, 5 μM final concentration.
3. Hoechst 33342, 1 μg/ml final concentration.

2.3 Control Compounds

1. CCCP, stock made in DMSO, diluted to 10 μM final concentration in fluorescent microscopy medium.
2. Rotenone, stock made in DMSO, diluted to 500 nM final concentration in neuronal medium and fluorescent microscopy medium.
3. DMSO.

2.4 Image Analysis

1. CellProfiler software available from www.cellprofiler.org.

3 Methods

3.1 Seeding Cells

1. Seed iPSC-derived neurons into plates suitable for live fluorescent microscopy at a density of 120,000 cells/cm^2 by preferred protocol (*see* **Note 1**).
2. At least 6 wells are needed to use the two control compounds plus DMSO in duplicate.
3. Incubate for 7 days, changing media every 2 days.

3.2 Addition of Control Compound Rotenone

1. Prepare neuronal medium in two tubes, to one tube add rotenone and to the other add DMSO to the equivalent amount as added to the rotenone tube.
2. Remove media from 4 wells and add either rotenone or DMSO solutions to 2 well each. Leave two wells for CCCP.
3. Incubate for 24 h (*see* **Note 2**).

3.3 Addition of Dyes and CCCP

1. Make up imaging media—to phenol red-free Neurobasal medium add TMRM (25 nM), Calcein AM (5 μM) and Hoechst 33342 (5 μg/ml).
2. Divide imaging media into three tubes, to tube 1 add CCCP (10 μM), to tube 2 add rotenone (500 nM), to tube 3 add equivalent amount of DMSO.
3. Remove neuronal growth media from wells and add the solution containing CCCP to the wells that are not used the day before, add rotenone and DMSO-containing solutions to the same wells as the previous day.
4. Incubate at 37 °C for 40 min.

3.4 Fluorescence Microscopy

1. Remove cells from incubator and transfer to fluorescent microscope for live imaging, ensure microscope is equipped with temperature control set to 37 °C and 5% CO_2.
2. Configure microscope to image TMRM (Ex552/Em578), Calcein AM (Ex495/Em516) and Hoechst 33342 (Ex350/Em461) and acquire images from multiple sites per well at 40× magnification (*see* **Note 3**).

3.5 Image Analysis Using CellProfiler Software

1. Settings such as diameter of objects and threshold correction factor will vary depending on image acquisition parameters. Below are suggested settings only.
2. Configure input modules:
   (a) On the Images module, drag images you wish to analyze into the File List box.
   (b) Metadata can be extracted from file or folder names to extract information describing the images. Regular

expression code can be entered to tell the software how the image names relate to information describing the images.

(c) In the NamesAndTypes module, assign names to the different channels using either file name or metadata.

(d) Groups can be used to split a list of images into subsets.

3.6 Configure Analysis Modules

1. IdentifyPrimaryObjects—Identifies nuclei from Hoechst image; these objects are named Nuclei. Use Otsu thresholding with three-class thresholding, minimize weighted variance, assign pixels in middle intensity class to foreground, with automatic smoothing for thresholding and a threshold correction factor of 3. Distinguish clumped objects by shape and draw dividing lines between clumped objects by intensity. Lower and upper bounds on threshold can be adjusted to avoid incorrect segmentation.
2. IdentifySecondaryObjects—Identifies cell soma from Calcein image by propagation from nuclei identified in module 1; these objects are named Cells. Use two-class thresholding and minimize weighted variance with automatic thresholding and a threshold correction factor of 3.5. Lower and upper bounds on threshold can be adjusted to avoid incorrect segmentation.
3. MeasureObjectSizeShape—Measures the size of Nuclei and Cells identified in modules 1 and 2.
4. CalculateMath—Divides the area of each Nuclei object by area of each related Cell to give CellNucleiRatio.
5. FilterObjects—Removes "Cells" with a CellNucleiRatio lower than 1.1 to remove any objects without cytoplasmic staining, that is, dead cells; the remaining objects are named FilteredCells.
6. EnhanceOrSuppressFeatures—Applies tophat filter to TMRM image to enhance mitochondria staining and reduce background, based on object size; the resulting image is called EnhancedMito (*see* **Note 4**).
7. IdentifyPrimaryObjects—Identifies mitochondria from EnhancedMito image, these objects are named Mitochondria. First analyze a selection of control images using three-class Otsu thresholding, minimize weighted variance, assign pixels in middle intensity class to background, with automatic smoothing and a threshold correction factor of 2. Then determine the median threshold used in test images and apply a manual threshold to analyze images. Lower and upper bounds on threshold can be adjusted to avoid incorrect segmentation.
8. ReassignObjectNumbers—Classifies mitochondria with 0 pixels between objects as a single object, to make objects that are

touching one object; these objects are called JoinedMito. This improves mitochondrial morphology measurements.

9. MeasureObjectSizeShape—Measures the size and shape of JoinedMito objects.
10. CalculateMath—Divide MinorAxisLength (in AreaShape category) of JoinedMito by MajorAxisLength (in AreaShape category) to give the aspect ratio of JoinedMito.
11. RelateObjects—Relates JoinedMito with associated FilteredCells and calculates per cell means for measurements from module 9 and 10.
12. ReassignObjectNumbers—Classifies all mitochondria within a cell as one object per cell, called UnifyMito; this enables measurement of the total area of mitochondria, and the average intensity of all mitochondria, in each cell.
13. MeasureObjectIntensity—Measures the intensity of TMRM image in Cell and UnifyMito objects.
14. MeasureObjectSizeShape—Measures the size and shape of UnifyMito objects.
15. CalculateMath—Divides the area (in AreaShape category) of UnifyMito, by area (in AreaShape category) of FilteredCells to give the proportion of the area of each cell occupied by mitochondria.
16. RelateObjects—Relates UnifyMito with associated FilteredCells and calculates per cell means for measurements from module 13 and 14.
17. ExportToSpreadsheet—Exports data from previous modules to spreadsheets for each object and per image. Add image metadata columns to object data file and calculate per-image mean values for object measurements. Select measurements to export in particular within the FilteredCells category: mean intensity, mean JoinedMito area measurements, area and major axis length, mean UnifyMito mean intensity and measurements from the calculate math modules. Within the Image category make sure to select appropriate metadata and final threshold of Nuclei, Cells, and Mitochondria. Also select count for nuclei and filtered cells to calculate the number of cells identified per well and the proportion of excluded cells (Nuclei count minus FilteredCells count) to determine cell viability.
18. Determine correct segmentation parameters for each object identification module using test mode. OverlayOutlines and SaveImages modules can be used to check accuracy of segmentation by saving an outline of the segmentation on top of the relevant image. Check settings using a subset of control images then run analysis on all images with manual thresholding for mitochondria identification.

4 Notes

1. It is recommended to seed iPSC-derived neurons at approximately 45 days from start of neuronal induction.
2. Rotenone is incubated for 24 h to cause an effect on mitochondrial morphology as well as on membrane potential. It may be added at the same time as the dyes to detect the effect on mitochondrial membrane potential alone.
3. Z-stacks can be acquired to ensure that the full extent of each cell is captured, but this is not essential for the accuracy of the assay. If stacks are acquired, then perform the image analysis on projections of the stacks.
4. Image enhancement uses a white tophat filter that reduces low intensity staining around brighter objects of a defined size. TMRM image enhancement is optional, depending on the acquired images and may not be necessary for z-stack projections.

References

1. Chaturvedi RK, Beal MF (2013) Mitochondria targeted therapeutic approaches in Parkinson's and Huntington's diseases. Mol Cell Neurosci 55:101–114. https://doi.org/10.1016/j.mcn.2012.11.011
2. Dhillon VS, Fenech M (2014) Mutations that affect mitochondrial functions and their association with neurodegenerative diseases. Mutat Res Rev Mutat Res 759:1–13. https://doi.org/10.1016/j.mrrev.2013.09.001
3. Hroudova J, Singh N, Fisar Z (2014) Mitochondrial dysfunctions in neurodegenerative diseases: relevance to Alzheimer's disease. Biomed Res Int 2014:175062. https://doi.org/10.1155/2014/175062
4. Keeney PM, Xie J, Capaldi RA, Bennett JP Jr (2006) Parkinson's disease brain mitochondrial complex I has oxidatively damaged subunits and is functionally impaired and misassembled. J Neurosci 26(19):5256–5264. https://doi.org/10.1523/JNEUROSCI.0984-06.2006
5. Grskovic M, Javaherian A, Strulovici B, Daley G (2011) Induced pluripotent stem cells—opportunities for disease modelling and drug discovery. Nat Rev Drug Discov:1–15. https://doi.org/10.1038/nrd3577
6. Khurana V, Tardiff DF, Chung CY, Lindquist S (2015) Toward stem cell-based phenotypic screens for neurodegenerative diseases. Nat Rev Neurol 11(6):339–350. https://doi.org/10.1038/nrneurol.2015.79
7. Guo X, Disatnik MH, Monbureau M, Shamloo M, Mochly-Rosen D, Qi X (2013) Inhibition of mitochondrial fragmentation diminishes Huntington's disease-associated neurodegeneration. J Clin Invest 123(12):5371–5388. https://doi.org/10.1172/JCI70911
8. Kalmar B, Innes A, Wanisch K, Kolaszynska AK, Pandraud A, Kelly G, Abramov AY, Reilly MM, Schiavo G, Greensmith L (2017) Mitochondrial deficits and abnormal mitochondrial retrograde axonal transport play a role in the pathogenesis of mutant Hsp27-induced Charcot Marie tooth disease. Hum Mol Genet 26(17):3313–3326. https://doi.org/10.1093/hmg/ddx216
9. Ludtmann MH, Angelova PR, Ninkina NN, Gandhi S, Buchman VL, Abramov AY (2016) Monomeric alpha-Synuclein exerts a physiological role on brain ATP synthase. J Neurosci 36(41):10510–10521
10. Mortiboys H, Thomas KJ, Koopman WJ, Klaffke S, Abou-Sleiman P, Olpin S, Wood NW, Willems PH, Smeitink JA, Cookson MR, Bandmann O (2008) Mitochondrial function and morphology are impaired in parkin-mutant fibroblasts. Ann Neurol 64(5):555–565. https://doi.org/10.1002/ana.21492
11. Little D, Luft C, Mosaku O, Lorvelle M, Yao Z, Paillusson S, Kriston-Vizi J, Gandhi S, Abramov AY, Ketteler R, Devine MJ, Gissen P (2018) A single cell high content assay detects

mitochondrial dysfunction in iPSC-derived neurons with mutations in SNCA. Sci Rep. https://doi.org/10.1038/s41598-018-27058-0 SREP-18-02363

12. Brand MD, Nicholls DG (2011) Assessing mitochondrial dysfunction in cells. Biochem J 435(2):297–312. https://doi.org/10.1042/BJ20110162
13. Koopman WJ, Distelmaier F, Esseling JJ, Smeitink JA, Willems PH (2008) Computer-assisted live cell analysis of mitochondrial membrane potential, morphology and calcium handling. Methods 46(4):304–311
14. Nicholls DG (2012) Fluorescence measurement of mitochondrial membrane potential changes in cultured cells. Methods Mol Biol 810:119–133. https://doi.org/10.1007/978-1-61779-382-0_8
15. Chan DC (2012) Fusion and fission: interlinked processes critical for mitochondrial health. Annu Rev Genet 46:265–287. https://doi.org/10.1146/annurev-genet-110410-132529
16. Friedman JR, Nunnari J (2014) Mitochondrial form and function. Nature 505 (7483):335–343. https://doi.org/10.1038/nature12985
17. Devine MJ, Kittler JT (2018) Mitochondria at the neuronal presynapse in health and disease. Nat Rev Neurosci 19(2):63–80. https://doi.org/10.1038/nrn.2017.170
18. Carpenter AE, Jones TR, Lamprecht MR, Clarke C, Kang IH, Friman O, Guertin DA, Chang JH, Lindquist RA, Moffat J, Golland P, Sabatini DM (2006) CellProfiler: image analysis software for identifying and quantifying cell phenotypes. Genome Biol 7(10):R100. https://doi.org/10.1186/gb-2006-7-10-r100
19. Kamentsky L, Jones TR, Fraser A, Bray MA, Logan DJ, Madden KL, Ljosa V, Rueden C, Eliceiri KW, Carpenter AE (2011) Improved structure, function and compatibility for CellProfiler: modular high-throughput image analysis software. Bioinformatics 27 (8):1179–1180. https://doi.org/10.1093/bioinformatics/btr095
20. Kitami T, Logan DJ, Negri J, Hasaka T, Tolliday NJ, Carpenter AE, Spiegelman BM, Mootha VK (2012) A chemical screen probing the relationship between mitochondrial content and cell size. PLoS One 7(3):e33755. https://doi.org/10.1371/journal.pone.0033755

Chapter 17

Evaluating the Effect of Drug Compounds on Cardiac Spheroids Using the Cardiac Cell Outgrowth Assay

Jonas Christoffersson, Florian Meier, Henning Kempf, Kristin Schwanke, Michelle Coffee, Mario Beilmann, Robert Zweigerdt, and Carl-Fredrik Mandenius

Abstract

The ideal cell culture model should mimic the cell physiology and the mechanical and the chemical cues that are present in specific tissues and organs, within a convenient high-throughput format. A possible key feature for such models is to recapture the cell polarity, the interactions between cells, and the interactions between the cells and the elastic extracellular matrix (ECM) by orienting the cells in a three-dimensional (3D) matrix. A common method to create 3D cell environments is to let the cells aggregate into spheroids with a diameter of around 200 μm. A major challenge for 3D cell cultures is to perform quick and easy imaging of the dense cell population, especially noninvasively. This protocol explains how to take advantage of the number of cells growing out from cell spheroids over time as a readout of the effect of a drug. The assay is compatible with standard imaging techniques and can be performed noninvasively using light microscopy or as a complement to other fluorescent imaging assays.

Key words Drug testing, Organ-on-a-chip, Cardiac spheroids, Cardiomyocytes, Induced pluripotent stem cells (iPSCs), 3D cell culture

1 Introduction

The effects of a drug on cell cultures in vitro and animals in vivo are examined before testing its impact on humans. Standard cell culture models are suitable for high-throughput screenings but lack a physiological relevance as the cells often are cultured on a two-dimensional (2D) surface under static conditions. Recently, perfused three-dimensional (3D) cell culture models, also referred to as microphysiological systems or organs-on-chips, have emerged as alternatives to conventional 2D models [1–4]. These models strive to better recapitulate the complex heterogenic environment within tissues and organs in order to obtain highly functional cells that respond to a drug in vitro as they would do in vivo. Several

Carl-Fredrik Mandenius and James A. Ross (eds.), *Cell-Based Assays Using iPSCs for Drug Development and Testing*, Methods in Molecular Biology, vol. 1994, https://doi.org/10.1007/978-1-4939-9477-9_17,
© Springer Science+Business Media, LLC, part of Springer Nature 2019

recent publications show 3D cell culture models of, for example, liver cells [5, 6], heart cells [7, 8], kidney cells [9, 10], and tumor cells [11, 12]. 3D cell environments can be created by encapsulating cells in ECM-mimicking scaffolds such as hydrogels [13, 14] and porous microcarriers [15], or by letting the cells aggregate into spheroids with defined sizes, typically in the range of 150–250 μm in diameter [16–18]. One advantage of using spheroids is that they are discrete entities that can be seeded into perfused microfluidic devices [19]. Concentration gradients of oxygen, nutrients, and metabolic waste products arise within cell spheroids due to the distance of the central portions from the surrounding cell culture medium, which in human tissue physiology is referred to as zonation [20–23]. A range of methods to analyze the impact of a drug on the target cells for both 2D and 3D cell cultures have been developed, including live/dead assays, quantification of biomarkers in the supernatant, and fluorescent labeling for microscopy of fixed cells. For cardiac cells, other common approaches are the recording of the beating frequency, and the recording of electrocardiograms using microelectrode arrays, which can be conducted with minimal interference with the cells. Noninvasive approaches are the analysis of cell growth and cell morphology as conducted in the neuronal network formation assay for neurites [24] or the wound healing assay for endothelial cells [25]. We recently reported that the outgrowth of cardiac cells from cardiac spheroids can be used to determine the impact of drug compounds [26]. We demonstrated the method with the drug compounds endothelin-1 and doxorubicin which both decreased the number of cells in the immediate surrounding of the spheroids, and with phenylephrine and amiodarone which stimulated an increase of the cell number around each spheroid. Thus, the assay could distinguish between positive and negative effects on the proliferation and/or migration of cells induced by drug compounds. This method can be implemented as a noninvasive assay using a standard phase contrast microscope, or by staining the cell nuclei and quantifying the number of surrounding cells using fluorescence microscopy. Both approaches are detailed in the protocol described below and exemplified with the three drug compounds doxorubicin, acetylsalicylic acid, and amiodarone. The outgrowth assay is suggested to serve as a complement to existing cell analysis methods.

2 Materials

1. Cardiac spheroids containing approximately 2500 cells (~250 μm in diameter) (*see* **Note 1**).
2. Cell culture medium (RPMI1640 supplemented with B27 minus insulin, Life Technologies, CA, USA).

3. μ-Slide $VI^{0.4}$ untreated (Ibidi GmbH, Germany) (*see* **Note 2**),
4. Laminin (100 μg/ml, Sigma-Aldrich, MO, USA).
5. Phosphate buffered saline (PBS, Sigma-Aldrich).
6. Dimethyl sulfoxide (DMSO, Sigma-Aldrich).
7. Doxorubicin (Sigma-Aldrich) dissolved in DMSO and diluted in cell culture medium (final DMSO concentration 0.25%) to 0.11 nM, 1.1 nM, and 10 nM.
8. Acetylsalicylic acid (Sigma-Aldrich) dissolved in DMSO and diluted in cell culture medium (final DMSO concentration 0.25%) to 0.11 nM, 1.1 nM, and 10 nM.
9. Amiodarone (Sigma-Aldrich) dissolved in DMSO and diluted in cell culture medium (final DMSO concentration 0.25%) to 0.11 nM, 1.1 nM, and 10 nM.
10. Motorized rocking table (Mixerlab 440, Swelab Instrument, Sweden).
11. Paraformaldehyde (PFA 4%, Sigma-Aldrich).
12. Hoechst 33342 (ThermoFisher, MA, USA).
13. Light microscope (Zeiss Axiovert 40C, Zeiss, Germany).
14. High content imaging system (Opera Phenix, PerkinElmer, MA, USA).
15. Image analysis software (Harmony, PerkinElmer).

3 Method

The principle of the assay is summarized in Fig. 1. The assay is here exemplified with the three drug compounds doxorubicin, acetylsalicylic acid, and amiodarone, each at three concentrations (0.11 nM, 1.1 nM, and 10 nM).

1. Incubate the μ-Slide $VI^{0.4}$, the cell culture medium and the drug supplemented cell culture medium at 37 °C and 5% CO_2 overnight to minimize the risk of air bubbles forming during the experiments (*see* **Note 3**).
2. Using a 1 ml syringe, coat the surfaces of the channels of a μ-Slide $VI^{0.4}$ by adding 30 μl laminin (100 μg/ml) for 1 h at room temperature.
3. Wash the channels with PBS 3 times.
4. Wash the channels with cell culture medium 3 times.
5. Seed 20–40 cardiac spheroids in 30 μl cell culture medium to one well of the channel and quickly remove 20 μl from the well on the opposite side of the channel (*see* **Note 4**).

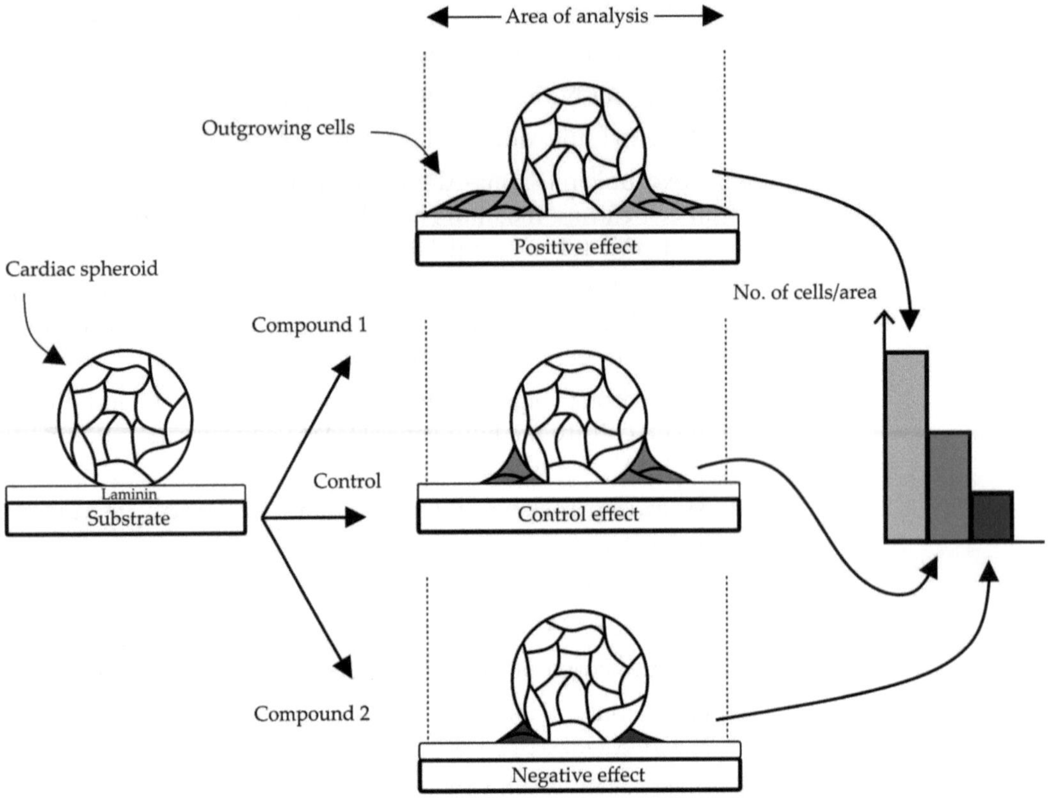

Fig. 1 Overview of the cardiac cell outgrowth assay. A certain time after adding drug compounds to cardiac spheroids on a cell-adhesion treated surface, the number of cells within a limited area (area of analysis) around each spheroid can be categorized to either have a positive effect (increased cell number), a negative effect (decreased cell number) or an insignificant compared to a control. Reprinted with permission [26]

6. Incubate the μ-Slide $VI^{0.4}$ for 16 h at 37 °C and 5% CO_2 to allow the spheroids to attach to the bottom of the channels (*see* **Note 5**).
7. Prepare the motorized rocker. Use a μ-Slide $VI^{0.4}$ filled with 90 μl of a sterile liquid and fix it in the middle of the rocker. Observe the height of the liquid in each well as the rocker turn. There should always be enough liquid in each well to ensure no air enters the channel. Adjust the movement of the rocker if needed (*see* **Note 6**).
8. Count the number of spheroids in each channel in a light microscope (*see* **Note** 7).
9. Replace the medium in each channel with 90 μl drug supplemented cell culture medium or control medium (cell culture medium supplemented with 0.25% DMSO).
10. Place the μ-Slide $VI^{0.4}$ on the motorized rocker moving at 0.5 Hz and incubate at 37 °C and 5% CO_2 (*see* **Note 8**).

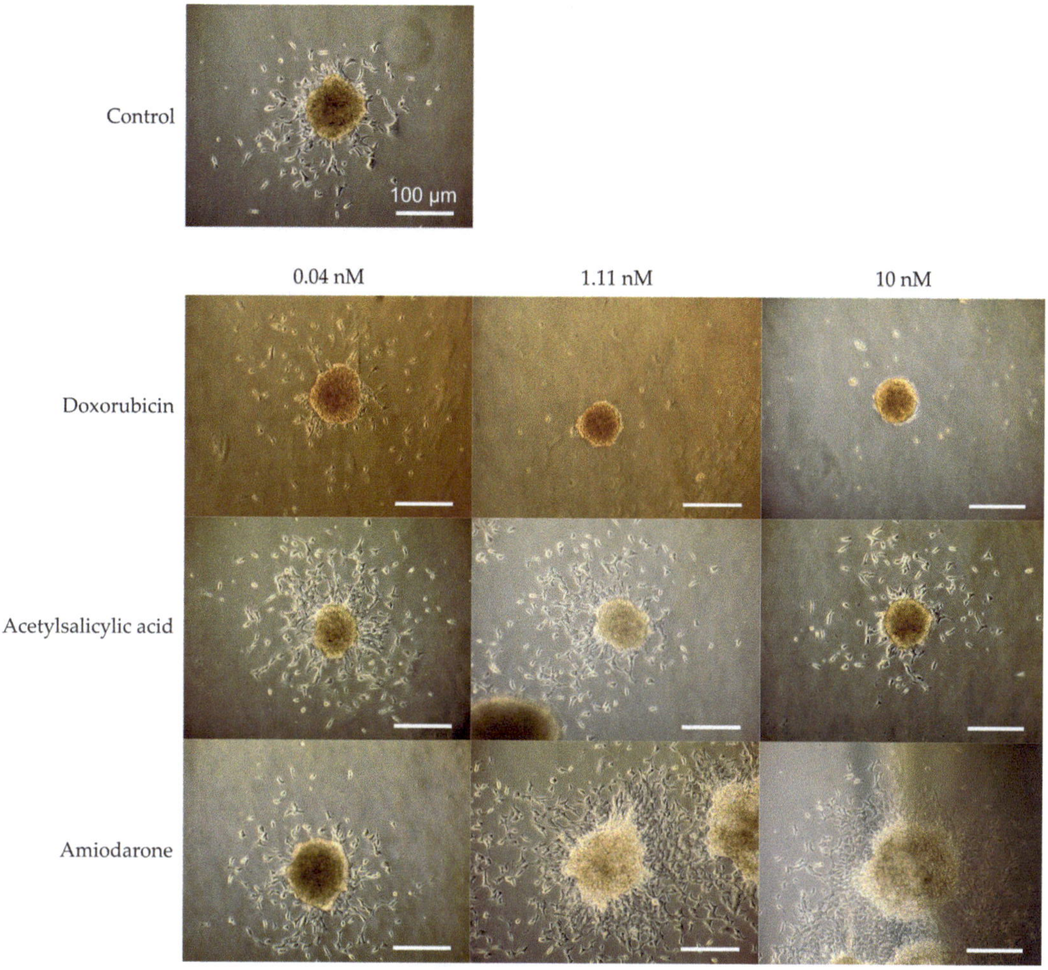

Fig. 2 Cardiac cells growing out from cardiac spheroids after 48 h of culture. Note the number of cells surrounding each spheroid at the different drug compound and concentration conditions. Scale bars represent 100 μm. Modified from [26] with permission

11. After 48 h, cells growing out from the spheroids should be visible in the control samples and at low concentrations of all the drug compound samples (Fig. 2) (*see* **Note 9**).
12. Wash the spheroids with PBS 3 times.
13. Fix the spheroids in PFA for 1 h at room temperature.
14. Wash the spheroids with PBS 3 times.
15. Stain the nuclei of the cells with Hoechst 33342.
16. Wash the spheroids with PBS 3 times.
17. Image the nuclei surrounding the spheroids using a high content imaging system with a 5× objective.

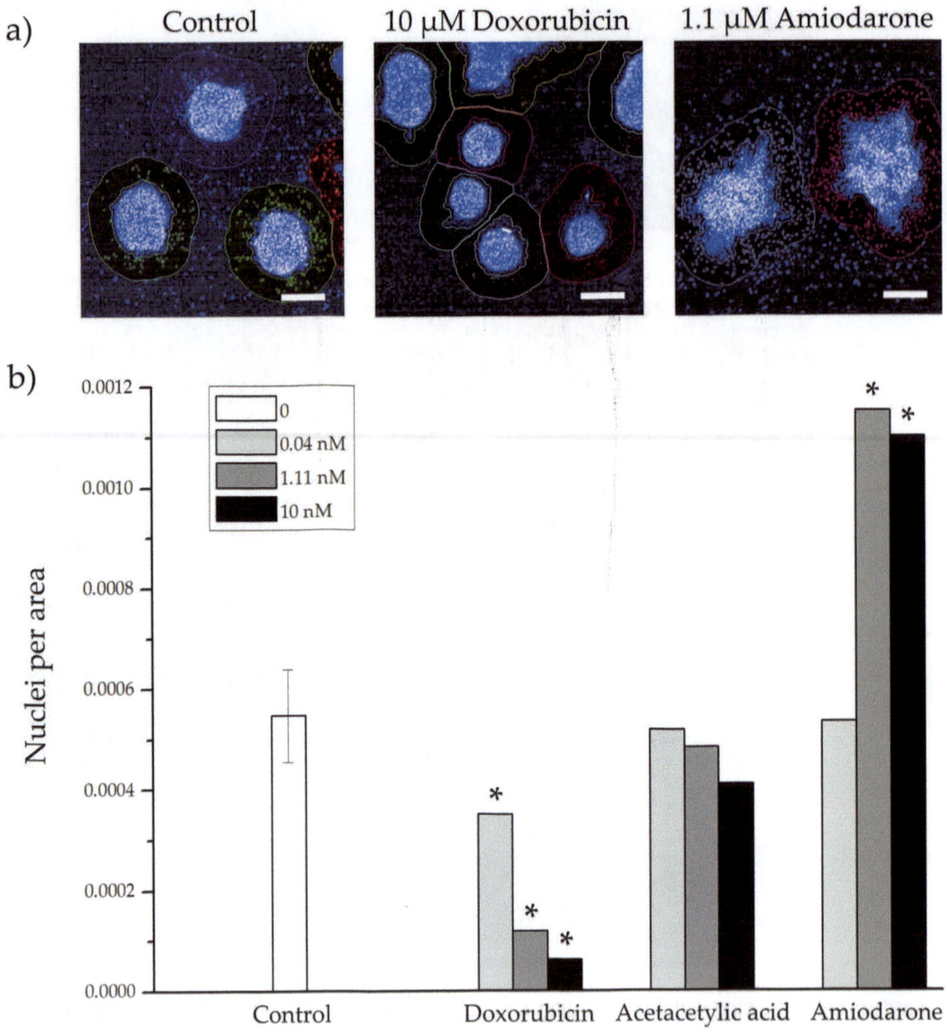

Fig. 3 Quantifying the number of cells surrounding the spheroids using fluorescence imaging. (**a**) The area of analysis was determined by two contours applied outside each spheroid. Cell nuclei were stained with Hoechst 33342 and the number of nuclei within each area of analysis was counted by the high content imaging software. Scale bars represent 100 μm. (**b**) The number of nuclei per area after 48 h of culture with cells exposed to doxorubicin, acetylsalicylic acid, amiodarone, or a DMSO control. Statistical difference between drug compound and control is noted by a *. Modified from [26] with permission

18. Using the image analysis software, locate each spheroid and apply two contours 5 and 50 pixels outside the boundary of each spheroid (Fig. 3a) (*see* **Note 10**).
19. Quantify the number of nuclei within the two contours.
20. Determine the area within the two contours (area of analysis).
21. Normalize the number of nuclei within one contour to the constrained area within the contour.
22. Compare the number of nuclei per area of the spheroids exposed to drug compounds to the controls (Fig. 3b).

4 Notes

1. The cardiac spheroids can be generated as described in [27] or cultured and differentiated using previously established protocols [28–30].
2. For these experiments, devices consisting of six parallel channels, each $0.4 \times 17 \times 3.8$ mm (30 μl), were purchased from a commercial vendor. Similar devices are also possible to create using photolithography and soft lithography as described in [31].
3. Bubbles that arise within microfluidic systems is a frequent problem and are difficult to remove and can disturb the flow of liquids and destroy or disrupt cells and cell aggregates. As gasses are released from liquids when they are heated, bubbles tend to form over time. The risk of bubble formation is minimized by preheating all liquids and equipment that is in contact with the cell culture medium prior to the experiments.
4. Make sure the cardiac spheroids are beating under a microscope before seeding. The beating might stop or slow down immediately after pipetting but should recover within 1 h. Tilt the pipette tip toward the channel entrance at the bottom of the well when seeding the spheroids.
5. The time for cell attachment can be both increased or decreased to fit into a certain schedule. However, a few hours are recommended for the cell outgrowth to begin before adding drug compounds. Furthermore, the outgrowth from closely neighboring spheroids might merge if the time for cell attachment is too long which can make the analysis more difficult.
6. Alternatively, make the wells taller by gently inserting cutoff 1 ml syringes and adding more cell culture medium. Note however that this can increase the velocity of the cell culture medium through the channels and therefore the shear stress experienced by the cells.
7. This step is especially important if, for example, biomarkers in the supernatant should be quantified to have an estimate of the number of cells within the channel.
8. The frequency of the rocking motion can be altered to fit the design of other devices and the volume of the wells. It is also possible to attach a syringe pump or a peristaltic pump to drive the perfusion through the channels for a more uniform and user-determined flow velocity.
9. For non-invasive analysis, the same spheroids can be imaged at several time-points using a phase contrast microscope and the cells can be counted manually or using standard imaging software (e.g., ImageJ). However, under conditions where many

cells are populating the surface (e.g., amiodarone at 1.11 nM and 10 nM in Fig. 2), fluorescence imaging might be necessary.

10. The 5 μm contour is applied to move the area of analysis away from the high density of cells in the spheroid.

Acknowledgments

The research leading to these results has received support from the Innovative Medicines Initiative Joint Undertaking under (grant no. 115439), resources of which are composed of financial contribution from the European Union's Seventh Framework Programme (FP7/2007-2013) and EFPIA companies. This publication reflects only the author's views, and neither the IMI JU nor EFPIA nor the European Commission is liable for any use that may be made of the information contained therein.

References

1. Bhatia SN, Ingber DE (2014) Microfluidic organs-on-chips. Nat Biotechnol 32 (8):760–772
2. Esch EW, Bahinski A, Huh D (2015) Organs-on-chips at the frontiers of drug discovery. Nat Rev Drug Discov 14(4):248–260
3. Marx U, Andersson TB, Bahinski A et al (2016) Biology-inspired microphysiological system approaches to solve the prediction dilemma of substance testing. ALTEX 33 (3):272–321
4. Christoffersson J, van Noort D, Mandenius C-F (2017) Developing organ-on-a-chip concepts using bio-mechatronic design methodology. Biofabrication 9(2):025023
5. Ma L-D, Wang Y-T, Wang J-R et al (2018) Design and fabrication of a liver-on-a-chip platform for convenient, highly efficient, and safe in situ perfusion culture of 3D hepatic spheroids. Lab Chip
6. Ehrlich A, Tsytkin-Kirschenzweig S, Ioannidis K et al (2018) Microphysiological flux balance platform unravels the dynamics of drug induced steatosis. Lab Chip
7. Jackman CP, Carlson AL, Bursac N (2016) Dynamic culture yields engineered myocardium with near-adult functional output. Biomaterials 111:66–79
8. Lemoine MD, Mannhardt I, Breckwoldt K et al (2017) Human iPSC-derived cardiomyocytes cultured in 3D engineered heart tissue show physiological upstroke velocity and sodium current density. Sci. Rep 7(1):5464
9. Wang L, Tao T, Su W et al (2017) A disease model of diabetic nephropathy in a glomerulus-on-a-chip microdevice. Lab Chip 17 (10):1749–1760
10. DesRochers TM, Suter L, Roth A et al (2013) Bioengineered 3D human kidney tissue, a platform for the determination of nephrotoxicity. PLoS One 8(3):e59219
11. Caballero D, Blackburn SM, de Pablo M et al (2017) Tumour-vessel-on-a-chip models for drug delivery. Lab Chip 17(22):3760–3771
12. Toh Y-C, Raja A, Yu H et al (2018) A 3D microfluidic model to recapitulate cancer cell migration and invasion. Bioengineering 5 (2):29
13. Prodanov L, Jindal R, Bale SS et al (2016) Long-term maintenance of a microfluidic 3D human liver sinusoid. Biotechnol. Bioengineering 113(1):241–246
14. Lanz HL, Saleh A, Kramer B et al (2017) Therapy response testing of breast cancer in a 3D high-throughput perfused microfluidic platform. BMC Cancer 17(1):709
15. Bergström G, Nilsson K, Mandenius CF et al (2014) Macroporous microcarriers for introducing cells into a microfluidic chip. Lab Chip 14(18):3502–3504
16. Urich E, Patsch C, Aigner S et al (2013) Multicellular self-assembled spheroidal model of the blood brain barrier. Sci Rep 3:1500
17. Bell CC, Hendriks DFG, Moro SML et al (2016) Characterization of primary human hepatocyte spheroids as a model system for

drug-induced liver injury, liver function and disease. Sci Rep 6:25187
18. Bergstrom G, Christoffersson J, Schwanke K et al (2015) Stem cell derived in vivo-like human cardiac bodies in a microfluidic device for toxicity testing by beating frequency imaging. Lab Chip 15(15):3242–3249
19. Christoffersson J, Bergström G, Schwanke K et al (2016) A microfluidic bioreactor for toxicity testing of stem cell derived 3D cardiac bodies. In: Turksen K (ed) Bioreactors in stem cell biology: methods and protocols. Springer Verlag, London, pp 159–168
20. Lin R-Z, Chang H-Y (2008) Recent advances in three-dimensional multicellular spheroid culture for biomedical research. Biotechnol J 3(9–10):1172–1184
21. Wrzesinski K, Fey JS (2018) Metabolic reprogramming and the recovery of physiological functionality in 3D cultures in micro-bioreactors. Bioengineering 5(1):22
22. Materne EM, Tonevitsky AG, Marx U (2013) Chip-based liver equivalents for toxicity testing-organotypicalness versus cost-efficient high throughput. Lab Chip 13 (18):3481–3495
23. Oinonen T, Lindros OK (1998) Zonation of hepatic cytochrome P-450 expression and regulation. Biochem J 329(1):17–35
24. Frimat J-P, Sisnaiske J, Subbiah S et al (2010) The network formation assay: a spatially standardized neurite outgrowth analytical display for neurotoxicity screening. Lab Chip 10 (6):701–709
25. van der Meer AD, Vermeul K, Poot AA et al (2009) A microfluidic wound-healing assay for quantifying endothelial cell migration. Am J Phys Heart Circ Phys 298(2):H719–H725
26. Christoffersson J, Meier F, Kempf H et al (2018) A cardiac cell outgrowth assay for evaluating drug compounds using a cardiac spheroid-on-a-chip device. Bioengineering 5(2):36
27. Halloin C, Coffee M, Manstein F, Zweigerdt R, Production of cardiomyocytes from human pluripotent stem cells by bioreactor technologies, Chapter 5 in Cell-Based Assays Using iPSCs for Drug Development and Testing, Methods in Molecular Biology, vol. 1994, Springer Nature 2019
28. Kempf H, Olmer R, Kropp C et al (2014) Controlling expansion and cardiomyogenic differentiation of human pluripotent stem cells in scalable suspension culture. Stem Cell Reports 3(6):1132–1146
29. Kempf H, Kropp C, Olmer R et al (2015) Cardiac differentiation of human pluripotent stem cells in scalable suspension culture. Nature Protocol 10(9):1345–1361
30. Kempf H, Olmer R, Haase A et al (2016) Bulk cell density and Wnt/TGFbeta signalling regulate mesendodermal patterning of human pluripotent stem cells. Nat Commun 7:13602
31. Christoffersson J, Mandenius C-F (2018) Fabrication of a microfluidic cell culture device using photolithographic and soft lithographic techniques. this volume:xxx–xxx

Part IV

Electrophysiology Assays

Chapter 18

Assay Procedures for Compound Testing of hiPSC-Derived Cardiomyocytes Using Multiwell Microelectrode Arrays

Udo Kraushaar and Elke Guenther

Abstract

The cardiac action potential requires a precise timing of activation and inactivation of ion channel subtypes. Deviations, for example, due to blockage of specific voltage-gated potassium channels, can result in live-threatening arrhythmias. Due to the limitations of standard cellular assays based on cells which artificially express only single ion channel subtypes, many potentially interesting compounds are discarded during drug development. More predictive functional assays are required. With the upcoming of human stem-cell derived cardiomyocytes (hiPS-CM) these assays are available, supporting even the design of patient-derived disease models. Microelectrode array systems allow to noninvasively record and evaluate cardiac field action potentials. In this chapter we describe how to cultivate hiPS-CM on two parallelized MEA systems and suggest an experimental strategy for compound tests.

Key words Microelectrode array, MEA, Human stem cell-derived cardiomyocytes, hiPS, Compound testing, Safety pharmacology

1 Introduction

Unexpected cardiotoxicity has been the most common cause of market drug withdrawals or developmental termination of promising drug candidates in recent years. An early detection of drug-induced cardiac side effects is desired as it prevents cost-intensive late-stage withdrawals and ensures the patients' safety. Especially drug induced ventricular arrhythmia triggered by the blockage of specific ion channels like the human-ether-a-gogo-related-gene (hERG) channel are, although rare side effects, life-threatening [1, 2]. The cytosolic domain of the hERG channel is known for its molecular affinity to cardiac and noncardiac drugs [3]. Blockage leads to a prolongation of the cardiac action potential. In the electrocardiogram (ECG) this translates into a prolonged QT interval, a major indicator of an increased risk of arrhythmia [4].

Carl-Fredrik Mandenius and James A. Ross (eds.), *Cell-Based Assays Using iPSCs for Drug Development and Testing*, Methods in Molecular Biology, vol. 1994, https://doi.org/10.1007/978-1-4939-9477-9_18,

© Springer Science+Business Media, LLC, part of Springer Nature 2019

Microelectrode arrays (MEA) allow for the extracellular recording of cardiac action potentials (fAP) with high spatial and temporal resolution [5–7]. The heart of an MEA chip consists of substrate-integrated microelectrodes (diameters between 10 and 100 μm, depending on chip and manufacturer) that are arranged in a two-dimensional array. This array is embedded in a recording chamber, which also serves as a culture dish. Electrogenic cells like cardiomyocytes, neurons, and others are plated directly on top of the electrodes, and electrical activity is recorded extracellularly as field action potentials from all electrodes simultaneously. The defined geometry of the 2D array provides information of not only the activity at the individual electrodes but also the 2D flow of excitation of the cellular network under investigation. Optional lids on top of the chamber ensure sterile conditions, a requirement for long-term recordings. Academic research experiments typically use single-well MEA chips that, due to the high number of electrodes, permit extensive investigation of network activity even over a long period of time. In areas where higher throughputs are of interest, chips with six to nine wells are available. For even higher throughput, multi-well plates are available in which up to 96 parallel recordings can be achieved. A limitation of this increase in number of experimental wells is the concomitant decrease in number of electrodes per well. For most cardiac applications, though, a so-called "golden electrode" is picked, which is the electrode with the best signal-to-noise ratio and most typical fAP waveform. The composition of cardiac fAPs is well known.

The effects of sodium, calcium, and potassium channels are reflected in the depolarization, plateau, and repolarization phase (for example *see* [8]). With the rise of stem-cell technologies human induced pluripotent stem cell-derived cardiomyocytes (hiPSC-CMs) together with the MEA are becoming an efficient and sensitive model for the investigation of drug-induced cardiotoxicity [9].

In this chapter we will address functional cardiotoxicology on hiPSC-derived cardiomyocytes using MEA chips with multiple wells.

2 Materials

hiPSC-derived cardiomyocytes can either be generated in-house or obtained from one of several commercial manufacturers. As the field is highly dynamic at the time of preparation of this chapter, a list of manufacturers will not necessarily reflect the current market status. Therefore, no recommendation can be given about the best cell source. When using cells obtained from commercial sources, best results are achieved when the recommended protocols and culturing conditions are obeyed. Special attention should be given to the cell thawing procedure, since there are several critical steps that can make the difference between a high yield of healthy cells and a major loss of cells.

MEA systems are available from several companies including Multi Channel Systems MCS (Germany), Axion BioSystems (USA), and AlphaMed (Japan), to mention just a few. They all are based on the same recording principle (i.e., extracellular recordings from substrate-integrated electrodes) and are all suitable for experiments with hiPSC-derived cardiomyocytes. Differences are subtle and include the software integrated for analysis, maximal sampling frequency, the capability to be used in cell culture incubators (not all of them are resistant to the humidity inside the incubator) and material and durability of the MEA electrodes.

Two systems are described here in this chapter: The low-throughput MEA (LTM) system consists of an MEA2100 system (MCS) attached to a PC running the software MC_Rack which comes with the system (Fig. 1a). The very same software can be used for both recording and analysis. The MEA chips used here

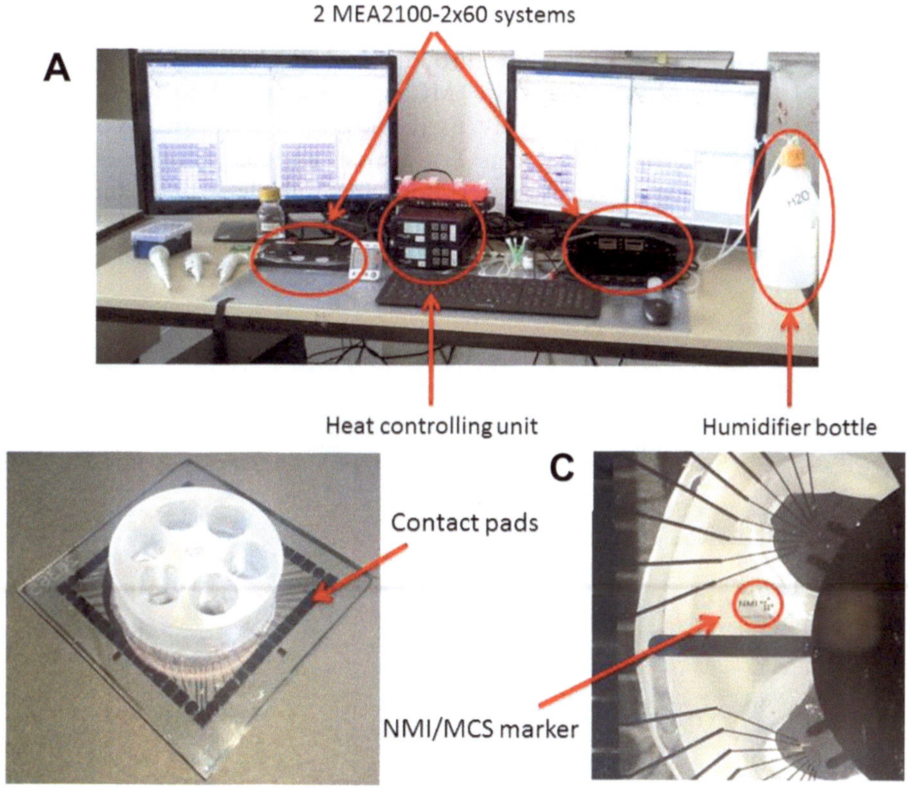

Fig. 1 (**a**) MEA recording system. The system shown comprises of 2 MEA2100–2 $\times$ 60 systems. (**b**, **c**) 6-well MEA chip. (**b**) top view on the six wells of an MEA chip (chamber in white). Black squares are the contact pads of each electrode of the array. These should be cleaned with 2-propanol prior to the experiment. (**c**) Magnified bottom view of the chip. Note the small NMI/MCS logo indicating the location of the grounding. This marker should point to the left when the MEA chip is inserted into the MEA2100 system

are 6-well chips (60-6wellMEA200/30iR-Ti). This type of MEA chip comes in two varieties: one with small round chambers (60-6wellMEA200/30iR-Ti-rcr) comparable to the wells of a 96-well plate (Fig. 1b), and one with triangular chambers (60-6wellMEA200/30iR-Ti-tcr) that support a larger media volume. In both MEA chips, nine electrodes per well (arranged as a 3 × 3 array) are integrated with an electrode diameter of 30 μm and a pitch of 200 μm. The material of the electrodes is sputtered titanium nitrate (TiN) which provides low electrode impedance resulting in very low noise. Sterilization of the chips prior to cultivation can be achieved either by autoclaving procedures (unless recommended differently by the chip manufacturer) or submersion of the chips for approximately 30–45 min in 70% (nondenatured) ethanol followed by air drying in a laminar flow hood/biosafety cabinet. The bottom of the MEA system has an integrated heating plate which needs to be attached to a heat controlling unit (TC02) which should be set to 37 °C. Last but not least, it is useful to place a climate chamber (MEA2100-CO2-C) connected to a carbogen outlet (5% CO_2, 95% O_2) on top of the MEA chip to ensure that the pH of the medium within the wells remains constant.

The higher throughput MEA (HTM) system consists of a Multiwell system (MCS), also attached to a PC (Fig. 2a). The software that is used in this system for recording is called Multiwell-Screen. For analysis another software is needed named Multiwell-Analyzer. Both are available at the manufacturer's website free of charge. Multiwell plates are available in 24- and 96-well format. For higher throughput, typically experiments are

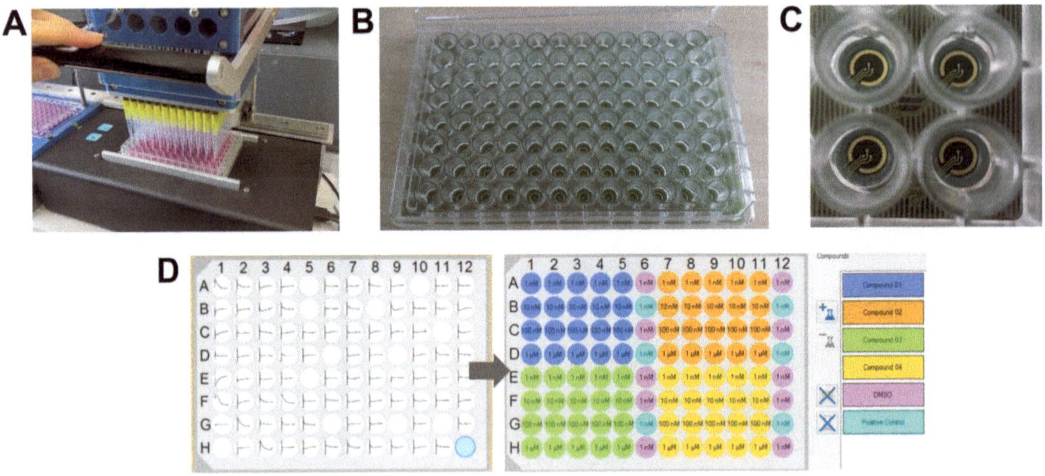

Fig. 2 Multiwell system (MCS). (**a**) Multiwell with optional 96-well pipettor. (**b**) 96-well multiwell plate. (**c**) Magnification of 4 wells. Note the three recording electrodes in the middle surrounded by a round golden reference electrode. (**d**) Screenshot of a compound test. Left: Activity is visible in almost every well. Right: Layout of a compound test (4 compounds, positive and negative control. Single concentration per well, $c = 4$ concentrations, $n = 5$ repeats)

performed with the 96-well plates (Fig. 2b). Each well has three golden recording electrodes integrated, surrounded by a golden reference electrode (Fig. 2c). The Multiwell plates are shipped sterile and cannot be autoclaved. Also, the bottom of the Multiwell system is heated, here the controller already in integrated into the device. The temperature is set within the software. With the system also comes a transparent box which, when hooked up to carbogen, keeps the pH of the medium stable.

Several different cell attachment substrates are available, including nitrocellulose (NC solution, *see* **Note 1**) and polyethylenimide (PEI solution, *see* **Note 2**) mostly used as coating for cardiomyocytes. Unless an alternative is specifically recommended by the cell manufacturer, these are suitable provided the pH is maintained within the range of 4–9. A pH outside of this range tends to destroy the insulation of the electrode leads, which in turn leaves the chip damaged.

Cleaning of the chips for the LTM system is best done by rinsing the chips with a powerful (deionized) water jet from a squirt bottle, with a subsequent submersion in a solution containing 1% Tergazyme (Alconox, USA) in distilled water over night at room temperature. Following a careful rinsing with distilled water, any remaining residue can be washed off by placing the chips overnight in large beaker with distilled water (approx. 1 L for four to six chips). Air drying completes the cleaning cycles, and the chips can then be sterilized for further use. Multiwell plates for the HTM system are meant to be for single use only.

Test compounds often cannot be dissolved directly in aqueous solutions but require a solvent. Typically, either ethanol or dimethylsulfoxide (DMSO) is used to prepare stock solutions. In most cell systems, final concentrations of up to 0.1% are well tolerated by the cells. However, running vehicle control tests in parallel to the experiments is highly recommended.

Cell cultivation for MEA does not require special equipment compared to that recommended in other typical cell culture procedures. A horizontal laminar flow hood is beneficial when a stereomicroscope is used during plating of the cells (e.g., when plating small droplets). When seeding the cells in small droplets, low adhesion surface 10 μl pipette tips are ideal for forming a defined droplet.

3 Methods

3.1 Cell Seeding on MEA Chips

The surface of a MEA chip is to some extent hydrophobic, depending on both storage condition and duration. The manufacturer MCS recommends to make the surface hydrophilic to improve cell adhesion and an optimal signal-to-noise ratio. While this is true for neuronal cultures this can be ignored for cardiomyocyte-

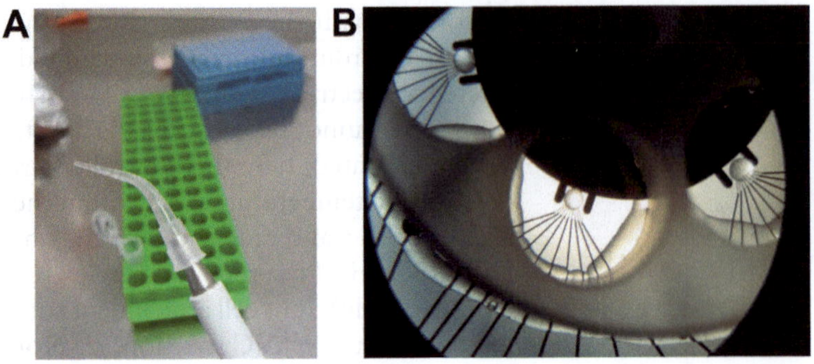

Fig. 3 (**a**) Pipette tip bend to an angle of approx. 40°. (**b**) Microscopic view on 3 out of 6 wells of a 6-well MEA chip. Note the drops on top of the electrode array

based experiments. The adhesion typically is very good, and the signals are much larger compared to neurons. In fact, a hydrophobic surface has the advantage that coating and cell suspension drops stay in place and do not flatten, which would make keeping the intended cell density more complicated, resulting in unstable assays [10]. Coating of the MEA chip requires a steady hand. To optimize the signal-to-noise ratio of recording, plating of the cells onto the reference electrode should be avoided. For MEAs obtained from MCS the reference electrode can easily be identified because for LTM system chips it is a very large solitary electrode placed some distance from the array of recording electrodes (Fig. 3b), while for the Multiwell plates of the HTM system it is defined by the round golden circular electrode (Fig. 2c). To avoid the reference electrode, only small droplets directly spotted on top of the electrode array are optimal. Depending on the number and spacing of the electrodes within the area, droplet volumes of 1–3 μL can easily be achieved without moistening the reference electrode (Fig. 3b). When placing the droplet, it is critical NOT to touch the electrodes of the array of the LTM system chips, since the TiN is very fragile and will easily be damaged by the mechanical stress. Electrodes of the Multiwell HTM system are more robust because they consist of gold. A stereomicroscope is a very helpful tool for both coating and plating of the cells. When using the stereomicroscope, it is difficult to insert the pipet tip with the cells (or coating substrate) for steric reasons. A solution is to kink the pipet tip to an angle of approx. 45–60° (Fig. 3a, *see* **Note 3**). When plating these small amounts of liquid, evaporation is a major issue. To minimize this, it is advised to (1) place the MEA chips into a sufficiently large petri dish containing a moist filter paper at the bottom, and (2) perform the chip coating and cell plating steps rapidly and return the MEA chip back to the cell culture incubator as quickly as possible. The optimal cell density depends on the type of cells under investigation and can range from 1000–12,000 cells/μL. Usually cells should be allowed to rest in the incubator for 1–3 h in order to attach to the surface of the MEA chip within the droplet before each well of the is slowly filled with medium (typically 200 μL) with the pipet touching the inner wall of the well to avoid disturbing the plated cells. Unless stated differently by the manufacturer's manual, a 50% media should be performed 1 day post-plating and twice a week thereafter.

3.2 Recording of Spontaneous Electrical Activity

Spontaneous electrical activity can already be detected 2–4 days postplating. To record this activity the cells should be transferred to the MEA setup for a minimum of 10 min prior to the recordings. This allows the cells to recover from mechanical stress introduced by vibrations and movement from incubator to the setup and ensures that the activity stabilizes. Additionally, the heat system located below the MEA chip in the platform typically is set to 37 °C. When experiments are performed in bicarbonate buffered solutions or media, stabilization of recordings for long-time periods can be achieved by covering the top of the MEA with a small chamber in which humidified carbogen (typically 5% CO_2 and 95% O_2, depending on the bicarbonate concentration in the medium) is flowing constantly. The carbogen stabilizes the pH of the solution and the humidified environment minimizes evaporation of the recording solution which would otherwise result in an increased osmolarity. This could be suboptimal for cell survival and increase the effective test compound concentration. To ensure that the contact pins are clean it is recommended to both carefully wipe clean the contact pins of the respective MEA system and the contact pads of the MEA chip for the LTM system. The correct orientation of the 6-well MEA chips in case of the LTM system is a bit tricky to identify. At the back side of the MEA chip locate the tiny NMI/MCS logo (Fig. 1c). Carefully transfer the MEA chip into the MEA2100 system, with the NMI/MCS logo pointing to the left of the experimenter. For the Multiwell plates of the HTM system the orientation is much easier, as the plate will only fit in one direction (with the curved edges towards the blue buttons which trigger the closing mechanism of the system). To ensure that the field action potential (fAP) amplitude is recorded in full, the sampling frequency of the MEA setup should be sufficiently high to avoid undersampling effects [11]. A sampling frequency set to 10–20 kHz is sufficient for most preparations. Please note that the data file size depends on the sampling frequency and easily can reach sizes of several tens of GBytes. Unless required differently in the experimental design, a hardware bandpass filter of 0.1 Hz–3 kHz is a good choice.

Figure 4 depicts a typical experiment for the 6-well MEA. Each of the boxes shown on the left is the oscillographic voltage representation of a single electrode. Each of the wells of the 6-well MEA used in this experiment contains nine electrodes. Each red box in Fig. 4 indicates the nine electrodes comprising one well. It can be

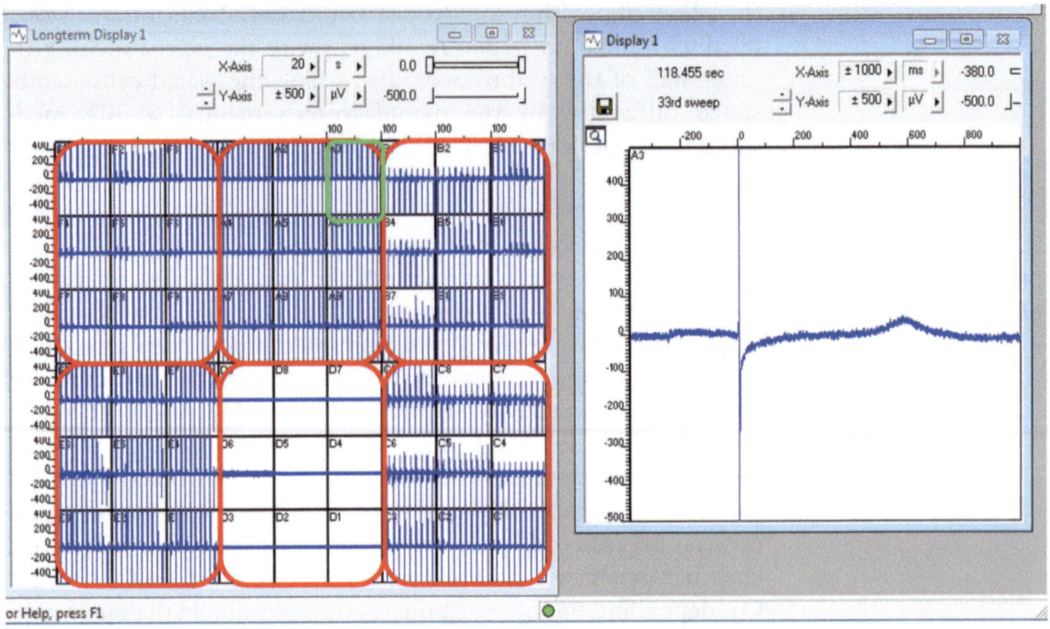

Fig. 4 Sample recording of spontaneously beating cardiomyocytes in a 6-well MEA. Left: overview over all electrodes. Red boxes indicate the 6 wells, while green box represents the representation of a single well. Right: time-resolved magnification of the field action potential from the electrode marked by the green box on the left

seen that spontaneous and regularly occurring fAPs could be recorded on most of the electrodes. On the right-hand side a recording from one electrode depicted in green is shown. The magnitude and deflection (upward or downward) of a single fAP can vary depending on the localization of the respective cardiomyocyte relative to the recording electrodes [12, 13].

Using the abovementioned setup allows to investigate different parameters as visualized in Fig. 5b. One of the most relevant parameters during compound tests is the duration of the field action potential which indicates if the compound of interest has an effect on the I_{Kr} potassium ion channel, also known as hERG (human-ether-a-gogo-related-gene) channel. Blockage of this channel can induce severe and live-threatening arrhythmia of the torsade-de-pointes type [1, 14, 15]. In vitro equivalents of this arrhythmia can be seen in Fig.5c, d.

3.3 Suggestions for Running an Experiment in 6-Well Format

Let us consider we want to perform an experiment with four concentrations of a test compound. As MC_Rack allows us full freedom for the file name. Each concentration should be recorded in a separate file. A suggested pattern could be as follows:

YYYYMMDD_MEA**a**_**bbb**_**C**x.mcd
YYYY–4 digits year (e.g., 2018)
MM–2 digits month (e.g., 08 for August)

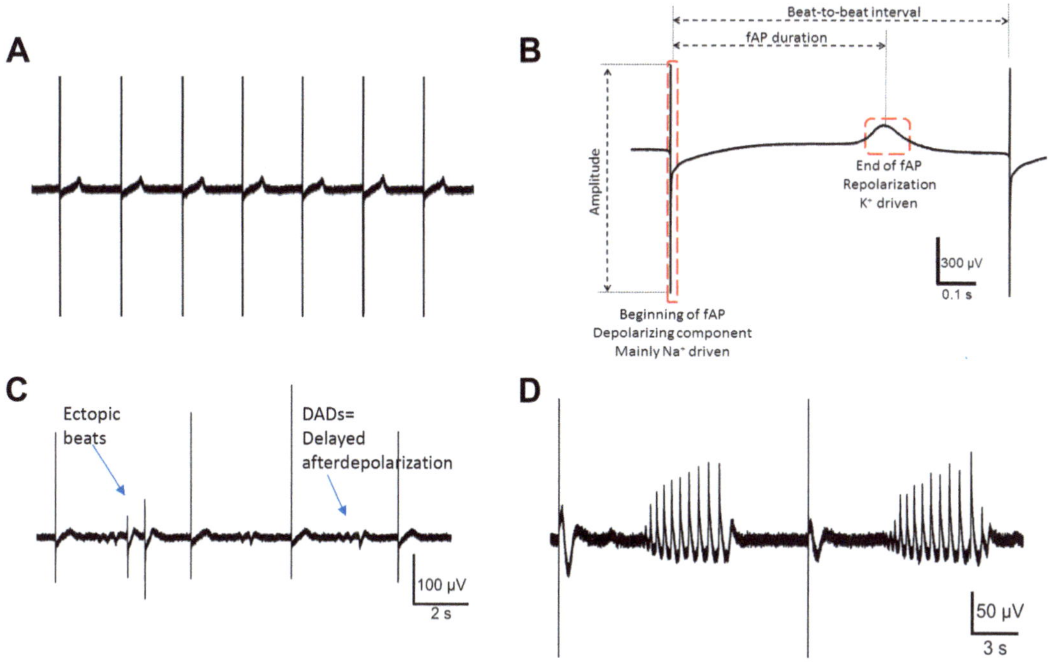

Fig. 5 Field action potential (fAP) recordings of cardiomyocytes on MEA. (**a**) regular beating of spontaneously active cells. (**b**) Selection of most relevant parameters with can be obtained from the cardiac field action potential. (**c**) proarrhythmic activity induced by a cardiotoxic compound. (**d**) tachycardic events

DD–2 digits day (e.g., 04 for 4th)
a–consecutive MEA number 1–n
bbb–three-digit internal compound code
x–concentration number (0 for control, 1 for 1st, 2 for 2nd, ...)

Example: 20180804_MEA1_001_C2.mcd: experiment from 4 August, 2018; first MEA of compound 001, 2nd concentration.

3.4 Preparation of the Test Compound for a Final Concentration Series of 0.01, 0.1, 1 and 10 μM

Prepare a 10 mM stock solution in diemethylsulfoxide (DMSO). Make sure the compound dissolves completely (*see* **Note 4**). From this stock solution prepare a serial dilution: 1000 μM; 100 μM; 10 μM; 1 μM, 0.1 μM in culture medium. Therefore, dilute the solution 1:10 from the respective higher concentration. Shake each concentration well. Prepare a serial dilution of DMSO in medium at the same concentrations as with the compound serial dilutions.

3.5 Experimental Procedure

1. Place a 10 cm petri dish next to the MEA chip (Fig. 6a).
2. After equilibration of the cells in the MEA system record control conditions for 2 min.
3. Lift climate chamber from MEA chip and set aside.
4. Reserve one well for vehicle control experiment. Note in protocol sheet.
5. Apply lowest compound concentration of 0.01 μM (*see* **Note 5**).

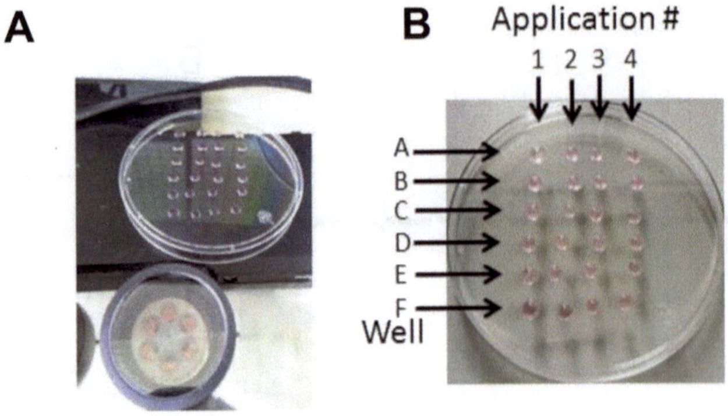

Fig. 6 Drop collection of discarded medium during compound application. (**a**) Overview during experiment. Top: petri dish with drops, Bottom: MEA chip with climate chamber sitting on top. (**b**) organization of the drops. Each line represents drops from one well (A–F), each column represents discarder prior to compound application

6. Slowly remove 20 μL of medium from each well within the MEA chip. Make sure to pipette on the side wall of the well to not disturb the cells.
7. Discard drop on the petri dish (Fig. 6b).
8. Keep the drops in a top-to-bottom order. This helps tracking if all six wells are used (Fig. 6b).
9. From stock solution of 0.1 μM slowly add 20 μL of compound solution to the side wall of each well but the vehicle control well. Pipette up and down 2× to mix (*see* **Note 4**).
10. Add same amount of volume from DMSO serial dilution to the vehicle-control well.
11. Place climate chamber on top of the MEA chip. Avoid vibrations.
12. Set timer to 5 min to allow wash in. During this time change filename to the correct name (here: ..._C0 to ..._C1).
13. After 5 min record for 2 min.
14. Apply next compound concentration of 0.1 μM.
15. Slowly remove 18 μL of medium from each well within the MEA chip.
16. Discard drop on the petri dish.
17. Place each drop right to the previous drop.
18. Look for the colour of the drop. The colour should be less purple than the previously pipetted drop. If not, carbogen supply to the MEA chip might be insufficient.
19. From stock solution of 1 μM slowly add 18 μL compound solution to the side wall of each well. Pipet up and down 2× to mix.

20. Repeat **steps 5–7**.
21. Apply next compound concentration of 1 μM and 10 μM accordingly (*see* **Note 5**).

4 Notes

1. To prepare NC solution, 2 cm^2 is dissolved in 40 mL methanol and stored at 4 °C. For coating add a droplet (ideally 1–2 μL) of NC solution to the well and let evaporate the methanol under the sterile hood. No further wash process is needed.
2. To prepare a PEI solution, a solution containing 50 mM boric acid and 12.5 mM borax are dissolved in distilled water. After stirring overnight add 0.1 mL of a 5% polyethylenimide solution. Sterile filter and store at 4 °C. For coating add a droplet (ideally 1–2 μL) of PEI solution to the well and leave at room temperature for 1 h. Discard the solution and wash 3× with distilled water. Allow well to dry until use.
3. To kink a pipet tip for the seeding of the cells simply move the pipet tip rapidly through a small flame (e.g., a crème brûlée torch) until the hot area becomes a bit more transparent compared to the cold plastic material. Bend the tip by application of force using a sterile forceps until the material has cooled down. Make sure that the tip is still hollow after this procedure (Fig. 3A).
4. To minimize the mechanical stress of the cardiomyocytes during application of a compound, an optimal media exchange is on the order if 10% of the well volume. This implies that the compound to be added consists of a stock solution with the respective high concentration. To simplify the calculations, it is advised to keep the volume within the well constant by withdrawing the pipetting volume prior to adding the compound.
5. If the test compound does not dissolve, there are several strategies to facilitate it:
   - Sonicate for 5 min.
   - Vivid shaking at temperatures up to 50 °C.
   - Dissolve in other liquids, e.g. distilled water, DMSO, ethanol. Note that for the latter two the final concentration of the solvent in the cellular environment usually should not exceed 0.1%.
   - Acidify with HCl.
   - Alkalinify with NaOH.
   - Lower the concentration of the stock solution. Adapt the serial dilution accordingly.

Acknowledgments

The research leading to these results has received support from the Innovative Medicines Initiative Joint Undertaking under (grant no. 115439), resources of which are composed of financial contribution from the European Union's Seventh Framework Programme (FP7/2007–2013) and EFPIA companies. This publication reflects only the author's views, and neither the IMI JU nor EFPIA nor the European Commission is liable for any use that may be made of the information contained therein.

References

1. Gintant GA, Su Z, Martin RL, Cox BF (2006) Utility of hERG assays as surrogate markers of delayed cardiac repolarization and QT safety. Toxicol Pathol 34:81–90. https://doi.org/10.1080/01926230500431376
2. Antzelevitch C (2007) Ionic, molecular, and cellular bases of QT-interval prolongation and torsade de pointes. Europace 9(Suppl 4):iv4–i15. https://doi.org/10.1093/europace/eum166
3. Mitcheson JS, Chen J, Lin M et al (2000) A structural basis for drug-induced long QT syndrome. Proc Natl Acad Sci U S A 97:12329
4. Morissette P, Hreiche R, Turgeon J (2005) Drug-induced long QT syndrome and torsade de pointes. Can J Cardiol 21:857–864
5. Stett A, Egert U, Guenther E et al (2003) Biological application of microelectrode arrays in drug discovery and basic research. Anal Bioanal Chem 377:486–495. https://doi.org/10.1007/s00216-003-2149-x
6. Kraushaar U, Buckenmaier S, Guenther E (2009) MEA-based biosensors to study cellular differentiation and integration. Tissue Eng Part A 15:726–726
7. Kraushaar U, Meyer T, Hess D et al (2011) Cardiac safety pharmacology: from human ether-a-gogo related gene channel block towards induced pluripotent stem cell based disease models. Expert Opin Drug Saf:1–14. https://doi.org/10.1517/14740338.2012.639358
8. Halbach M, Egert U, Hescheler J, Banach K (2003) Estimation of action potential changes from field potential recordings in multicellular mouse cardiac myocyte cultures. Cell Physiol Biochem 13:271–284. https://doi.org/10.1159/000074542
9. Millard D, Dang Q, Shi H et al (2018) Cross-site reliability of human induced pluripotent stem cell derived cardiomyocyte based safety assays using microelectrode arrays: results from a blinded CiPA pilot study. Toxicol Sci 164(2):550–562. https://doi.org/10.1093/toxsci/kfy110
10. Mulder P, de Korte T, Dragicevic E et al (2018) Predicting cardiac safety using human induced pluripotent stem cell-derived cardiomyocytes combined with multi-electrode array (MEA) technology: a conference report. J Pharmacol Toxicol Methods 91:36–42. https://doi.org/10.1016/j.vascn.2018.01.003
11. Nyquest H (1928) Certain topics in telegraph transmission theory. Transact Am Inst Electr Eng 47:617–644
12. Holt GR, Koch C (1999) Electrical interactions via the extracellular potential near cell bodies. J Comput Neurosci 6:169–184. https://doi.org/10.1023/A:1008832702585
13. Buzsáki G, Anastassiou CA, Koch C (2012) The origin of extracellular fields and currents—EEG, ECoG, LFP and spikes. Nat Rev Neurosci 13:407–420. https://doi.org/10.1038/nrn3241
14. Bowlby MR, Peri R, Zhang H, Dunlop J (2008) hERG (KCNH2 or Kv11.1) K+ channels: screening for cardiac arrhythmia risk. Curr Drug Metab 9:965–970
15. Joshi A, Dimino T, Vohra Y et al (2004) Preclinical strategies to assess QT liability and torsadogenic potential of new drugs: the role of experimental models. J Electrocardiol 37 (Suppl):7–14

Chapter 19

Multielectrode Array (MEA)-Based Detection of Spontaneous Network Activity in Human iPSC-Derived Cortical Neurons

Valeria Kizner, Sandra Fischer, and Maximilian Naujock

Abstract

Multielectrode arrays enable the detection of spontaneous cellular network activity, which can be utilized for the characterization of a neuronal culture. Here, we describe the detection of spontaneous neuronal activity in iPSC-derived cortical neurons using a 24-well plate for a multiwall-MEA system.

Key words MEA, iPSC, Neuron, Neuronal network activity, Neurophysiology

1 Introduction

Self-organization of neurons leads to neuronal network formation during early neurodevelopment. These rising networks generate spontaneous neuronal activity. Different developmental processes like neurogenesis, cellular differentiation, migration, and programmed cell death can be affected by neuronal activity [1]. Therefore, spontaneous electrophysiological activity in neuronal networks of a cell population during development can be used to understand the structural and functional formation of neuronal circuits and their implication in network plasticity and adaptability [2, 3].

Multielectrode array (MEA) technology is a neurophysiological method that provides recordings of spontaneous electrical activity from neuronal cultures in vitro. By this method, proper development of the neuronal activity of a culture can be investigated [4]. The cell culture surface of an MEA plate is composed of multiple extracellular electrodes per well, allowing the detection of local field potentials displayed as spikes (arising from single action potentials) and burst activity (arising from groups of action potentials) at different positions within a neuronal network.

Carl-Fredrik Mandenius and James A. Ross (eds.), *Cell-Based Assays Using iPSCs for Drug Development and Testing*, Methods in Molecular Biology, vol. 1994, https://doi.org/10.1007/978-1-4939-9477-9_19,
© Springer Science+Business Media, LLC, part of Springer Nature 2019

Besides, this technology can be used to determine the effects of various neuromodulators on a human neuronal network [5].

Neuronal network dysfunction appears in various neuropsychiatric and neurodegenerative disorders [6–8]. Human induced pluripotent stem cells (iPSCs)-derived neurons can be target of MEA recordings to study neurological disorders, since these neurons display basic electrophysiological properties [9]. In a traditional disease modeling setting the electrophysiological properties of iPSC-derived neurons can be described and compared with patient or genetically engineered cell lines [10, 11]. The measurement of spontaneous activity, plasticity and responsiveness to a neurostimulant can also be used to provide in-depth characterization of the functionality and viability of a respective iPSC-derived neuronal culture. In summary, MEA recordings are used to study neurophysiology, neuropharmacology, and neurotoxicology in cell cultures.

Here, we describe a protocol based on 24 multiwell MEA plates to record spontaneous network activity in human iPSC-derived cortical neurons, tailored to enable pharmacological studies of human iPSC-derived neuronal networks.

2 Materials

2.1 Cell Culture Media

Equilibrate media at room temperature every time before working with cells. Store all cell culture medium at 4 °C and use it within 3 weeks.

1. Neurobasal™ medium (Thermo Fisher Scientific).
2. DMEM/F-12 GlutaMAX™ (Thermo Fisher Scientific).
3. B-27™ Supplement (50×) (Thermo Fisher Scientific).
4. N-2 Supplement (100×) (Thermo Fisher Scientific).
5. MEM nonessential amino acids solution (100×) (Thermo Fisher Scientific).
6. L-glutamine (200 mM) (Thermo Fisher Scientific).
7. Penicillin–streptomycin (10,000 U/ml) (Thermo Fisher Scientific).
8. 2-mercaptoethanol (50 mM) (Thermo Fisher Scientific)
9. Sodium pyruvate (100 mM) (Merck, Darmstadt, HE, Germany).
10. Human insulin solution (10mg/ml) (Merck, Darmstadt, HE, Germany).

    Prepare standard medium by mixing 250 ml Neurobasal™ medium, 250 ml DMEM/F-12 GlutaMAX™, 5 ml B-27™ Supplement (50×), 2.5 ml N-2 Supplement (100×), 2.5 ml MEM nonessential amino acids solution (100×), 2.5 ml L-glutamine (200 mM), 1.25 ml penicillin–streptomycin (10,000 U/ml), 500 μl 2-mercaptoethanol (50 mM), 2.5 ml sodium pyruvate (100 mM), and 125 μl human Insulin solution.

11. Human FGF (rhFGF, R&D Systems, Wiesbaden, HE, Germany). Neuronal expansion medium (NEM) prepared by adding 20 ng/ml rhFGF into standard medium.
12. BDNF, GDNF, and dibutyryl cAMP (PeproTech, Rocky Hill, NJ, USA). Add 20 ng/ml BDNF, 20 ng/ml GDNF, and 50 μM dibutyryl cAMP to the standard medium to get maturation medium (MATU).
13. *N*-[*N*-(3,5-difluorophenacetyl)-l-alanyl]-*S*-phenylglycine *t*-butyl ester (DAPT, Stemcell Technologies, Vancouver, BC, Canada). Prepare predifferentiation medium (PREDIFF) by adding 10 μM DAPT to MATU medium.

2.2 Coating Solutions

1. Matrigel™ (Matrigel™ hESC-qualified Matrix, BD Biosciences, Franklin Lakes, NJ, USA). Thaw Matrigel in an ice box at 4 °C overnight. Ice-cold DMEM/F-12 GlutaMAX™ is 1:1 diluted with Matrigel™ to prepare an intermediate Matrigel solution. This intermediate Matrigel solution is aliquoted in 1 ml and stored at −20 °C. For the final Matrigel solution dilute 1 ml intermediate Matrigel solution in 25 ml ice-cold DMEM/F-12 GlutaMAX™. This final Matrigel solution can be stored for approximate 10 days at 4 °C.
2. Laminin from Engelbreth-Holm-Swarm murine sarcoma basement membrane (L2020, Merck, Darmstadt, HE, Germany). Thaw laminin at room temperature or overnight at 4 °C. For the laminin coating, dilute laminin 1:100 in Hanks' Balanced Salt Solution (HBSS, no calcium, no magnesium), immediately before use.

2.3 Polyethyleneimine (PEI) Solution

1. Borate buffer solution prepared by diluting 20× borate buffer 1:20 with sterile water.
2. PEI solution. Prepare a 7% PEI solution with 1 ml of 50% PEI and 6 ml 1× borate buffer. Vortex briefly. Aliquot and store for 1 month at −20 °C. Prepare the final 0.07% PEI solution by diluting 500 μl of 7% PEI solution in 49.5 ml 1× borate buffer. Sterile filter through a 0.22 μm filter unit.

2.4 Cell Culture

1. Human iPSC-derived neuronal progenitor cells (NPCs) (*see* **Note 1**).
2. 6-well plate, sterile, tissue-culture treated.
3. 24-well plate for Multiwell-MEA-system (Multi Channel Systems, Reutlingen, BW, Germany).

4. Hardware Multiwell-MEA workstation (Multi Channel Systems, Reutlingen, BW, Germany).
5. Software Multiwell Screen (Multi Channel Systems, Reutlingen, BW, Germany).
6. Software Multiwell Analyzer (Multi Channel Systems, Reutlingen, BW, Germany).

3 Methods

3.1 Cell Culture

The timeline for cell culture is shown in Fig. 1.

1. Prepare a 6-well plate with 1 ml Matrigel solution at room temperature for at least 30 min.
2. Add 5 ml standard medium into a 15 ml falcon tube.
3. Use a water bath preheated at 37 °C to thaw up NPCs until an ice crystal is left (*see* **Note 2**).
4. Resuspend cells carefully by adding dropwise medium from the 15 ml falcon tube.
5. Add the cell suspension into the 15 ml falcon tube.
6. Pellet the cells at 280 × *g* for 5 min at room temperature.
7. Discard the supernatant and resuspend the cell pellet carefully with 1 ml NEM medium.
8. Discard the Matrigel solution from the 6-well plate and add 2 ml NEM medium.
9. Add the cell suspension into a well ($\geq 2 \times 10^6$ per well).
10. Use a humidified incubator for maintenance of cells containing 5% CO_2 at 37 °C.

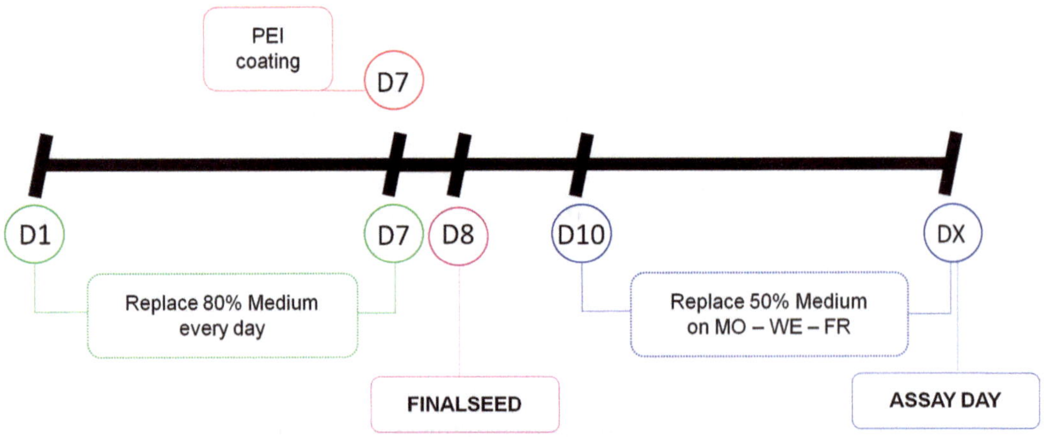

Fig. 1 Workflow of finale iPSC neuronal differentiation and MEA measurement. Assay day X (DX) is the day of measurement

11. Feed confluent cells with 3 ml PREDIFF medium.
12. Carefully feed cells with 2/3 of medium exchange every day for 6 more days. Do not aspirate the whole medium, because cells may detach from the wells during the course of differentiation.

3.2 PEI Coating

1. Prepare the 24-well MEA plate on day 7.
2. Pipette 100 μl of 0.07% PEI solution directly covering electrodes on each well.
3. Incubate at 37 °C for 1 h.
4. Aspirate the PEI solution and wash twice with sterile 1× PBS and once more with sterile ddH_2O immediately.
5. Air-dry plates with open lid in a sterile hood overnight without UV light. Critical: aqueous residues will interfere with proper coating of laminin.
6. Coat the plates with 100 μl laminin 1:100 in HBSS for 2 h on the next day.

3.3 Final Cell Seeding

1. Remove cells from the plate with 1 ml MATU media using a P1000 pipette.
2. Gently resuspend cells less than 5 times to get single cells and transfer cells to a 15 ml falcon tube.
3. Use a hemocytometer with trypan blue to count cells.
4. Adjust cell concentration to 1 mio/ml living cells with MATU media.
5. Aspirate laminin before seeding the cells on a PEI/laminin double coated 24-well MEA plate with a density of 150,000 alive cells per well (*see* **Notes 3** and **4**).
6. Add 500 μl fresh MATU medium after 30 min.
7. Only change ½ of the medium every Monday, Wednesday, and Friday until assay day. Laminin is added to MATU media (1:100) to prevent detaching of the cells once per week. The final seeding density of the cells is shown in Fig. 2.

3.4 MEA Measurement

Typical MEA measurements are illustrated in Fig. 3.

1. Test electrical activity or start your measurement of the neurons (*see* **Note 5**).
2. Open the software Multiwell Screen and set the temperature of the instrument to 37 °C to record the MEA plate (*see* **Note 6**).
3. Apply a bandwidth with a filter for 100 Hz to 5 kHz cutoff frequencies and at a sampling rate of 20 kHz.
4. Equilibrate plates for 5 min.
5. Record for 10 min.

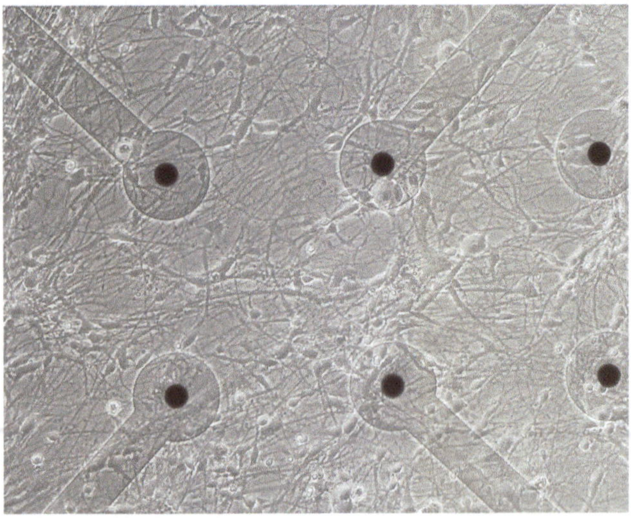

Fig. 2 Bright-field image recording with a 10× objective of an MEA well with six electrodes and iPSC-derived neurons

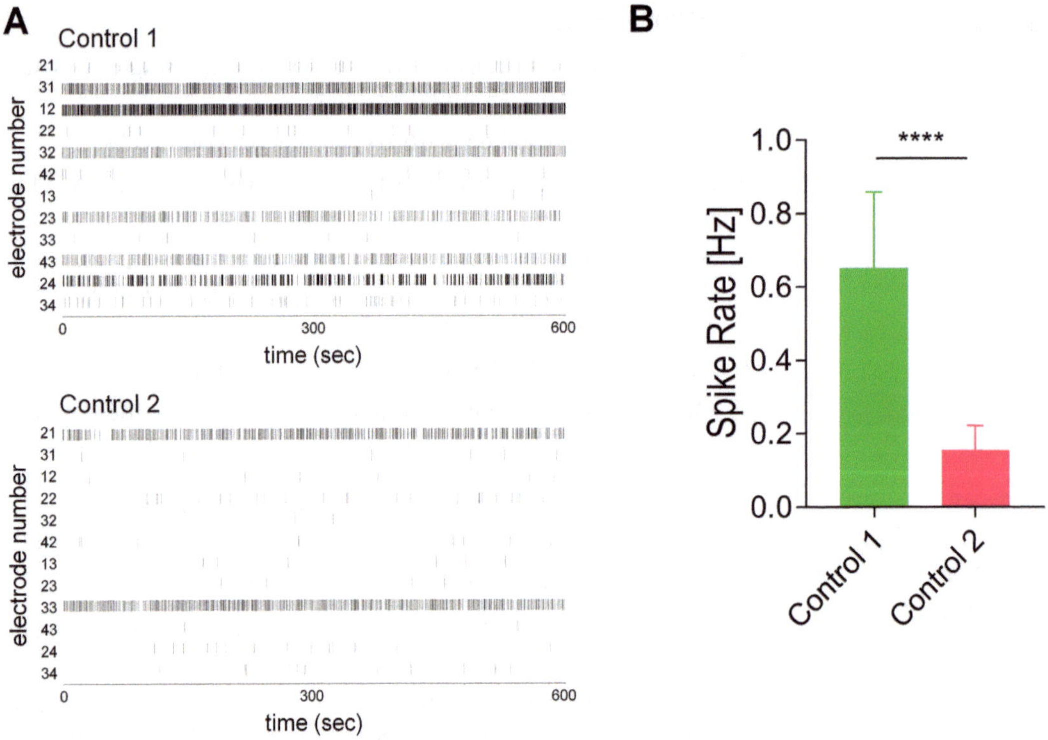

Fig. 3 Example of an analysis of spontaneous neuronal activity with MEA 1 week after final seed. (**a**) Detection of neuronal activity of control 1 and control 2 from one well. (**b**) The control 1 has a higher mean spike rate compared to the control 2 after the first week. $p < 0.0001$; Mann–Whitney test

6. Open the recordings in Multiwell Analyzer. Apply the Spike Detector to all wells by changing Baseline duration to 10,000 ms.
7. Select all electrodes and export results from the 'Dose Response' tab.
8. Include electrodes with greater or equal 0.01 Hz spike rate in the analysis.

4 Notes

1. Use your protocol to generate iPSC derived NPCs or use this modified version of a monolayer based differentiation protocol [12]. Briefly, confluent iPSCs are fed every day with standard medium, 1 μM dorsomorphin (Tocris, Bristol, UK), and 10 μM SB431542 (Merck, Darmstadt, HE, Germany) for neuronal induction. Cells are replated as cluster on Matrigel-coated 6-well plates by breaking up the neuroepithelial sheet with 1 mg/ml Dispase between days 8 and 12. Neuronal rosette formation is promoted by feeding the cells with standard medium and 20 ng/ml rhFGF2. Continue feeding cells with standard medium on day 17. Background differentiation is removed with multiple Dispase steps until day 25. Next, use Accutase to get single NPC culture. The culture can be expanded or frozen.
2. Use your standard protocol to thaw cells.
3. The density can range between 100,000 till 150,000 per well depending on your cell culture and the proportion of proliferative cells.
4. Use a parallel culture on a PEI/laminin-coated 96-well plate with the same density to have a look on cell culture and for immunofluorescence staining to characterize your cell culture (e.g., for ratio of glutamatergic vs. GABAergic neurons).
5. The first spike signals of neuronal cell cultures can be observed after 1 week of seeding depending on neuronal maturation.
6. Do not change the medium before recording. Optional: Add compounds in prewarmed conditioned maturation media after a baseline recording to modulate neuronal spike activity. Normalize data to the vehicle application.

References

1. Luhmann HJ, Sinning A, Yang JW et al (2016) Spontaneous neuronal activity in developing neocortical networks: from single cells to large-scale interactions. Front Neural Circuits 10:40
2. Marom S, Shahaf G (2002) Development, learning and memory in large random networks of cortical neurons: lessons beyond anatomy. Q Rev Biophys 35(1):63–87

3. van Pelt J, Vajda I, Wolters PS et al (2005) Dynamics and plasticity in developing neuronal networks in vitro. Prog Brain Res 147:173–188
4. Amin H, Maccione A, Marinaro F et al (2016) Electrical responses and spontaneous activity of human iPS-derived neuronal networks characterized for 3-month culture with 4096-electrode arrays. Front Neurosci 10:121
5. Gross GW, Harsch A, Rhoades BK et al (1997) Odor, drug and toxin analysis with neuronal networks in vitro: extracellular array recording of network responses. Biosens Bioelectron 12 (5):373–393
6. Braak H, Del Trecidi K (2015) Neuroanatomy and pathology of sporadic Alzheimer's disease. Adv Anat Embryol Cell Biol 215:1–162
7. Holtkamp M, Buchheim K, Elsner M et al (2011) Status epilepticus induces increasing neuronal excitability and hypersynchrony as revealed by optical imaging. Neurobiol Dis 43 (1):220–227
8. Uhlhaas PJ, Singer W (2010) Abnormal neural oscillations and synchrony in schizophrenia. Nat Rev Neurosci 11(2):100–113
9. Shi Y, Kirwan P, Smith J et al (2012) Human cerebral cortex development from pluripotent stem cells to functional excitatory synapses. Nat Neurosci 15(3):477–486. S1
10. Du X, Parent JM (2015) Using patient-derived induced pluripotent stem cells to model and treat epilepsies. Curr Neurol Neurosci Rep 15 (10):71
11. MacLaren EJ, Charlesworth P, Coba MP et al (2011) Knockdown of mental disorder susceptibility genes disrupts neuronal network physiology in vitro. Mol Cell Neurosci 47(2):93–99
12. Shi Y, Kirwan P, Livesey FJ (2012) Directed differentiation of human pluripotent stem cells to cerebral cortex neurons and neural networks. Nat Protoc 7(10):1836–1846

Part V

Miscellaneous Cell-Based Assays

Chapter 20

Assessing Neuronal Excitability on a Fluorometric Imaging Plate Reader (FLIPR) Following a Defined Electrostimulation Paradigm

Valeria Kizner, Sandra Fischer, Benjamin Jähnke, and Maximilian Naujock

Abstract

FLIPR-based calcium assay enables the detection and characterization of neuronal excitability by using electrical field stimulation to evoke and record action potential-driven calcium transients in induced pluripotent stem cell (iPSC)-derived cortical forebrain neurons. Here we describe high throughput measurement of neuronal excitability with a defined electrostimulation paradigm in a 384-well plate format using FLIPR.

Key words FLIPR, Calcium imaging, iPSC, Neuron, Excitability

1 Introduction

Differences in neuronal excitability can indicate deficits in the functionality of voltage-gated ion channels, which finally are important for the initiation and conduction of action potentials [1, 2]. Various psychiatric and neurological disorders such as schizophrenia or amyotrophic lateral sclerosis are associated with impaired neuronal excitability [3–6], which can be investigated using a variety of preclinical research tools such as patch-clamp, multielectrode arrays, or calcium imaging.

In our experimental calcium imaging setting, the physiological form of neuronal excitability is simulated by the application of an external electrical stimulus to change the electric gradient on the cell membrane surface to finally elicit the rise and propagation of action potentials. This electrostimulation in turn causes dose-dependent calcium influx while following defined paradigm. For the recording of neuronal excitability we estimated the

Carl-Fredrik Mandenius and James A. Ross (eds.), *Cell-Based Assays Using iPSCs for Drug Development and Testing*, Methods in Molecular Biology, vol. 1994, https://doi.org/10.1007/978-1-4939-9477-9_20,
© Springer Science+Business Media, LLC, part of Springer Nature 2019

measurement of intracellular changes in calcium concentration via a calcium indicator dye and external electrical stimulus [7].

Calcium imaging visualizes calcium signaling in living neuronal cultures. Fluorescent dye indicators are small molecules that can chelate calcium ions. They are used for the visualization and measurement of intracellular calcium. The fluorescent calcium indicator finally changes its brightness by binding to calcium ions [8]. Calcium ions generate intracellular signals that regulate different cellular functions including contraction, secretion, and embryonic development [9]. In the nervous system, intracellular elevations in calcium concentration play a pivotal role in the regulation of neurite outgrowth, neuronal migration, and also neuronal cell death. Furthermore, calcium signaling alters gene expression, neurotransmitter release from synaptic vesicles and plays a role in the induction of memory- and learning-related forms of neuronal plasticity [10].

The FLIPR Tetra High Throughput Cellular Screening System (FLIPR Tetra System) is a kinetic plate-based, cellular assay screening system which uses calcium fluorescent dyes to detect neuronal excitability by measuring fluctuations of intracellular calcium levels occurring in a large number of cells [11, 12].

Here we describe a simple and robust fluorescence-based calcium assay to measure network excitability by a defined electrostimulation paradigm using human iPSC-derived cortical neurons in a 384-well format on a FLIPR device.

2 Materials

2.1 Cell Culture Medium

Store all cell culture medium at 4 °C for approx. 3 weeks. Equilibrate medium at room temperature before starting the cell culture work.

1. Standard medium: 250 ml DMEM/F-12 GlutaMAX™, 250 ml Neurobasal™ medium, 500 μl 2-mercaptoethanol (50 mM), 2.5 ml MEM nonessential amino acids solution (100×), 1.25 ml penicillin–streptomycin (10,000 U/ml), 2.5 ml N-2 Supplement (100×), 5 ml B-27™ Supplement (50×), 2.5 ml L-glutamine (200 mM) (all Thermo Fisher Scientific, Waltham, MA, USA), 125 μl insulin solution human (10mg/ml), 2.5 ml sodium pyruvate (100 mM) (both Merck, Darmstadt, HE, Germany).
2. Neuronal expansion medium (NEM): Standard medium +20 ng/ml Recombinant Human FGF (rhFGF, R&D Systems, Wiesbaden, HE, Germany).
3. Maturation medium (MATU): Standard medium +20 ng/ml BDNF, 20 ng/ml GDNF, 50 μM dibutyryl cAMP (all PeproTech, Rocky Hill, NJ, USA).

4. Predifferentiation medium (PREDIFF): MATU medium +10 μM *N*-[*N*-(3,5-difluorophenacetyl)-l-alanyl]-*S*-phenylglycine *t*-butyl ester (DAPT, Stemcell Technologies, Vancouver, BC, Canada).

2.2 Preparing Coating Solutions

1. For the preparation of the intermediate Matrigel solution allow the Matrigel™ (Matrigel™ hESC-qualified Matrix, BD Biosciences, Franklin Lakes, NJ, USA) to thaw up overnight at 4 °C in an ice box.
2. Dilute Matrigel™ 1:1 with ice-cold DMEM/F-12 GlutaMAX™.
3. Make 1 ml aliquots and store at −20 °C.
4. Prepare the Matrigel solution by diluting 1 ml intermediate Matrigel solution in 25 ml ice-cold DMEM/F-12 GlutaMAX™. Keep it for up to approximate 10 days at 4 °C.
5. Thaw Laminin overnight at 4 °C or at room temperature. Prepare 1:100 laminin from Engelbreth-Holm-Swarm murine sarcoma basement membrane (L2020, Merck, Darmstadt, HE, Germany) in Hank's Balanced Salt Solution (HBSS), no calcium, no magnesium, immediately before use.

2.3 FLIPR Reagents

1. Ringer buffer: 130 mM NaCl, 5 mM KCl, 1 mM $CaCl_2*2H_2O$, 1 mM $MgCl_2*6H_2O$, 2 mM KH_4PO_4, 20 mM HEPES, 5 mM glucose, pH 7.4.
2. FLIPR Calcium 4 Assay Kit (Molecular Devices, San Jose, CA, USA).

2.4 Equipment

1. Prepare human iPSC-derived neuronal progenitor cells (NPCs) according to Shi et al. [13] or use the modified protocol below for the generation of NPCs (*see* **Note 1**).
2. 6-well tissue-culture treated (TC) plate, sterile.
3. 384-well TC plate, Poly-D-Lysine, Flat Clear Bottom.
4. FLIPR Tetra System High Throughput Cellular Screening System (Molecular Devices, San Jose, CA, USA).
5. Platinum electrodes.
6. Screen Works 4.0 Software (Molecular Devices, San Jose, CA, USA).
7. Stimulator 1/8, Typ:215/1 (Hugo Sacks Elektronik, March-Hugstetten, BW, Germany).

3 Methods

3.1 Cell Culture (Fig. 1)

1. Coat a 6-well plate with 1 ml Matrigel solution at room temperature for at least 30 min.

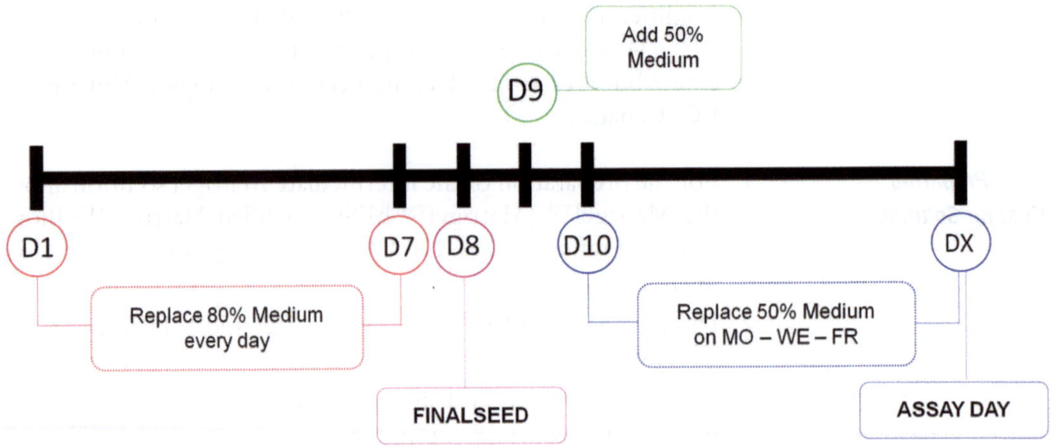

Fig. 1 Workflow of iPSC neuronal differentiation and FLIPR calcium imaging. Assay day X (DX) is on the time point of interest

2. Prepare a 15 ml falcon tube with 5 ml standard medium.
3. Thaw NPCs in a preheated 37 °C water bath until no ice crystal is left (*see* **Note 2**).
4. Transfer some medium from the 15 ml falcon tube dropwise into the cryovial and resuspend the cells carefully.
5. Transfer the cell suspension into the 15 ml falcon tube.
6. Centrifuge the cells at 280 × *g* at room temperature for 5 min.
7. Aspirate supernatant and resuspend cell pellet carefully with 1 ml NEM medium.
8. Aspirate the Matrigel solution and add 2 ml NEM medium. Add the cell suspension into a well ($\geq 2 \times 10^6$ per well).
9. Maintain cells in a humidified incubator containing 5% CO_2 at 37 °C.
10. Cells are feed with 2 ml NEM medium every second day.
11. Feed cells with 3 ml PREDIFF medium after reaching 80–90% cell confluence.
12. Change carefully 2/3 of medium every day for 6 more days. Critical: cells may detach from the wells during the course of differentiation.

3.2 Final Seed

1. Coat plates with 30 μl laminin 1:100 in HBSS for 2 h on day 8.
2. Wash off cells with 1 ml MATU media using a P1000 pipette.
3. Resuspend cells gently to get single cells, but less than 5 times.
4. Transfer the cells to a 15 ml falcon tube.

5. Count the cells with a hemocytometer and adjust the cell concentration to 1 mio/ml alive cells by adding MATU media to cell suspension.
6. Aspirate laminin and seed the cells on laminin coated 384-well plate with a density of 30,000 cells per well (*see* **Note 3**).
7. Add 30 μl fresh MATU medium on top next day.
8. Change ½ the medium every Monday, Wednesday, and Friday until assay day. Add 1:100 laminin to MATU media to prevent detaching of the cells once time per week.

3.3 Assay Day—Preparation of FLIPR Measurement

1. Use the cells for the assay after 6 weeks of neuron maturation (*see* **Note 4**).
2. Aspirate medium and wash 1× with 30 μl Ringer buffer.
3. Add 20 μl Ringer solution and 20 μl Calcium 4 dye loading solution.
4. Incubate at 37 °C for 1 h.

3.4 Assay Day—FLIPR Measurement

1. Turn on the FLIPR device.
2. Perform the "Signal Test" and save it as an screenshot (*see* **Note 5**).
3. Configure your data set for 450 reads at a temperature 37 °C.
4. Record a baseline of 1 min before stepwise electrostimulation.
5. Apply the stimulation paradigm in Table 1 (*see* **Notes 6–8**).
6. Evaluate your results (Figs. 2 and 3) by measuring $\Delta F/F_0$ values by calculating the ratio between the change in maximal relative light units (ΔF) and minimal fluorescent light units (F_0).

Table 1
Electrostimulation paradigm for FLIPR application

Time (min)	Volt	Hz	Stimulation time (s)
1	12	5	5
3	12	10	5
5	12	30	5
7	12	50	5

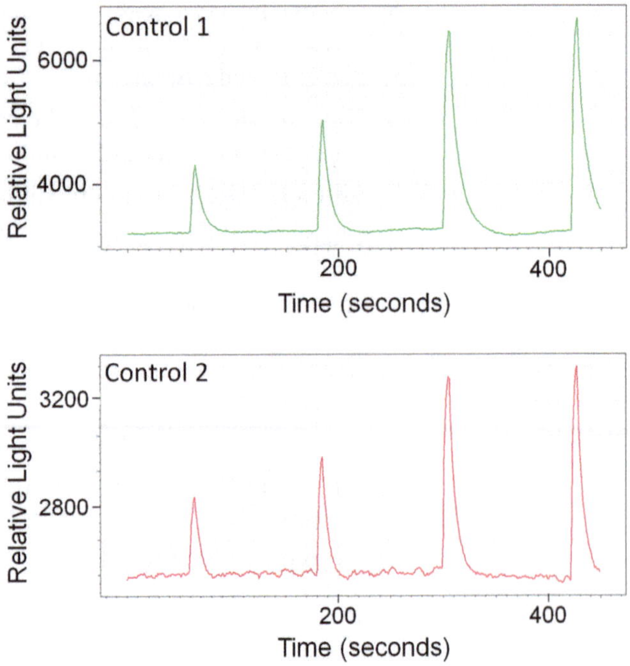

Fig. 2 Each graph shows individual traces of electrostimulation paradigm with four pulses of increasing frequency of two different iPSC-derived neuron cell lines showing an increase in calcium responses

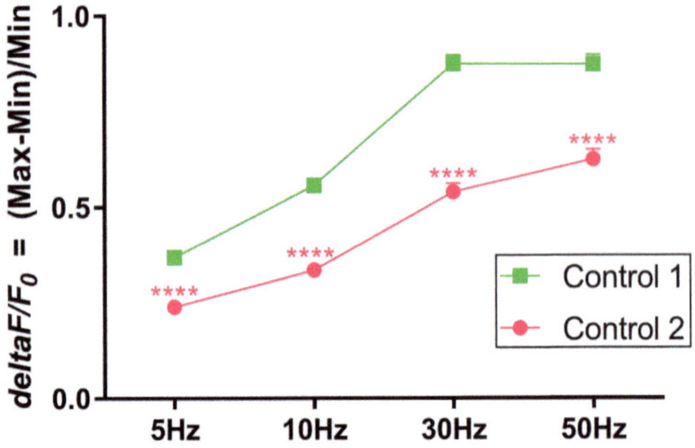

Fig. 3 The iPSC-derived neuron control 2 showed hypoexcitability compared to control 1. ****$p < 0.0001$; 2-way ANOVA

4 Notes

1. iPSC-derived NPCs are generated according to a modified version of a monolayer based differentiation protocol [13]. Briefly, neuronal induction is started when iPSCs reach a confluence between 80% and 90%. Therefore, cells are fed every day with Standard medium +1 μM dorsomorphin

(Tocris, Bristol, UK) and 10 μM SB431542 (Merck, Darmstadt, HE, Germany). The neuroepithelial sheet is broken up with 1 mg/ml Dispase in standard medium between day 8 and 12 and replated as clusters on Matrigel-coated 6-well plates. On day 13 cells are fed with Standard medium +20 ng/ml rhFGF2 to promote neuronal rosette formation. Cells are fed with standard medium at day 17. Multiple Dispase dissociation steps are included to get rid of background differentiation until day 25. Then, cells are passed with Accutase to get single cell culture and subsequently expanded or frozen.

2. Thaw cells according to your standard protocol.
3. Seeding density is 30,000 cells per well but may be different depending on your cell culture.
4. Measurement of neuron culture is also depending on your cell culture, maturation, and time point of interest.
5. The Signal test provide information of cell density, which can be compared between wells or cell lines.
6. Optional: After the stimulation paradigm you can add neurostimulants such as bicuculline to provoke spontaneous firing or test other compounds of interest.
7. After FLIPR measurement, cells can be used for calcium imaging by confocal laser scanning microscopy to analyze individual cells.
8. Cells can be fixed and stained for, for example, neuronal marker microtubule-associated protein 2 (MAP2), GABA- or glutamatergic markers (vesicular GABA transporter marker 1 (VGAT1) and vesicular glutamate transporter 1 (VGLUT1)), or presynaptic and postsynaptic markers. Critical: control for possibly interfering signals from Calcium 4 dye loading buffer.

Reference

1. Eijkelkamp N et al (2012) Neurological perspectives on voltage-gated sodium channels. Brain 135(Pt 9):2585–2612
2. Hedrich UB et al (2014) Impaired action potential initiation in GABAergic interneurons causes hyperexcitable networks in an epileptic mouse model carrying a human Na(V)1.1 mutation. J Neurosci 34(45):14874–14889
3. Frazzini V et al (2016) Altered Kv2.1 functioning promotes increased excitability in hippocampal neurons of an Alzheimer's disease mouse model. Cell Death Dis 7:e2100
4. Lasarge CL, Danzer SC (2014) Mechanisms regulating neuronal excitability and seizure development following mTOR pathway hyperactivation. Front Mol Neurosci 7:18
5. Mertens J et al (2015) Differential responses to lithium in hyperexcitable neurons from patients with bipolar disorder. Nature 527 (7576):95–99
6. Wainger BJ et al (2014) Intrinsic membrane hyperexcitability of amyotrophic lateral sclerosis patient-derived motor neurons. Cell Rep 7 (1):1–11
7. Virdee JK et al (2017) A high-throughput model for investigating neuronal function and synaptic transmission in cultured neuronal networks. Sci Rep 7(1):14498
8. Grynkiewicz G, Poenie M, Tsien RY (1985) A new generation of Ca2+ indicators with greatly improved fluorescence properties. J Biol Chem 260(6):3440–3450

9. Berridge MJ, Lipp P, Bootman MD (2000) The versatility and universality of calcium signalling. Nat Rev Mol Cell Biol 1(1):11–21
10. Berridge MJ (1998) Neuronal calcium signaling. Neuron 21(1):13–26
11. Mezler M et al (2012) Development and validation of a fluorescence-based HTS assay for the identification of P/Q-type calcium channel blockers. Comb Chem High Throughput Screen 15(5):372–385
12. Smetters D, Majewska A, Yuste R (1999) Detecting action potentials in neuronal populations with calcium imaging. Methods 18 (2):215–221
13. Shi Y, Kirwan P, Livesey FJ (2012) Directed differentiation of human pluripotent stem cells to cerebral cortex neurons and neural networks. Nat Protoc 7(10):1836–1846

Chapter 21

Fabrication of a Microfluidic Cell Culture Device Using Photolithographic and Soft Lithographic Techniques

Jonas Christoffersson and Carl-Fredrik Mandenius

Abstract

Photolithography and soft lithography are two common methods for fabrication of microfluidic cell culture devices. Well-defined microstructures are created by exposing a photoresist to UV-light under a photolithographic mask in which the intended patterns are transparent. The subsequent cross-linking of UV-exposed photoresist generates a reusable master that serves as a template for an elastomer, commonly polydimethylsiloxane (PDMS), that reciprocally recaptures the structures of the master in an optically clear and oxygen-permeable rubber-like material. Connections to the cell culture-containing channels of the device for perfusion of culture medium can be established by inserting tubing through the elastomer. In this protocol, the basic steps for making a standard microfluidic device for cell-based assays from photolithography and soft lithography techniques are outlined.

Key words Photolithography, Soft lithography, PDMS, Microfluidics

1 Introduction

Photolithography originates from the production of microelectronics and the manufacturing of integrated circuits [1]. The technique was developed in order to create well-defined microstructures on a substrate such as silicon. In photolithography, a silicon wafer is covered by a viscous photoresist by spin coating. The thickness of the photoresist (i.e., the height of the subsequent microstructures) is determined by the spin speed. When exposed to high energy UV light, the photoresist starts to cross-link and become soluble or insoluble to a developer depending on whether it is of a positive or negative sort. Microstructures of the photoresist can therefore be created by exposing it to UV-light through an opaque mask with transparent patterns. The cross-linking is completed by heat and the structures are revealed by removing the UV-exposed (positive) photoresist or by removing the unexposed (negative) photoresist, using a developer solution. The resulting

Carl-Fredrik Mandenius and James A. Ross (eds.), *Cell-Based Assays Using iPSCs for Drug Development and Testing*, Methods in Molecular Biology, vol. 1994, https://doi.org/10.1007/978-1-4939-9477-9_21,
© Springer Science+Business Media, LLC, part of Springer Nature 2019

photoresist-patterned silicon wafer, also known as a master, with a resolution in the micrometer range, is a reusable template for the replica molding process known as soft lithography [1–3]. The most commonly used elastomer for soft lithography is polydimethylsiloxane (PDMS). PDMS is a transparent, gas permeable, and non-cytotoxic material that covers the surface of the master and solidifies after 2 h at 60 °C. The cured PDMS is then easy to remove from the master and can be bonded to a glass slide or a coverslip by creating reactive hydroxyl groups on the surfaces using oxygen plasma. PDMS is a very hydrophobic material which means that air bubbles are easily created when filling the channels of such a device. A way to overcome this problem is to first rinse the device with ethanol before extensive washing in water and cell culture medium. There are further potential drawbacks of PDMS such as the absorption of small (<500 Da) hydrophobic molecules into the bulk of the PDMS, the risk of un-cross-linked monomers contaminating the cell culture medium if the ratio of the prepolymer and the curing agent is too high, as well as the gas permeability that can lead to water vapor passing through the device creating air bubbles or a change in pH or salt concentration [4]. However, for creating prototypes of microfluidic cell culture devices, PDMS is a very convenient material with simple and relatively inexpensive fabrication procedures [5–7].

The procedure described in this protocol is applicable to small-scale fabrication of microdevices for iPSC derived cells to be used for drug testing assays. Other protocols in this volume (*see* [8, 9]) can be applied for microfluidic devices fabricated with this protocol and the fabrication protocol has previously been applied by us in other microfluidic assays [10–12].

2 Materials

2.1 Photolithography Reagents and Equipment

1. Silicon wafer, 4″ (SWI, Taiwan) (*see* **Note 1**).
2. Hydrogen peroxide (H_2O_2, VWR International, PA, USA).
3. Ammonia solution (NH_3, Merck, NJ, USA).
4. SU8–3035 negative photoresist (MicroChem Corp., MA, USA).
5. Wafer spin coater (Spin150, APT, Germany) (*see* **Note 2**).
6. Hotplate (RCT basic, IKA-Werke GmbH & Co. KG, Germany).
7. Photolithographic mask (Coated Screens Scandinavia AB, Sweden) (*see* **Note 3**).
8. Mask aligner with a UV source (Kari Süss, Germany) (*see* **Note 4**).

9. Resist developer (Micro Resist Technology, Germany).
10. Isopropanol (VWR International).
11. Dektak stylus profiler (Bruker Corp., MA, USA).

2.2 Silanization Reagents

1. Tetramethylsilane (Sigma-Aldrich, MO, USA).
2. Ethanol (99.5%, Solveco AB, Sweden).

2.3 Soft Lithography Reagents and Equipment

1. PDMS prepolymer and curing agent (Sylgard 184, Dow Corning, MI, USA).
2. Microscope slide (VWR International).
3. Punch (Syneo, TX, USA).
4. Steel tubing (0.64 mm OD, 0.4 mm ID, 12 mm long, New England Small Tube Corp., NH, USA).
5. Tygon tubing (1.52 mm OD, 0.5 mm ID, VWR International).
6. Plasma oven (Diener Electronic GmbH & Co. KG, Germany).

3 Methods

The overall process steps in photolithography and soft lithography are summarized in Fig. 1. It is recommended to perform photolithography in a clean room environment to minimize the risk of contamination of the wafer by micro particles. The following protocol will create microstructures with a height of 50 μm on a silicon wafer using a negative photoresist (*see* **Note 5**). By altering the spinning speed of the photoresist, other heights can be generated as shown in the guidelines of the manufacturer of the photoresist.

3.1 Photolithography

Photolithography is the process of creating a master with structures of photoresist in defined patterns which have been cross-linked by UV-light under a photolithographic mask (Fig. 1).

1. Clean the wafer by immersing it in a washing solution containing H_2O, H_2O_2, and NH_3 (5:1:1 ratio by volume) at 85 ° C for 10 min.
2. Rinse the wafer with water, blow dry, and put it in the oven for 10 min at 100 ° C to make sure all the water has evaporated.
3. Pour about 4 ml (3 cm in diameter) of SU8–3035 in the center of the wafer. Spin the wafer using the spin coater at 500 rpm ramped at 150 rpm/s for 10 s followed by 2000 rpm ramped at 150 rpm/s for 35 s. The photoresist should now be evenly distributed on the wafer (*see* **Note 4**).
4. Prebake the wafer at 65 ° C for 1 min on a hotplate followed 100 °C for 40 min in an oven.

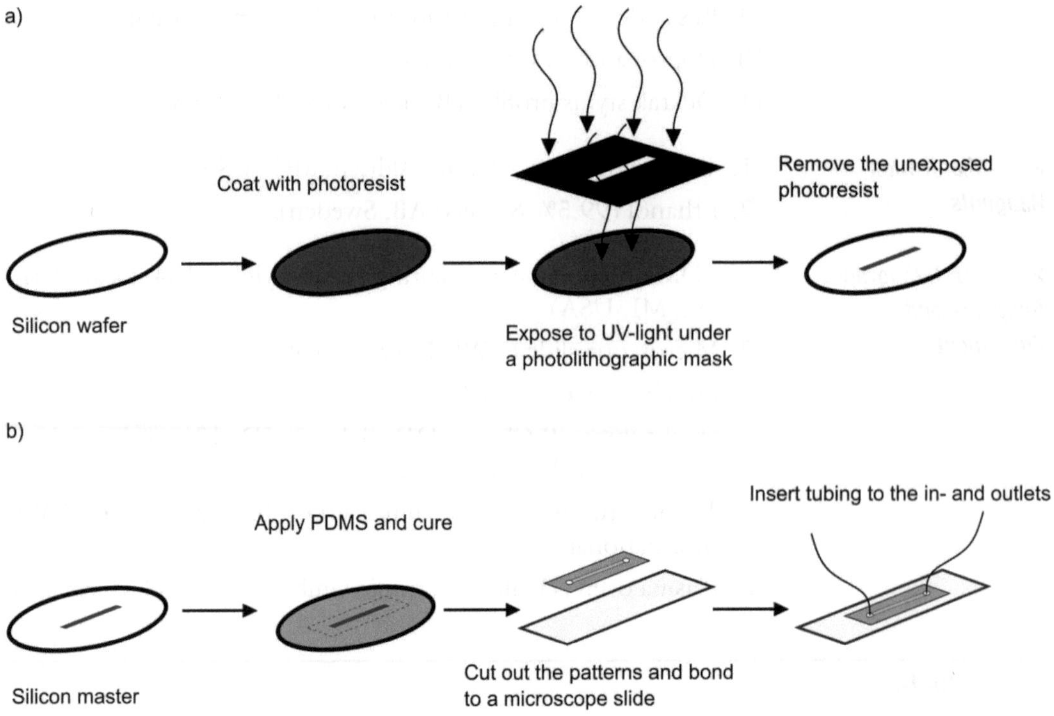

Fig. 1 An overview of the major process steps during photolithography (**a**) and soft lithography (**b**)

5. Expose the photoresist on the wafer to UV-light at 187 mJ/cm under the photolithographic mask in the mask aligner.
6. Post-bake the wafer at 65 °C for 1 min followed by 95 °C for 5 min (*see* **Note 6**).
7. Remove the unexposed photoresist by immersing the wafer in negative resist developer for 10 min followed by a rinse with isopropanol. Check under a microscope to make sure all of the unexposed photoresist has been removed resulting in patterns with sharp edges. Remove further photoresist residues by immersing the wafer in negative resist developer for 2 more min followed by a rinse with isopropanol.
8. Use the Dektak stylus profiler to check the height of the patterns on the finalized master (*see* **Note 7**).

3.2 Silanization

Silanization is done to make PDMS easier to remove from the master.

1. Immerse the master in 3% tetramethylsilane in ethanol for 10 min.
2. Wash the wafer with ethanol and incubate at 120 °C for 30 min.

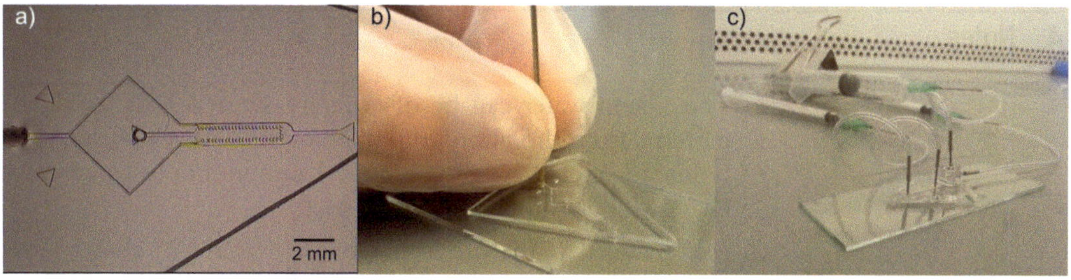

Fig. 2 Example of microfluidic channels and structures created by photolithography and molded in PDMS by soft lithography (**a**). Connections to the channels can be created by manual punching (**b**) and by inserting tubing (**c**) for inlet and outlet

3.3 Soft Lithography

Soft lithography is the technique for replica molding of microstructures on a master using an elastomer such as PDMS (Fig. 2).

1. Mix PDMS prepolymer and curing agent at a 10:1 ratio (w/w) by stirring the mixture for at least 2 min. Degas in vacuum in a desiccator to remove bubbles.
2. Clean the master twice in water and ethanol to remove microparticles. Blow the master dry. Pour the degassed PDMS mixture over the master and cure the PDMS in an oven at 60 °C for at least 2 h (*see* **Notes 8** and **9**).
3. Cut rectangles around the patterns (leave at least 0.5 cm between the pattern and the cut) using a scalpel and gently remove the PDMS from the master.
4. Make sure all channels and structures are intact using a microscope (Fig. 2a).
5. Use a punch to make holes for inlets and outlets in the PDMS (Fig. 2b).
6. Use tape to remove any residues on the patterned side of the PDMS piece. Rinse in water and ethanol and blow dry.
7. Oxidize the surface of the patterned side of the PDMS piece and the surface of a clean microscope slide in an oxygen plasma oven for 1 min.

7. Bond the oxidized surfaces of the PDMS piece and the microscope slide. Make sure no air bubbles are trapped in between and incubate in the oven at 60 ° C for 10 min for irreversible bonding.

8. Attach the Tygon tubing to the steel tubing and insert the steel tubing into the inlet and the outlet of the PDMS piece. Seal the steel tubing with uncured PDMS and cure in the oven at 60 ° C for at least 10 min (Fig. 2c) (*see* **Note 10**).

9. Autoclave for 20 min at 121 ° C.

4 Notes

1. The silicon wafer can be of different sizes as long as all the patterns from the photolithographic mask fit well within their boundaries. However, the required amount of photoresist will change (approximately 1 ml per inch).
2. Make sure that the properties of the wafer spin coater (size, speed, and acceleration) comply with the specific photoresist being used as per the instructions of the manufacturer.
3. A photolithographic mask is an opaque stencil with transparent patterns for UV-light to pass through. The mask can be custom-made in many vector graphics software (Inkscape, Illustrator, Sketch, etc.) and printed on a plastic sheet with a high-resolution printer, or, for the best results, on quartz.
4. Check the effect of the light source prior to exposure and adjust the exposure time accordingly.
5. A negative photoresist cross-links under UV-light and the remaining unexposed negative photoresist is removed using a developer. With a positive photoresist, the exposed parts can be removed using a developer, while the unexposed parts will cross-link. Hence, if a positive photoresist would be used, the patterns of the photolithographic mask must be inverted (i.e., the intended structures should be opaque and the rest of the mask transparent), to achieve the same result.
6. Structures of different heights can be created on the same wafer by repeating Subheading 3.1, **steps 3** and **6**. However, when additional photoresist is spin-coated on top of a photoresist-covered wafer, the spin speed might be adjusted as the friction on the two materials is different. This step should be revised if the height of the structures is not as expected.
7. Make sure that the heights of the structures are even all over the wafer by taking measures on several locations from the periphery to the center.
8. If there are bubbles on the in the PDMS after pouring it over the master, these must be removed. This can be done either by preservation 5 min in a vacuum desiccator or by leaving the wafer at room temperature for an hour until all the bubbles have disappeared.
9. PDMS cures in room temperature after about 48 h (depending on the volume). Uncured PDMS can be stored in a freezer and used for at least a week.
10. Uncured PDMS that has been stored in the freezer for a few days is appropriate to use for this step as it is more viscous and will not go into the channels.

Acknowledgments

The research leading to these results has received support from the Innovative Medicines Initiative Joint Undertaking under (grant no. 115439), resources of which are composed of financial contribution from the European Union's Seventh Framework Programme (FP7/2007-2013) and EFPIA companies. This publication reflects only the author's views, and neither the IMI JU nor EFPIA nor the European Commission is liable for any use that may be made of the information contained therein.

References

1. Whitesides GM (2006) The origins and the future of microfluidics. Nature 442:368–373
2. Xia Y, Whitesides GM (1998) Soft lithography. Angew Chem (Intern Ed) 37:550–575
3. del Campo A, Greiner C (2007) SU-8: a photoresist for high-aspect-ratio and 3D submicron lithography. J Michromech Microeng 17:R81–R95
4. Halldorsson S, Lucumi E, Gomez-Sjoberg R, Fleming RM (2015) Advantages and challenges of microfluidic cell culture in polydimethylsiloxane devices. Biosens Bioelectron 63:218–231
5. Booth R, Kim H (2012) Characterization of a microfluidic in vitro model of the blood-brain barrier (μBBB). Lab Chip 12:1784–1792
6. Huh D, Matthews BD, Mammoto A, Montoya-Zavala M, Hsin HY, Ingber DE (2010) Reconstituting organ-level lung functions on a chip. Science 328:1662–1668
7. Schimek K et al (2013) Integrating biological vasculature into a multi-organ-chip microsystem. Lab Chip 13:3588–3598
8. Christoffersson J, Mandenius CF (2018) Using a microfluidic device for culture and drug toxicity testing of 3D cells. Chapter 22 in Cell-Based Assays Using iPSCs for Drug Development and Testing, Methods in Molecular Biology, vol. 1994, Springer Nature, 2019
9. Christoffersson J et al (2018) Evaluating the effect of drug compounds on cardiac spheroids using the cardiac cell outgrowth assay. Chapter 17 in Cell-Based Assays Using iPSCs for Drug Development and Testing, Methods in Molecular Biology, vol. 1994, Springer Nature, 2019
10. Christoffersson J, Bergström G, Schwanke K, Kempf H, Zweigerdt R, Mandenius C-F (2016) A microfluidic bioreactor for toxicity testing of stem cell derived 3D cardiac bodies. In: Turksen K (ed) Bioreactors in stem cell biology: methods and protocols. Springer Verlag, London
11. Bergström G, Christoffersson J, Schwanke K, Zweigerdt R, Mandenius C-F (2015) Stem cell derived in vivo-like human cardiac bodies in a microfluidic device for toxicity testing by beating frequency imaging. Lab Chip 15:3242–3249
12. Bergström G, Nilsson K, Mandenius CF, Robinson ND (2014) Macroporous microcarriers for introducing cells into a microfluidic chip. Lab Chip 14:3502–3504

Chapter 22

Using a Microfluidic Device for Culture and Drug Toxicity Testing of 3D Cells

Jonas Christoffersson and Carl-Fredrik Mandenius

Abstract

Microfluidic devices provide convenient assays tools for testing cell cultures in three-dimensional (3D) formats. These devices have significant potential for establishing assays that are better at predicting drug toxicity and efficacy effects compared to assays in conventional two-dimensional (2D) cultures. Microfluidic cell culture devices consist of perfused cell culture chambers with inlets and outlet for seeding, culturing, sampling, and assaying the cells. This protocol describes how to prepare and seed cells in a microfluidic cell culture device for drug toxicity testing on cells in a 3D structure. The protocol exemplifies the use of a basic microfluidic device with HepG2 hepatoma cells but can be transferred and optimized for other cells and cell types, including iPSC-derived tissue and organ cells.

Key words Microfluidic devices, Microbioreactors, 3D cell-based assays, In vitro models, Drug testing

1 Introduction

Organs-on-chips, or microphysiological tissues, are cell culture platforms that intend to mimic the functions of a specific tissue or organ to serve as a model for drug testing [1, 2]. These cell culture models capture the cells in a three-dimensional (3D) scaffold such as hydrogels [3, 4] and microcarriers [5] or by self-aggregation into spheroids or tissue constructs [6–8]. The 3D orientation of cells has been shown to increase the expression of several markers of hepatocyte functionality including cytochrome P450 enzymes, albumin, and urea, compared to conventional two-dimensional (2D) cell cultures (i.e., cells cultured on flat substrates such as glass and plastic) [9, 10]. Furthermore, by integrating capillary-like fluidic channels into the cell culture model, several properties of the blood flow within organs, including continuous renewal of nutrients and gasses, removal of toxic wastes, and an applied shear stress, can be recapitulated in the microfluidic devices. This has

Carl-Fredrik Mandenius and James A. Ross (eds.), *Cell-Based Assays Using iPSCs for Drug Development and Testing*, Methods in Molecular Biology, vol. 1994, https://doi.org/10.1007/978-1-4939-9477-9_22,
© Springer Science+Business Media, LLC, part of Springer Nature 2019

been shown to have a positive effect on several cell types in microfluidic models [11, 12]. For hepatocytes, a common approach to culture cells in microfluidic devices is to seed the cells in a cell culture chamber built up by pillars, with perfusion of cell culture medium in adjacent channels [13–15]. These devices can be created by photolithography and soft lithography as described in another chapter of this book [16]. The cells are seeded at high density and form aggregates that can be used for drug testing of cell in 3D. In this protocol, the steps required to prepare the microfluidic device, seed the cells, and start the perfused cell culture for a hepatic cell culture (Hep G2) are described. An assay protocol in a similar device with 3D cardiac cells for testing drugs is described elsewhere in this book [17].

2 Materials

1. Microfluidic cell culture device (*see* **Note 1**).
2. Blunt-ended cannulas with female Luer connectors (21G) (VWR International, PA, USA).
3. 1 ml syringes (Becton, Dickinson and Company, NJ, USA).
4. 70% ethanol (Solveco AB, Sweden)
5. Phosphate buffered saline (PBS, Sigma-Aldrich, MO, USA).
6. BSA solution (1% bovine serum albumin in PBS) (Sigma-Aldrich).
7. Cell culture medium: Dulbecco's modified Eagle's medium (DMEM) supplemented with 10% fetal bovine serum (FBS), 1% nonessential amino acids (NEAA), and 1% penicillin–streptomycin (Pen/Strep) (all from Life Technologies, CA, USA).
8. HepG2 cells (ATCC, VA, USA) (*see* **Note 2**).
9. Syringe pump (Fusion100, KR Analytical LTD).
10. Drug substance for toxicity testing.
11. Live/Dead assay (1 μM Calcein AM and 4 μM ethidium homodimer-1 diluted in PBS) (Life Technologies).

3 Methods

The microfluidic device can be created in the elastomer polydimethylsiloxane (PDMS) using photolithography and soft lithography as described in reference [14]. The design of the device used for this protocol is outlined in Fig. 1a. The protocol can also be used as a general guidance for preparing and seeding cells in microfluidic devices.

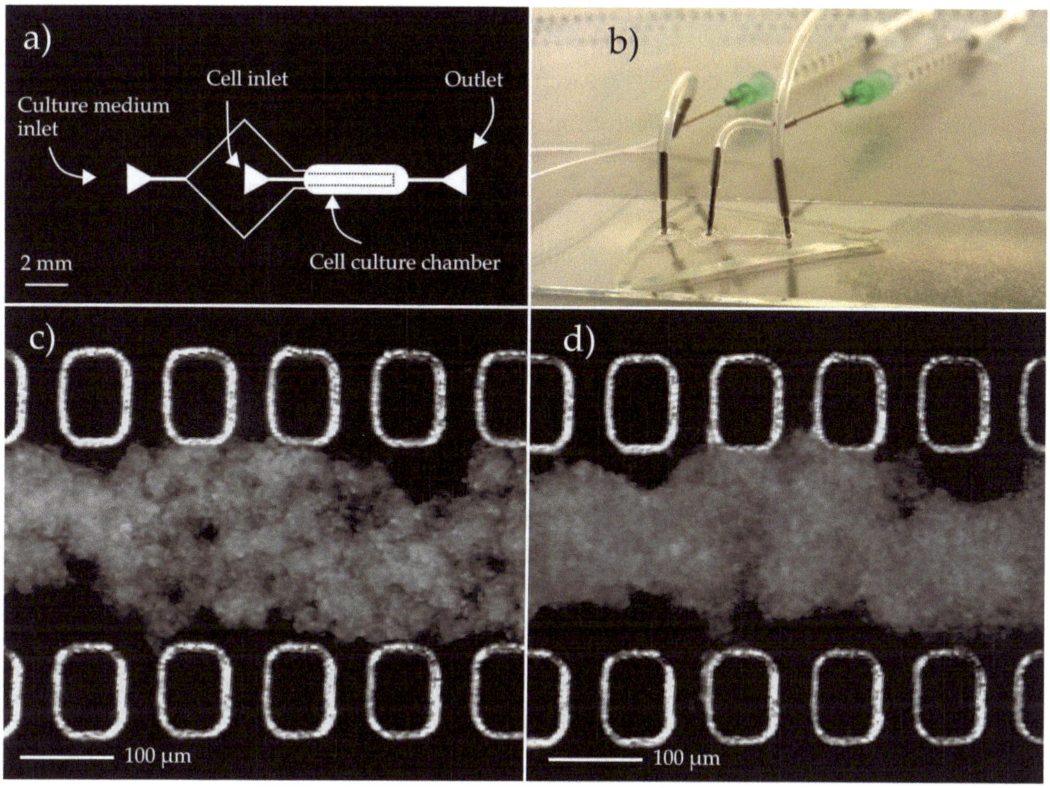

Fig. 1 Preparing a microfluidic device for cell-based assays. (**a**) The design of the device used to describe this protocol with (**b**) the device constructed in PDMS using photolithography and soft lithography, and HepG2 cells seeded into the cell culture chamber of the device after (**c**) 24 h and (**d**) 48 h of cell culture

3.1 Preparing the Device for Cell Culture (HepG2)

1. Incubate the microfluidic device, PBS, 1% BSA solution, and cell culture medium at 37 ° C and 5% CO_2 overnight before setting up the microfluidic cell culture to minimize the risk of bubbles forming within the microchannels.
2. Insert blunt-ended cannulas into the tubing of the microfluidic device (Fig. 1b).
3. Connect the cannulas at the culture medium inlet and the cell inlet to 1 ml syringes filled with ethanol.
4. Gently fill the channels of the device with ethanol and flush out all residing air bubbles (*see* **Note 3**).
5. Switch the ethanol filled syringe to a 1 ml syringe filled with PBS and replace the ethanol within the microfluidic device with PBS. Be careful not to introduce any bubbles to the device (*see* **Note 4**).
6. Switch the syringe to a 1 ml syringe filled with 1% BSA in PBS. Coat the channels of the device with BSA for 1 h at room temperature (*see* **Note 5**). Be careful not to introduce any bubbles to the device.

7. Switch the syringe to a 1 ml syringe filled with cell culture medium and replace the BSA solution within the microfluidic device. Be careful not to introduce any bubbles to the device.
8. Fill the tubing of the cell inlet with cell culture medium by closing the outlet with a 1 ml syringe (set the piston of the syringe around the 100 μl mark), removing the 1 ml syringe from the cell inlet, and gently pushing cell culture medium through the culture medium inlet.
9. The microfluidic device should at this point be filled with cell culture medium without any air bubbles within the channels or the tubing.
10. Prepare a cell suspension with 1–10 × 10^6 cells/ml and pipet approximately 100 μl into the Luer connector of the blunt ended cannula in the cell inlet (*see* **Note 6**).
11. Gently withdraw cell culture medium with the syringe connected to the outlet. Check under a microscope to see the cells filling the cell culture chamber. Make sure that there is sufficient amount of liquid in the tubing and Luer connector to not let any air enter the device. If needed, add more cells or cell culture medium to the Luer connector (*see* **Note** 7).
12. Close the cell inlet by inserting a 1 ml syringe to the Luer connector.
13. Place the microfluidic device with syringes connected to the inlet and outlet in the incubator at 37 ° C and 5% CO_2 for 2 h to let the cells start aggregating to each other (*see* **Note 8**).
14. Place the syringe connected to the culture medium inlet on a syringe pump and set the pumping speed to 1 μl/min.
15. Gently remove the syringe at the outlet and place the tubing from the outlet in a waste reservoir (*see* **Note 9**).

3.2 Infusion of Test Drugs into the Microfluidic Device and Assaying by Fluorescence Imaging

1. Change the syringe at the culture medium inlet to a drug supplemented cell culture medium for toxicity testing. Be careful not to introduce any bubbles to the device. Use a larger syringe depending on the time intended for cell culturing and assaying.
2. At the end of the experiment, replace the syringe at the culture medium inlet to a syringe filled with assay solution (e.g., a Live/Dead assay) and gently replace the cell culture medium by hand and incubate for 30–45 min at room temperature.
3. Image the cells with fluorescence microscopy (Fig. 2).

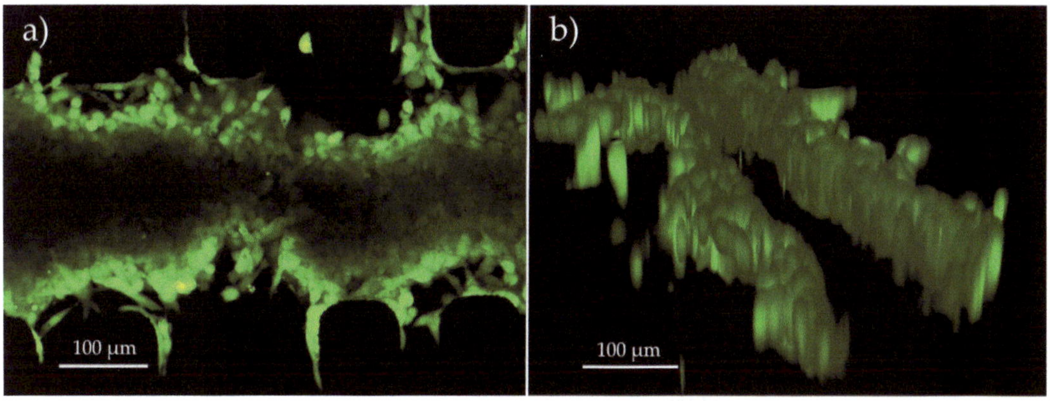

Fig. 2 Fluorescence imaging of living HepG2 cells (Calcein AM, green). The cells reside behind the pillars in the cell culture chamber of the microfluidic device and are here imaged using confocal microscopy and presented with (**a**) a 2D (from the bottom) view and (**b**) a 3D rendered z-stack

4 Notes

1. The microfluidic device used to describe this protocol is based on a design where cells are seeded through a dedicated cell inlet into a channel where micro pillars (75 × 100 μm) make up a cell culture chamber. Cell culture medium is then perfused through the channels adjacent to the cells. The device can be constructed using photolithography and soft lithography as previously described [16] and similar designs have previously been reported by others [13–15].
2. HepG2 cells are used as an example of this protocol but can be substituted to cells of different types (i.e., tumor cells, primary cells, or stem cell-derived cells). It is however recommended to start with a robust cell type to get familiar with the protocol before using more sensitive and expensive cells.
3. The hydrophobic properties of PDMS make air bubbles a frequent problem and can be difficult to remove. Therefore, to first fill the channels with ethanol and subsequently replace the liquid with PBS and cell culture medium, without introducing new air bubbles, is often a necessary step.
4. To switch syringes without introducing air to the tubing, make sure a meniscus of liquid is present at both the Luer connector and the syringe. Combine the meniscuses before fixing the syringe in the Luer connector.
5. Coating the microfluidic device with BSA minimizes the risk of cells attaching to the glass bottom outside of the cell culture chamber in the early phase after cell seeding.

6. Individual cells can often pass between the pillars of the cell culture chamber. A high cell concentration makes cell seeding faster as the cells clump together which makes them more likely to stay behind the pillars.
7. A slightly easier method to seed the cells is to push them into the cell culture chamber using the syringe. However, the force experienced by the cells is considerably lower when withdrawn from the outlet. Depending on the sensitivity of the cell type, this step can be revised.
8. If the cells have not properly aggregated, there is a considerable risk of cells escaping between the pillars when the perfusion starts.
9. The waste reservoir can also be a used as a sample reservoir for subsequent analysis of biomarkers in the supernatant.

Acknowledgments

The research leading to these results has received support from the Innovative Medicines Initiative Joint Undertaking under (grant no. 115439), resources of which are composed of financial contribution from the European Union's Seventh Framework Programme (FP7/2007-2013) and EFPIA companies. This publication reflects only the author's views, and neither the IMI JU nor EFPIA nor the European Commission is liable for any use that may be made of the information contained therein.

References

1. Bhatia SN, Ingber DE (2014) Microfluidic organs-on-chips. Nat Biotechnol 32 (8):760–772
2. Marx U, Andersson TB, Bahinski A et al (2016) Biology-inspired microphysiological system approaches to solve the prediction dilemma of substance testing. ALTEX 33 (3):272–321
3. Prodanov L, Jindal R, Bale SS et al (2016) Long-term maintenance of a microfluidic 3D human liver sinusoid. Biotechnol. Bioengineering 113(1):241–246
4. Lanz HL, Saleh A, Kramer B et al (2017) Therapy response testing of breast cancer in a 3D high-throughput perfused microfluidic platform. BMC Cancer 17(1):709
5. Bergström G, Nilsson K, Mandenius CF et al (2014) Macroporous microcarriers for introducing cells into a microfluidic chip. Lab Chip 14(18):3502–3504
6. Christoffersson J, Bergström G, Schwanke K et al (2016) A microfluidic bioreactor for toxicity testing of stem cell derived 3D cardiac bodies. In: Turksen K (ed) Bioreactors in stem cell biology: methods and protocols. Springer Verlag, London, pp 159–168
7. Christoffersson J, Meier F, Kempf H et al (2018) A cardiac cell outgrowth assay for evaluating drug compounds using a cardiac spheroid-on-a-chip device. Bioengineering 5(2)
8. Jackman CP, Carlson AL, Bursac N (2016) Dynamic culture yields engineered myocardium with near-adult functional output. Biomaterials 111:66–79
9. Ma L-D, Wang Y-T, Wang J-R et al (2018) Design and fabrication of a liver-on-a-chip platform for convenient, highly efficient, and safe in situ perfusion culture of 3D hepatic spheroids. Lab Chip

10. Miranda JP, Rodrigues A, Tostões RM et al (2010) Extending hepatocyte functionality for drug-testing applications using high-viscosity alginate–encapsulated three-dimensional cultures in bioreactors. Tissue Eng Part C Methods 16(6):1223–1232
11. Esch MB, Prot JM, Wang YI et al (2015) Multi-cellular 3D human primary liver cell culture elevates metabolic activity under fluidic flow. Lab Chip 15(10):2269–2277
12. Booth R, Kim H (2012) Characterization of a microfluidic in vitro model of the blood-brain barrier (μBBB). Lab Chip 12(10):1784–1792
13. Lee PJ, Hung PJ, Lee LP (2007) An artificial liver sinusoid with a microfluidic endothelial-like barrier for primary hepatocyte culture. Biotechnol Bioeng 97(5):1340–1346
14. Toh YC, Lim TC, Tai D et al (2009) A microfluidic 3D hepatocyte chip for drug toxicity testing. Lab Chip 9(14):2026–2035
15. Ong LJY, Chong LH, Jin L et al (2017) A pump-free microfluidic 3D perfusion platform for the efficient differentiation of human hepatocyte-like cells. Biotechnol. Bioengineering 114(10):2360–2370
16. Christoffersson J, Mandenius CF (2019) Fabrication of a microfluidic cell culture device using photolithographic and soft lithographic techniques. In: Mandenius C-F, Ross JA (eds) Chapter 21 in Cell-based assays using iPSCs for drug development and testing. Springer, New York
17. Christoffersson J et al (2019) Evaluating the effect of drug compounds on cardiac spheroids using the cardiac cell outgrowth assay. In: Mandenius C-F, Ross JA (eds) Chapter 17 in Cell-based assays using iPSCs for drug development and testing. Springer, New York

Chapter 23

hiPS Cell-Derived Neurons for High-Throughput Screening

Stefanie Traub and Ralf Heilker

Abstract

Human induced pluripotent stem (hiPS) cell-derived neurons promise to provide better model cells for drug discovery in the context of neurodegenerative and neuropsychiatric diseases. The neuronal differentiation protocol described encompasses a cellular amplification phase for hiPS-derived neural progenitor (NP) cells. Thus, the combination of growth factor-driven expansion and inhibition of notch (GRINCH) enabled the scalable production of neurons in sufficient numbers to meet the immense material needs of a high-throughput screening (HTS) campaign. These GRINCH cells matured in 384-well microplates display neuronal markers and electrophysiological activity. The differentiation protocol was applicable to various human hiPS cell clones. In a finding and profiling campaign for modulators of the tropomyosin receptor kinase B (TrkB), the GRINCH neurons were shown to be suitable for measuring the phosphorylation and downstream signaling of the endogenously expressed TrkB. The employed techniques in the amplified luminescent proximity homogeneous assay (Alpha) and the high-throughput reverse transcription polymerase chain reaction (RT-PCR) format are transferable to other pharmaceutical drug targets. Together with the GRINCH neurons, these detection technologies open new experimental routes with tremendous potential for early drug discovery.

Key words Human iPS cells, Neuronal differentiation, Neural progenitor cells, High-throughput screening

1 Introduction

In the past, there was frequently a disconnect between preclinical biological models and drug efficacy in human probands [1, 2]. The challenge to bridge between in vitro and in vivo is particularly grand in the context of neurodegenerative and neuropsychiatric diseases, where hardly any human primary neurons are available to translate simplistic early pharmacological studies into the physiological disease context. In recent years, the development of hiPS cell technology [3] has opened new routes toward human neuronal model cells for drug discovery. However, the high cell numbers required for HTS have hampered the application of hiPS cell-derived neurons in large drug candidate finding campaigns. Further roadblocks to the

Carl-Fredrik Mandenius and James A. Ross (eds.), *Cell-Based Assays Using iPSCs for Drug Development and Testing*, Methods in Molecular Biology, vol. 1994, https://doi.org/10.1007/978-1-4939-9477-9_23,
© Springer Science+Business Media, LLC, part of Springer Nature 2019

use of hiPS cell technology in this context derive from the typically lengthy neuronal differentiation protocols. In the here described method, these challenges have been met by FGF2-enhanced expansion [4] of the hiPS cell-derived NP cells and by an accelerated neuronal maturation based upon inhibited Notch signaling [5, 6]. To render the process suitable to the prerequisites of an industrial drug discovery environment, the last phase of the neuronal differentiation process was carried out in microplates of the 384-well format using an automated system for maintenance cell culture. The derived GRINCH neurons displayed characteristic synaptic markers. In addition, the functional activity of the GRINCH neurons was demonstrated by electrophysiological measurements.

The suitability of the GRINCH neurons as model cells for drug discovery was demonstrated by applying them in the search for and profiling of TrkB modulators [5, 7]. TrkB is the receptor for brain-derived neurotrophic factor (BDNF) [8, 9]. Lower expression levels of BDNF have been implicated with a multitude of CNS-related diseases [10] such as Alzheimer disease, Parkinson disease, schizophrenia, depression, bipolar disorder, anxiety, Huntington disease, stroke, epilepsy, eating disorders, and substance use. In consequence, the pharmacologic activation of TrkB provides a promising approach to the treatment of various CNS disorders [11]. To analyze TrkB stimulation on target, an Alpha assay was developed that measured the phosphorylation of TrkB (Tyr706).

2 Materials

2.1 Human iPS Cell Culture

1. hiPS cell line Ad3Cl1, derived from fibroblasts (Lonza CC-2511, tissue acquisition number 23447; Lonza Group, Ltd., Basel, Switzerland) and characterized within StemBANCC.
2. Minicircle (mc) hiPS cell line (catalogue number [cat#] SC301A-1; System Biosciences; Mountain View, CA), derived from adult human adipose stem cells [12].
3. 1× Dulbecco's Phosphate buffered saline (1× DPBS; cat# 4200-067; Thermo Fisher Scientific, Waltham, MA).
4. Cellstar™ 6-well plates (cat# 657160; Greiner Bio-One; Frickenhausen, Germany).
5. DMEM/F-12 GlutaMAX™ (cat# 10565-018; Thermo Fisher Scientific).
6. Accutase solution (cat# SCR005; Merck Millipore; Billerica, MA).

7. Matrigel™ (cat# 354277; Corning Inc., Corning, NY (*see* **Note 1**)).
8. Rho-associated, coiled-coil containing protein kinase (ROCK) inhibitor Y-27632 (cat# Y0503-5MG; Sigma-Aldrich; St. Louis, MO), 10 mM stock solution: Dissolve 5 mg of Y-27632 Rock inhibitor in 1.563 mL of water. Store aliquots of 100 μL at −20 °C or at 4 °C for up to 1 year. Once thawed, store aliquots at 4 °C and use within 4 weeks.
9. Ethylenediaminetetraacetic acid (EDTA; cat# 15575020; Thermo Fisher Scientific), 0.5 M stock solution: dilute to 0.5 mM in 1× DPBS. Store aliquots of 50 mL at 4 ° C for up to 4 weeks.
10. hiPS cell medium: mTeSR™1 complete medium (cat# 05850; STEMCELL Technologies; Vancouver, Canada), supplemented with 1× Antibiotic-Antimycotic (cat# 15240-062; Thermo Fisher Scientific).
11. hiPS cell cryopreservation medium: mFreSR (cat# 05854; STEMCELL Technologies), supplemented with 10 μM Y-27632.

2.2 Directed Differentiation to Neurons

1. 0.01% poly-L-ornithine (PLO; cat# P4957; Sigma-Aldrich).
2. Laminin (cat# L2020; Sigma-Aldrich).
3. T175 Nunclon Delta surface-treated EasYFlasks (cat# 178883; Thermo Fisher Scientific).
4. T-1000 flasks (cat# PFHYS1008, Merck Millipore).
5. CELLCOAT® μClear® poly-D-lysine 384-well microplate (cat# 781090; Greiner Bio-One).
6. Dispase solution (cat# 17105; Thermo Fisher Scientific), dissolve at 1 mg/mL in 1× DPBS. Store aliquots of 15 mL at −80 ° C for up to 1 year. Once thawed, filter dispase solution using a Steriflip® filter unit (cat# SE1M179M6; Merck Millipore), store aliquots at 4 °C and use within 4 weeks.
7. Noggin (cat# 3344-NG-050; R&D Systems; Minneapolis, MN), dissolved at 0.1 mg/mL in 1× DPBS. Store aliquots of 50 μL at −20 °C for up to 1 year. Once thawed, keep aliquots at 4 °C for up to 1 week.
8. SB431542 (cat# 1614/1; R&D Systems), 10 mM stock solution: Dissolve 10 mg of SB431542 in 2.602 mL of ethanol. Store at RT for up to 1 year.
9. Basic Fibroblast Growth Factor 2 (FGF2; cat# 4114-TC; R&D Systems (*see* **Note 2**), 0.16 μg/mL stock solution: Dissolve 1 mg of FGF2 in 6.25 mL of 1× DPBS, supplemented with 0.1% (w/v) bovine serum albumin (BSA; cat# A3059-100G; Sigma-Aldrich). Store aliquots of 50 μL at −20 °C for up to 1 year. Once thawed, keep aliquots at 4 °C for up to 1 week.

10. *N*-[*N*-(3,5-difluorophenacetyl)-l-alanyl]-*S*-phenylglycine *t*-butyl ester (DAPT; cat# 72082; STEMCELL Technologies), 10 mM stock solution: Dissolve 5 mg of DAPT in 1.16 mL of DMSO. Store aliquots of 50 μL at −20 °C for up to 1 year. Once thawed, keep aliquots at 4 °C for up to 1 week.
11. Neural induction medium (NIM): DMEM/F-12 with GlutaMAX™ (cat# 10565-018; Thermo Fisher Scientific) and Neurobasal™ Medium (cat# 12348-017; Thermo Fisher Scientific) mixed in a 1:1 (v/v) ratio, then supplemented with 0.5 × N-2 (cat# 17502-048; Thermo Fisher Scientific), 0.5 × B27 (cat# 17504-044; Thermo Fisher Scientific (*see* **Note 3**), 2.5 μg/mL insulin (cat# 19278; Sigma-Aldrich), 2 mM L-glutamine (cat# 25030-024; Thermo Fisher Scientific), 50 μM 2-mercaptoethanol (cat# 21985-023; Thermo Fisher Scientific), 0.5 × nonessential amino acids (NEAA; cat# 11140-050; Thermo Fisher Scientific), 100 U/mL penicillin and 100 μg/mL streptomycin (cat# 15140-122; Thermo Fisher Scientific), 500 ng/mL noggin (cat# 3344-NG-050; R&D Systems), and 10 μM SB431542 (cat# 1614; R&D Systems).
12. Neural progenitor medium (NPM): NIM omitting noggin and SB431542, supplemented with 20 ng/mL FGF2 (cat# 4114-TC; R&D Systems).
13. Neural progenitor cryopreservation medium (NPCM): NPM, supplemented with 10% (v/v) DMSO (cat# D2438-5S10mL; Sigma-Aldrich).
14. Neuronal differentiation medium (NDM): NPM omitting FGF2, supplemented with 6.6 μg/mL laminin and 10 μM *N*-[*N*-(3,5-difluorophenacetyl)-l-alanyl]-*S*-phenylglycine *t*-butyl ester (DAPT; cat# 72082; STEMCELL Technologies).

2.3 Immunofluorescence Staining

1. 4′,6-diamidino-2-phenylindole (DAPI; cat# D1306; Thermo Fisher Scientific), 10.8 μM stock solution: Dissolve 10 mg of DAPI in 2 mL of deionized water. Store aliquots of 50 μL at −20 °C for up to 1 year. Once thawed, store aliquots at 4 °C in the dark for up to 1 week.
2. 2× Fixing solution: 37% (v/v) paraformaldehyde solution (cat# 252549-500 mL; Sigma-Aldrich), diluted in deionized water to 8% (v/v).
3. Washing buffer: 10× Tris Buffered Saline (TBS; cat# T5912-1 L; Sigma-Aldrich), diluted in deionized water to 1× solution (v/v).
4. 2× Permeabilization solution: 100% Triton™X-100 (cat# T8787-100 mL; Sigma- Aldrich) diluted in 1× TBS to 0.6% (v/v).

5. 2× Blocking solution: 100% fetal bovine serum (FBS, cat# 26140-079; Thermo Fisher Scientific) diluted in in permeabilization solution to 20% (v/v).
6. STAINperfect Immunostaining Kit A (cat# SP-A-1000; Immusmol; Aquitaine, France).

2.4 RT-PCR Gene Expression Analysis

1. RealTime ready (RTR) Cell Lysis Kit (cat# 05943523001; Roche; Basel, Switzerland).
2. RTR Master Mix: 1× RTR Virus Master Mixture (cat# 05992877001; Roche) and 1× RTR Catalogue (cat# 05532957001; Roche) or Designer Assay (cat# 05583055001; Roche) solutions.
3. LightCycler® 480 Multiwell Plates 480 (cat# 04729749001; Roche).
4. LightCycler® 480 Sealing Foils (cat# 04729757001; Roche).

2.5 Whole-Cell Patch-Clamp Recordings

1. Artificial cerebrospinal fluid (ACSF) solution: 140 mM $NaCl_2$ (cat#31434-1 kg; Sigma-Aldrich), 5.3 mM KCl (cat# HC756324; Merck), 1.8 mM $CaCl_2$ (cat# 1.02382.0250; Merck), 0.8 mM $MgCl_2$ (M8530-500 g; Sigma-Aldrich), 27.8 mM D-glucose (cat# G5767-500 g; Sigma-Aldrich), and 10 mM Hepes (cat# 3375-1 kg (cat# 35256-1 L; Honeywell; Morris Plains, NJ)/NaOH (pH 7.4, 320 mOsmol/kg).
2. Patch pipette solution: 145 mM potassium D-gluconate (cat# G4500-100; Sigma-Aldrich), 1 mM EGTA (cat# 3054.1; Roth; Karlsruhe, Germany), 2 mM MgCl2, 4 mM Mg-ATP (cat# A9187-100MG; Sigma-Aldrich), and 0.4 mM Na-GTP (cat# G8877; Sigma-Aldrich) 10 mM HEPES/Tris (pH 7.2, 290–300 mOsmol/kg).

2.6 Amplified Luminescent Proximity Homogeneous Assay (ALPHA™)

1. Brain-derived neurotrophic factor (BDNF; cat# cyt-207; Biotrend Chemikalien GmbH; Cologne, Germany; molecular weight of 26,984 Da for the homodimer), 46.25 μg/mL stock solution: Dissolve 10 μg of BDNF in 0.38 mL of 1× DPBS, supplemented with 0.1% (w/v) BSA. Store aliquots of 10 μL at −20 °C for up to 1 year. Once thawed, store aliquots at 4 °C and use within 5 days.
2. TrkB stimulation buffer: 1% (w/v) BSA in 1× D-PBS.
3. TrkB cell lysis buffer: 1× CST Lysis Buffer (cat# 9803S; Cell Signaling Technology; Danvers, MA), supplemented with 2× Phosphatase inhibitor cocktail II and III (cat# P5726, P0044; Sigma-Aldrich) and 2× Complete protease inhibitor (cat# 11873580001; Roche).
4. AlphaLISA™ Assay Buffer (cat# AL000F; PerkinElmer; Waltham, MA), diluted in deionized water to 1× solution (v/v).

5. Phospho-TrkB ALPHA™ reagents: anti-TrkB antibody (cat# MAB3971; R&D Systems), conjugated with biotin (referred to as "TrkB biotinylated antibody"), and the anti-pTrkA/B (Tyr 706) antibody (cat# 4621BF; Cell Signaling Technology) immobilized to AlphaLISA™ acceptor beads (referred as to "TrkB Acceptor beads).
6. AlphaScreen™ Streptavidin Donor beads (cat# 6760002; PerkinElmer).
7. 384-well Small Volume Deep Well microplate (cat# 784201; Greiner Bio-One).

3 Methods

All incubations at 37 °C are carried out in a humidified, 5% CO_2 incubator.

3.1 hiPS Cell Maintenance

1. Prepare one Matrigel™-coated 6-well plate. Briefly, dilute Matrigel™ in DMEM/F-12 GlutaMAX™ according to the manufacturer's instructions. Employ 1 mL/well of Matrigel™ solution and incubate plates for 2 h at 37 °C.
2. (A) Ad3Cl hiPS cells: Detach cells for passaging at 80% confluence using 1 mL/well of 0.5 mM EDTA (diluted in 1× DPBS) for 4 min at 37 °C. Carefully aspirate EDTA and flush the loosened cells by adding 2 mL/well of hiPS cell medium and transfer cell suspension to a conical 50 mL tube. Add 10 mL of hiPS cell medium, resulting in a split ratio of 1:6 in **step 3**

   (B) mc hiPS cells: Detach cells for passaging at 80% confluence using 1 mL/well of Accutase solution (diluted in an equal volume of 1× DPBS) for 2 min at 37 °C. Add 9 mL/well of DMEM/F-12 GlutaMAX™. Collect the resulting clumps into a conical 50 mL tube, centrifuge at 200 × *g* for 5 min, then aspirate DMEM/F-12 GlutaMAX™ and resuspend cells in 10 mL of hiPS cell medium, supplemented with 10 μM Y-27632, resulting in a split ratio of 1:5 in **step 3**.
3. Aspirate the Matrigel™ solution from the above prepared 6-well plates and seed 2 mL/well of the cell suspension.
4. Let the cells settle in the laminar flow at RT for 10 more min to ensure even distribution.
5. Exchange medium daily with fresh hiPS cell medium without Y-27632 until next passaging step at 80% confluence.
6. For cryopreservation, detach Ad3Cl1 and mc hiPS cells as described in step 2 and distribute to cryovials in 1 mL-aliquots with approximately 4×10^6 cells/mL in hiPS cell cryopreservation medium. Store cryovials for 24 h at −80 °C in a freezing container, then transfer the cryovials to liquid nitrogen storage tank for long-term storage.

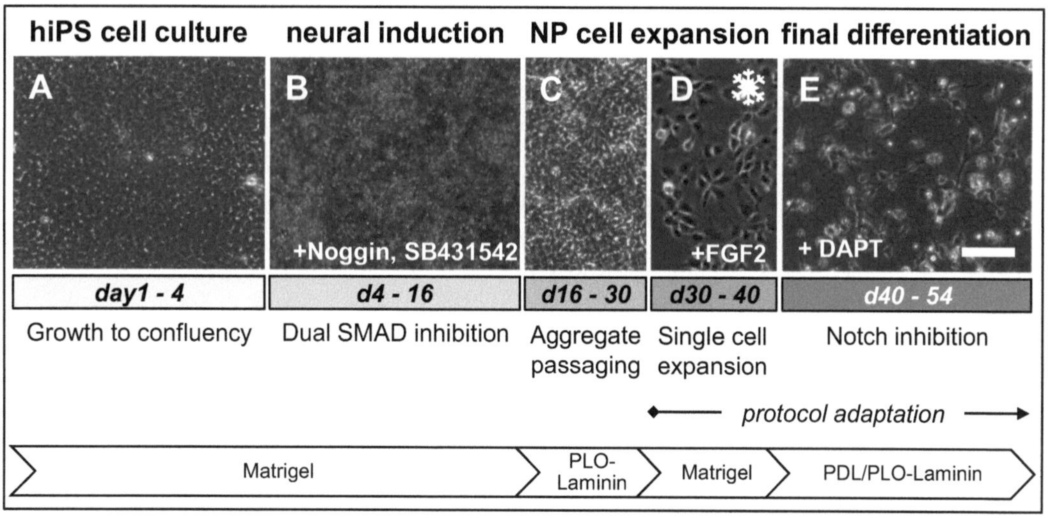

Fig. 1 Process overview: neuronal differentiation of hiPS cells. The neuronal differentiation protocol was based upon previously published work, adapted to high-throughput needs. Briefly, (**a**) a hiPS cell monolayer was formed; (**b**) neural induction was induced by dual SMAD inhibition via Noggin and SB431542; (**c**) NP cell expansion was carried out using aggregate passaging and (**d**) FGF2-driven single cell expansion. The expanded NP cells were frozen at this stage (indicated by star). For final neuronal differentiation in microplates, NP cells were incubated with Notch-signaling inhibitor DAPT (**e**). The duration of the individual phases and the respectively employed surface coatings are indicated below. Scale bar: 100 μm

3.2 Directed Differentiation into Neurons

The differentiation protocol [5], based upon previous publications [13–15], is divided up into three parts, (1) neural induction, (2) NP cell expansion and cryopreservation, and (3) final differentiation to neurons (Fig. 1). In the following method description, the days are counted from day 1, which corresponds to the seeding of hiPS cells.

3.2.1 Neural Induction

1. On day 1, prepare one well of a Matrigel™-coated 6-well plate. Briefly, dilute Matrigel™ in DMEM/F-12 GlutaMAX™ according to the manufacturer's instructions. Employ 1 mL/well of Matrigel™ solution and incubate plates for 2 h at 37 °C.
2. Detach Ad3Cl1 and mc hiPS cells from maintenance culture at 80% confluence using 1 mL/well of Accutase solution (undiluted) for 6 min at 37 °C.
3. Add 3 mL/well of DMEM/F-12 GlutaMAX™ and collect the resulting hiPS single-cell suspension into a conical 50 mL tube. Repeat this washing procedure three times, resulting in a total suspension volume of 10 mL.
4. Centrifuge the detached cells at 200 × *g* for 5 min.
5. Aspirate DMEM/F-12 GlutaMAX™ and resuspend cells at 2×10^6 cells/mL in hiPS cell medium, supplemented with 10 μM Y-27632 (*see* **Note 4**).

6. Aspirate the Matrigel™ solution from the above prepared 6-well plates and seed 2 mL/well of the single-cell suspension.
7. Let the cells settle in laminar flow at RT for 10 more minutes to ensure even distribution (*see* **Note 5**).
8. On days 2 and 3, replace medium with 2 mL/well of fresh hiPS cell medium without Y-27632 (*see* **Note 6**).
9. From days 4 to 15, replace medium with 2 mL/well of NIM daily (*see* **Note 7**).
10. On day 16, prepare one well of a PLO- and laminin-coated 6-well plate. Briefly, add 1 mL/well of PLO solution and incubate plates for 1 h at RT. Subsequently, wash plates with 1 mL/well of 1× DPBS. Next, employ 1 mL/well of laminin solution (diluted in DMEM/F-12 GlutaMAX™ at 20 μg/mL) and incubate plates for 2 h at 37 °C.
11. Detach the generated neuroectodermal sheet using 1 mL/well of dispase solution to for 5 min at 37 °C (*see* **Note 8**).
12. Carefully aspirate the dispase solution and replace by 2 mL/well of NPM.
13. Detach the neuroectodermal sheet from the bottom of the well using a cell scraper, and break it into smaller aggregates by gently pipetting up and down three times.
14. Collect the resulting neural aggregates into a conical 50 mL tube.
15. Allow the aggregates to settle for 5 min at RT. Carefully aspirate the supernatant, and add 10 mL/well of fresh NPM. Repeat this washing procedure three times.
16. Aspirate NPM and resuspend neural aggregates resulting from one well of a 6-well plate in 2 mL of NPM.
17. Aspirate the laminin solution from the above prepared 6-well plates and seed 2 mL/well of the neural aggregate solution. This corresponds to a split ratio of 1:1 (*see* **Note 9**).
18. From days 17 to 19, replace medium with 2 mL/well of NPM daily (*see* **Note 10**).
19. On day 20, repeat **steps 10–17**.
20. From days 21 to 24, replace medium with 2 mL/well of NPM daily.
21. On day 25, prepare two wells of a PLO- and laminin-coated 6-well plate as described in **step 10**.
22. Repeat **steps 11–15**.
23. Aspirate NPM and resuspend neural aggregates resulting from one well of a 6-well plate in 4 mL of NPM.

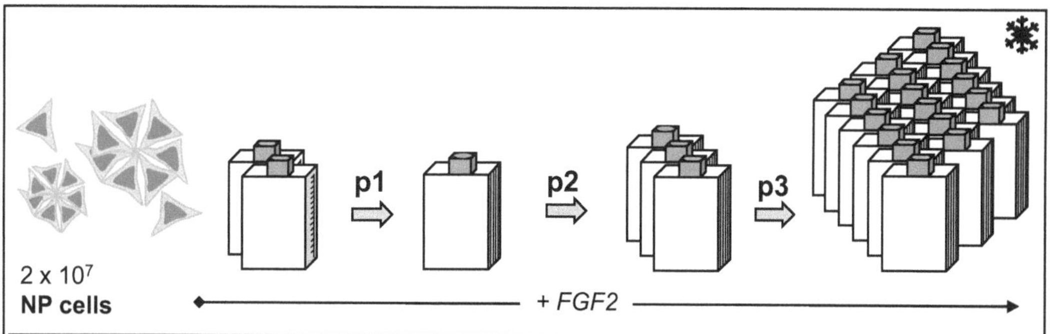

culture ware	2 x T-175	1 x T-1000	3 x T-1000	18 x T-1000
Neuronal differentiation [day]	30-32	32-34	34-37	37- 40
NP cell harvest [cells]	$6x10^7$	$1.8x10^8$	$1.1x10^9$	$6.4x10^9$

Fig. 2 Large-scale expansion of hiPS cell-derived NP cells. Scheme of NP cell large scale expansion. During FGF2-driven NP cell expansion, the number of cells increased over 10 days from 2×10^7 to 6.4×10^9 covering the scope of an HTS campaign. The expanded NP cells were frozen at this stage

24. Aspirate the laminin solution from the above prepared 6-well plates and seed 2 mL/well of the neural aggregate solution. This corresponds to a split ratio of 1:2.
25. From days 26 to 29, replace medium with 2 mL/well of NPM daily.

3.2.2 NP Cell Expansion and Cryopreservation

To obtain a sufficient number of cells for a large-scale screening campaign, the following expansion scheme was applied (Fig. 2).

1. On day 30, prepare two Matrigel™-coated T175 flasks. Briefly, dilute Matrigel™ in DMEM/F-12 GlutaMAX™ according to the manufacturer's instructions. Employ 12 mL/flask of Matrigel™ solution and incubate flasks for 2 h at 37 °C.
2. Detach neural aggregates using 1 mL/well of Accutase solution (undiluted) for 5 min at 37 ° C.
3. Add 3 mL/well of DMEM/F-12 GlutaMAX™ and collect the resulting NP cell suspension into a conical 50 mL tube. Repeat this washing procedure three times, resulting in a total suspension volume of 10 mL/well.
4. Centrifuge the detached cells at $360 \times g$ for 5 min.
5. Aspirate DMEM/F-12 GlutaMAX™ and resuspend NP cells resulting from two wells of a 6-well plate in 40 mL of NPM.
6. Aspirate the Matrigel™ solution from the above prepared T175 flasks and seed 20 mL/flask of NP cells suspension.
7. On day 31, replace medium with 30 mL/flask of NPM.

8. On day 32, prepare a Matrigel™-coated T-1000 flask as described in **step 1**. Employ 68 mL/flask of Matrigel™ solution and incubate flasks for 2 h at 37 °C.
9. Detach NP cells using 17 mL/flask of Accutase solution for 5 min at 37 ° C.
10. Add 8 mL/flask of DMEM/F-12 GlutaMAX™ and collect the resulting NP cell suspension into a conical 50 mL tube, resulting in a total suspension volume of 25 mL/flask.
11. Centrifuge the detached cells at 360 × *g* for 5 min.
12. Aspirate DMEM/F-12 GlutaMAX™ and resuspend the NP cells, resulting from two T175 flasks in 200 mL of NPM.
13. Aspirate the Matrigel™ solution from the above prepared T-1000 flask and seed 200 mL/flask of NP cell suspension.
14. On day 33, replace medium with 200 mL/flask of NPM.
15. On day 34, prepare three Matrigel™-coated T-1000 flasks as described in **step 8**.
16. Detach NP cells using 100 mL/flask of Accutase solution for 5 min at 37 °C.
17. Add 50 mL/flask of DMEM/F-12 GlutaMAX™ and collect the resulting NP cell suspension into a conical 175 mL tube.
18. Centrifuge the detached cells at 360 × *g* for 5 min.
19. Aspirate DMEM/F-12 GlutaMAX™ and resuspend NP cells resulting from one T-1000 flask in 600 mL NPM.
20. Aspirate the Matrigel™ solution from the above prepared T-1000 flasks and seed 200 mL/flask of NP cell suspension.
21. On days 35 and 36, replace medium with 200 mL/flask of NPM.
22. On day 37, prepare eighteen Matrigel™-coated T-1000 flask as described in **step 8**.
23. Repeat **steps 16–18**.
24. Aspirate DMEM/F-12 GlutaMAX™ and resuspend NP cells resulting from three T-1000 flasks in 3.6 L of NPM.
25. Aspirate the Matrigel™ solution from the above prepared T-1000 flasks and seed 200 mL/flask of NP cell suspension.
26. On days 38 and 39, replace medium with 200 mL/flask of NPM.
27. On day 40, detach NP cells as described in **steps 16–18**.
28. Aspirate DMEM/F-12 GlutaMAX™ and resuspend NP cells in NPCM.
29. Distribute NP cell solution to cryovials in 1-mL aliquots with approximately 5×10^7 cells/mL. Store cryovials for 24 h at −80 °C in a freezing container, then transfer cryovials to liquid nitrogen storage tank for long-term storage.

3.2.3 Final Differentiation into Neurons

1. Remove the NP cell cryovial from the liquid nitrogen storage tank.
2. Immerse the cryovial in a 37 °C water bath for 3 min.
3. Remove the cryovial from the water bath and gently transfer the NP cell solution to a conical 50 mL tube.
4. Rinse the cryovial with 1 mL of DMEM/F-12 GlutaMAX™, and transfer drop-wise to the conical 50 mL tube containing the NP cell suspension. Gently swirl the tube while adding the medium to mix the solution completely and minimize the osmotic shock on the thawed cells.
5. Slowly add 8 mL of DMEM/F-12 GlutaMAX™ to the conical 50 mL tube, resulting in a total suspension volume of 10 mL.
6. Centrifuge the NP cell suspension at 360 × *g* for 5 min.
7. Aspirate DMEM/F-12 GlutaMAX™ and resuspend cells at 0.15×10^6 cells/mL in NDM.
8. Seed 0.04 mL/well of the NP cell suspension into CELL-COAT® µClear® poly-D-lysine coated 384-well microplates (*see* **Note 11**).
9. From days 2 to 14 of final differentiation, replace half of the medium with 0.02 mL/well of NDM every other day.

3.3 Gene Expression Analysis

During directed differentiation into neurons, generate RT-PCR lysates of relevant differentiation stages (Fig. 3).

1. (A) For cellular lysis of hiPS cells, follow the protocol until Subheading 3.1, **step 5**.
   (B) For cellular lysis of the neuroectodermal sheet, follow protocol until Subheading 3.2.1, **step 9**.
   (C) For cellular lysis of neural aggregates, follow the protocol until Subheading 3.2.1, **step 25**.
   (D) For cellular lysis of NP cells, follow the protocol until Subheading 3.2.2, **step 27**. Aspirate DMEM/F-12 GlutaMAX™ and resuspend cells at 0.15×10^6 cells/mL in NPM. Seed 2 mL/well of the NP cell suspension into one well of a Matrigel™-coated 6-well plate and culture cells for 2 days.
2. Wash cells three times with 2 mL/well of 1× DPBS.
3. Aspirate 1× DPBS and add 2 mL/well of 1× RTR Cell Lysis Buffer, supplemented with 1× RNase inhibitor, of a 6-well plate.
4. Incubate cells for 5 min at RT (*see* **Note 12**).
5. Prepare RTR master mixture, composed of 1× RTR Virus Master and 1× RTR Assay (Table 1) according to the manufacturer's instructions.
6. Transfer 9 µL/well of the RTR master mixture into a 384-well LightCycler™ 480 plate.

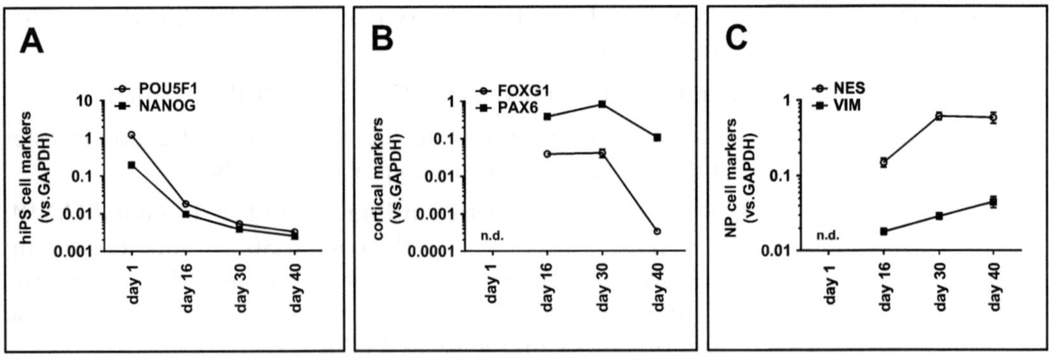

Fig. 3 Gene expression analysis during NP cell generation and expansion. Kinetic expression of hiPS cell markers POU5F1 and NANOG (**a**), cortical markers FOXG1, PAX6 (**b**) as well as NP cell markers NES and VIM (**c**) was measured in cellular lysates of directed differentiation day 1 (hiPS cells), day 16 (neuroectodermal sheet), day 30 (neural aggregates) and day 40 (NP cells, respectively. n.d. = not detected. RT-PCR data are given as relative expression [$2^{(-\Delta CP)}$] values. $n = 6$ replicates, error bars represent SEM

Table 1
Human RTR primer/probe assay sets

Gene	Supplier	Assay ID
FOXG1	Roche	138481
GAPDH	Roche	141139
NANOG	Roche	143702
NES	Roche	140436
PAX6	Roche	136139
POU5F1	Roche	113034
VIM	Roche	112941

Human RTR catalog or designer assays are listed according to gene name, supplier, and assay ID

7. Add 1 μL/well of the above generated RT-PCR lysates into the same 384-well LightCycler™ 480 plate.
8. Seal the LightCycler™ 480 plate and centrifuge for 4 min at 1500 × *g*.
9. Measure gene expression using a LightCycler™ 480 instrument (Roche). Apply the following RT-PCR program: (1) Reverse Transcription: 50 °C, 8 min, 1 cycle; (2) Preincubation: 95 °C, 30 s, 1 cycle; (3) Amplification: 95 °C, 1 s, followed by 60 °C, 20 s, 45 cycles; (4) Cooling: 40 °C, 30 s, 1 cycle.

3.4 Immunofluorescence Staining

During directed differentiation into neurons, perform immunofluorescence staining of relevant differentiation stages (Figs. 4 and 5).

1. (A) For immunofluorescence staining of NP cells, follow the protocol until Subheading 3.2.2, **step 27**. Aspirate

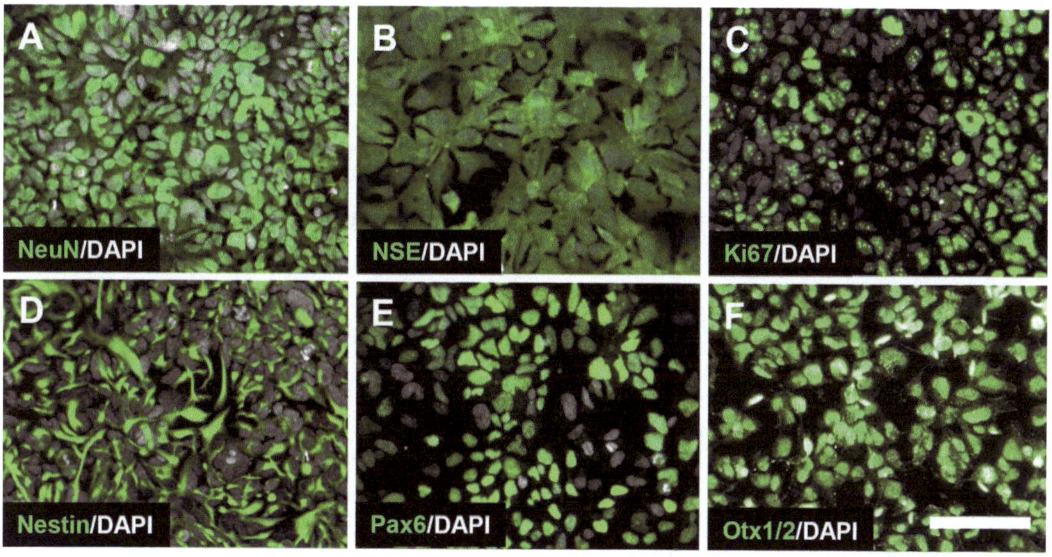

Fig. 4 Immunofluorescence characterization of hiPS cell-derived NP cells. (**a–f**) NP cell markers in immunofluorescence. The NP cells were stained using antibodies for NeuN (**a**, green), NSE (**b**, green), Ki67 (**c**, green), Nestin (**d**, green), Pax6 (**e**, green), and Otx1/2 (**f**, green). Nuclei are stained using DAPI (grey). Scale bar: 100 μm

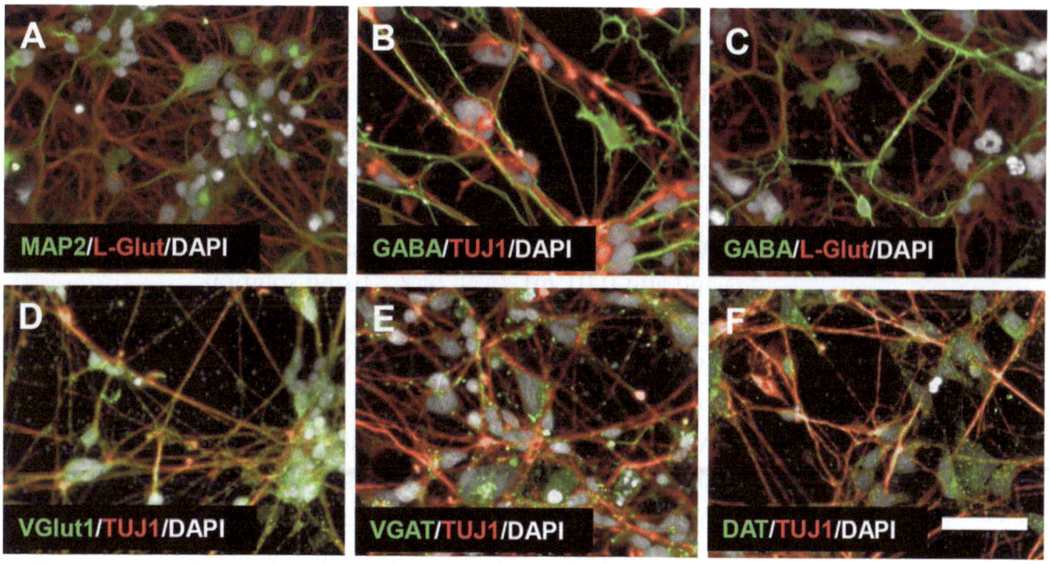

Fig. 5 Immunofluorescence characterization of hiPS cell-derived neurons. hiPS cell-derived neurons were characterized using synaptic antibodies for neurotransmitters L-Glut (**a**, **c**, red) and GABA (**b**, **c**, green), as well as transporter proteins VGlut1 (**d**, green), VGAT, (**e**, green), and DAT (**f**, green) were stained. Tuj-1 and MAP2 were used as neuronal cytoskeletal markers (red). Nuclei are stained using DAPI (grey). Scale bar: 50 μm

DMEM/F-12 GlutaMAX™ and resuspend cells at 0.15×10^6 cells/mL in NPM. Seed 0.04 mL/well of the NP cell suspension into a CELLCOAT® μClear® poly-D-lysine coated 384-well microplate and culture cells for 2 days.

Table 2
Primary antibodies

Antigen	Abbr.	Dilution	Supplier	Cat#
Nestin	Nes	1:150	Abcam, Cambridge, UK	ab22035
Paired-box protein 6	Pax6	1:150	Biolegend, San Diego, CA	PRB-278P
Homeobox protein OTX1/2	Otxl/2	1:150	Merck Millipore	AB9566
Neuron-specific enolase	NSE	1:200	Novus Biologicals, Littleton, CO	NB110-58870
Neuronal Nuclei	NeuN		Merck Millipore	MAB377X
Ki-67	Ki-67	1:200	Pharmingen, Franklin Lakes, NJ	550609
Beta-3-tubulin	TUJ1	1:500	Abcam	ab14545
Microtubule-associated protein 2	MAP2	1:1000	Abcam	ab5392
Vesicular glutamate transporter 1	VGlut1	1:150	SYSY, Goettingen, Germany	135503
Vesicular GABA transporter	VGAT	1:150	SYSY	131011
Gamma-aminobutyric acid	GABA	1:500	ImmuSmol, Aquitaine, France	IS1006
L-Glutamate	L-Glut	1:500	ImmuSmol	IS018

Primary antibodies are listed according to antigen, abbreviation (abbr.), dilution, supplier, and catalog number (cat#)

(B) For immunofluorescence staining of differentiated neurons, follow the protocol until Subheading 3.2.3, **step 9**.

2. Wash cells three times with 0.07 mL/well of washing buffer using a BioTek EL406 Washer Dispenser.
3. Aspirate washing buffer to a residual volume of 0.01 mL/well and add 0.01 mL/well of 2× fixing solution.
4. Incubate cells for 15 min at RT.
5. Wash cells three times with 0.07 mL/well of washing buffer (*see* **Note 13**).
6. Aspirate washing buffer to a residual volume of 0.01 mL/well and add 0.01 mL/well of 2× permeabilization solution.
7. Incubate cells for 15 min at RT.
8. Aspirate washing buffer to a residual volume of 0.01 mL/well and add 0.01 mL/well of 2× blocking solution.
9. Incubate cells for 60 min at RT.
10. Aspirate washing buffer to a residual volume of 0.01 mL/well and add 0.01 mL/well of primary antibodies (Table 2, *see* **Note 14**), diluted in 2× blocking solution.
11. Incubate cells over night at 4 °C.
12. Wash cells three times with 0.07 mL/well of washing buffer.

Table 3
Secondary antibodies

Target species	Fluorochrome	Dilution	Supplier	Cat#
Goat anti-mouse IgG	488	1:250	Thermo Fisher Scientific	A-11029
Goat anti-mouse IgG	546	1:250	Thermo Fisher Scientific	A-11030
Goat anti-rabbit IgG	488	1:250	Thermo Fisher Scientific	A-11034
Goat anti-rabbit IgG	546	1:250	Thermo Fisher Scientific	A-11010
Goat anti-chicken IgY	488	1:250	Thermo Fisher Scientific	A-11039

Secondary antibodies are listed according to target species, fluorochrome, dilution, supplier, and catalog number (cat#)

13. Aspirate washing buffer to a residual volume of 0.01 mL/well and add 0.01 mL/well of species-specific secondary Alexa Fluor™ antibodies (Table 3), diluted in 2× blocking solution.
14. Incubate cells for 120 min at RT in the dark.
15. Wash cells three times with 0.07 mL/well of washing buffer.
16. Aspirate washing buffer to a residual volume of 0.01 mL/well and add 0.01 mL/well of DAPI solution (600 nM) diluted in 2× blocking solution.
17. Incubate cells for 5 min at RT in the dark.
18. Wash cells three times with 0.07 mL/well of washing buffer.
19. Perform imaging using an Opera HCA reader (PerkinElmer) and a 405 nm, 488 nm, and a 561 nm laser, respectively.

3.5 Whole-Cell Patch-Clamp Recordings

During final differentiation into neurons, perform whole-cell patch-clamp recordings on glass coverslips to track neuronal maturation (Fig. 6).

1. Prepare PLO- and laminin-coated glass coverslips in 6-well plates. Briefly, add 1 mL/well of PLO solution and incubate plates for 1 h at RT. Subsequently, wash plates with 1 mL/well of 1× DPBS. Next, employ 1 mL/well of laminin solution (diluted in DMEM/F-12 GlutaMAX™ at 20 μg/mL) and incubate plates for 2 h at 37 °C.
2. Follow the protocol until Subheading 3.2.3, **step** 7.
3. Aspirate laminin solution from the above prepared 6-well plates and seed 2 mL/well of the NP cell suspension.
4. Culture the cells until day 20 of final differentiation and replace half of the medium with 0.02 mL/well of NDM every other day.
5. On days 11 and 20, perfuse each well at 1–2 mL/min with ACSF.

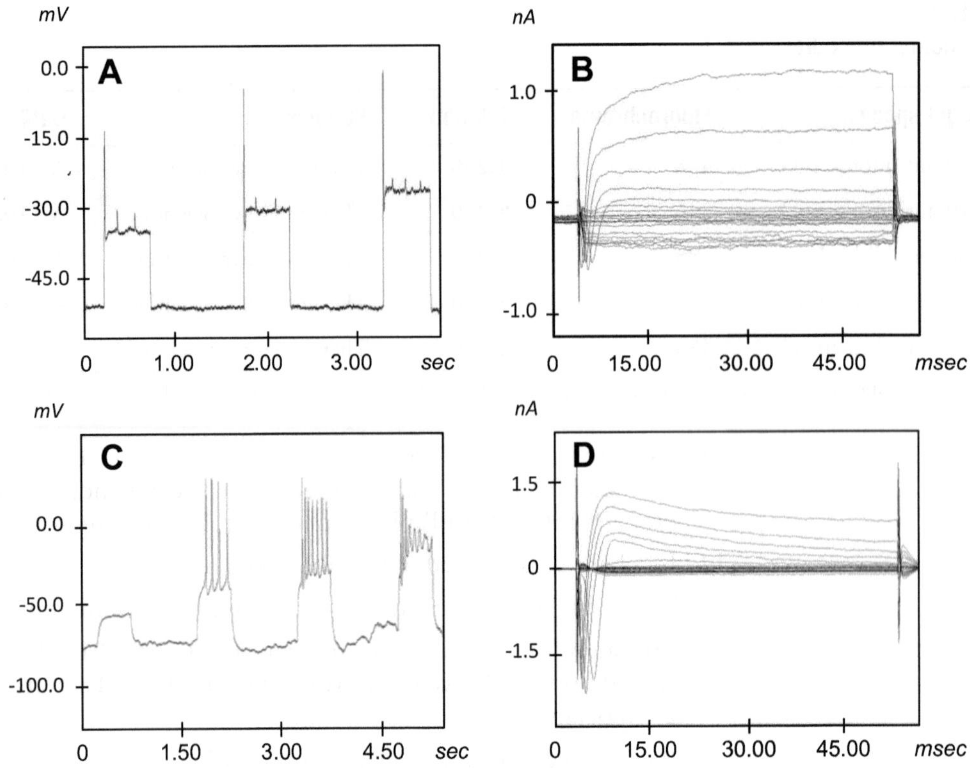

Fig. 6 Single-cell electrophysiological properties of the hiPS cell-derived neurons. (**a**, **c**) Action potentials measured in response to current injections on days 51 (**a**) and 60 (**c**) of neuronal differentiation. (**b**, **d**) Whole-cell sodium and potassium currents during stepwise depolarization from a holding potential of −70 to +20 mV on days 51 (**b**) and 60 (**d**) of neuronal differentiation

6. Perform whole cell current clamp and voltage-clamp recordings at RT using a Heka EPC 9 Patch Clamp Amplifier (Heka, Lambrecht, Germany) and the Tida software (Version 5.2.5, Heka).
7. Sample whole cell recordings with a rate of 20 kHz (10 kHz for recordings of synaptic activity) and low-pass filtered at 2.9 kHz.
8. Record sodium and potassium membrane currents in voltage clamp mode by performing a series of 10 depolarizing steps, followed by 10 hyperpolarizing steps from a holding potential of −70 mV to +20 mV, and −70 mV to −160 mV.
9. Record action potentials in current clamp mode with a holding current that leads to a membrane potential of −70 mV, by performing a series of stepwise increasing depolarizing current injections.

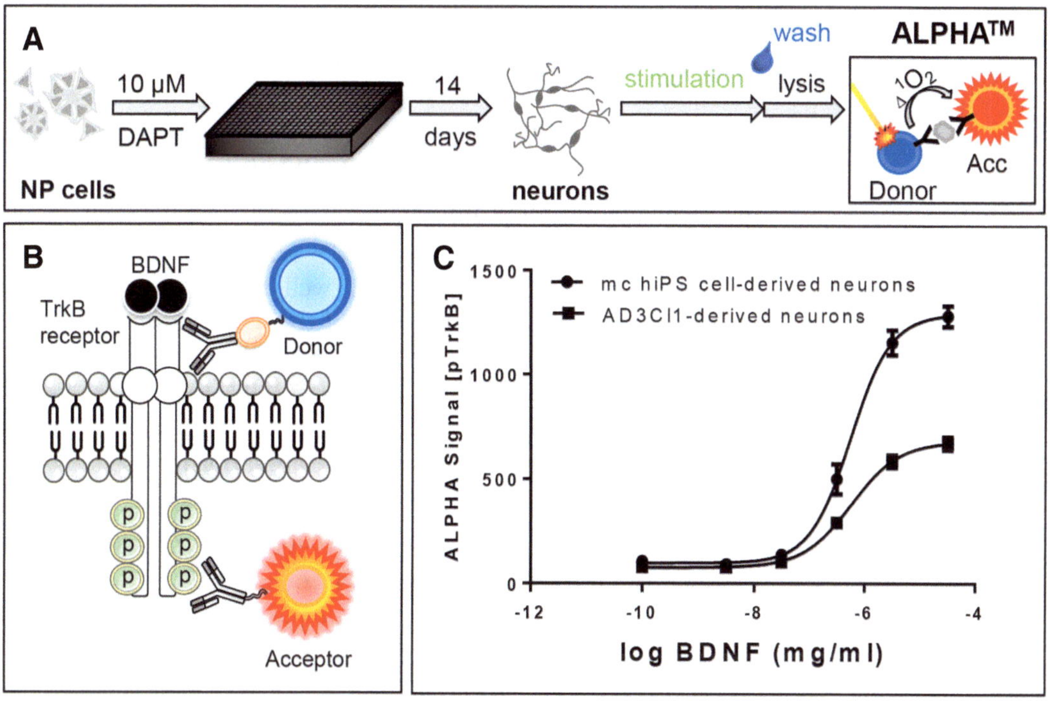

Fig. 7 TrkB phosphorylation in hiPS cell-derived neurons. (**a**) TrkB stimulation scheme: NP cells were seeded into 384-well plates. Final differentiation was induced by DAPT-treatment for 14 days. hiPS cell-derived neurons were stimulated with BDNF, washed, and lysed, respectively. Subsequently, pTrkB ALPHA™ was measured. (**b**) pTrkB ALPHA™ assay principle: Upon binding of the BDNF dimer to the TrkB binding site, receptor autophosphorylation occurs, triggering internal signalling cascades. Whereas the Acceptor bead detects the one of the C-terminally phosphorylated sites of TrkB (T706), the biotinylated Donor bead detects TrkB at an N-terminal site. The binding of the two antibodies brings donor and acceptor beads into proximity. Laser irradiation of donor beads at 680 nm generates singlet oxygen, triggering a cascade of chemical events in nearby acceptor beads, which results in a chemiluminescent emission at 615 nm. (**c**) TrkB dose-response experiments were carried out using the hiPS cell-derived neurons, generated from both hiPS cell lines, mc hiPS cells and AD3Cl1. The pTrkB Alpha™ signal was measured at the indicated concentrations of BDNF. $N = 4$ replicates. Error bars represent SEM

3.6 Automated Measurement of TrkB Phosphorylation in 384-Well Format

Activity of tropomyosin receptor kinase B (TrkB) in the hiPS cell-derived neurons was measured using an Alpha assay (Fig. 7).

1. Continue from Subheading 3.2.3, **step 9**, and measure TrkB phosphorylation (Tyr 706) on day 14 of final differentiation.
2. Prepare 2× concentrated BDNF in TrkB stimulation buffer.
3. Aspirate NDM to a residual volume of 0.01 mL/well using a BioTek EL406 Washer Dispenser (Biotek; Winooski, Vermont).
4. Add 0.01 mL/well of the indicated BDNF concentrations using a CyBi®-Well Vario 384 Channel Simultaneous Pipettor (Cybio; Jena, Germany).

5. Incubate cells for 15 min at 37 °C.
6. Wash cells three times with 0.07 mL/well of 1× DPBS using a BioTek EL406 Washer Dispenser.
7. Aspirate 1× DPBS to a residual volume of 0.01 mL/well and add 0.01 mL/ 384-well of TrkB cell lysis buffer.
8. Incubate cells for 5 min at RT.
9. Prepare phospho-TrkB ALPHA™ reagents in AlphaLISA™ Assay Buffer as follows: 2.5 μL/well of TrkB acceptor beads (f.c.: 10 μg/mL); 2.5 μL/well of the TrkB biotinylated antibody (f.c.: 1 nM); 2.5 μL/384-well of AlphaScreen™ streptavidin-coated donor beads (f.: 20 ng/mL).
10. Transfer 5 μL/well of the above-generated cellular lysates into a 384-well Small Volume Deep Well microplate.
11. Add 2.5 μL of the above prepared TrkB acceptor beads into each well of the 384-well Small Volume Deep Well microplate.
12. Incubate the cellular lysates for 45 min at RT.
13. Add 2.5 μL of the above prepared TrkB biotinylated antibody into each well of the 384-well Small Volume Deep Well microplate.
14. Incubate the cellular lysates for 45 min at RT.
15. Add 2.5 μL of the above prepared AlphaScreen™ streptavidin-coated donor beads under subdued light conditions into each well of the 384-well Small Volume Deep Well microplate.
16. Incubate the cellular lysates for 30 min at RT.
17. Measure AlphaLISA™ signals using an EnVision™ 2104 Multilabel Reader (PerkinElmer; excitation: 680 nm, emission: 615 nm).

4 Notes

1. Matrigel™ may clump upon thawing. This may impair cell attachment. To avoid clumping, add 0.5 mL of room temperature DMEM/F-12 GlutaMAX™ per frozen Matrigel™ aliquot and flick several times. Once liquefied in DMEM/F-12 GlutaMAX™, Matrigel™ can be stored for 1 week at 4 °C without clumping.
2. The bioactivity of cytokines such as FGF2 may vary between different suppliers. Do not change the supplier during neuronal differentiation as this may alter the efficiency of differentiation. Furthermore, the optimal concentration of cytokines may vary. Accordingly, compare various concentrations of cytokines.

3. The quality of the B27 supplement can vary between different batches. Compare several lots in terms of neural induction, and reserve the optimal lot for further experiments.
4. hiPS cells should be seeded at 80–90% confluence. Do not use overgrown cultures as this may impair efficiency of differentiation.
5. A Single cells solution of hiPS cells will attach rapidly. If the cells are not dispersed evenly after being plated, the uneven density can alter the growth and subsequent differentiation.
6. If the hiPS cell cultures do not reach 100% confluence within 2 days after initial seeding for neuronal differentiation, cell density has to be adjusted. The density of the hiPS is a key factor in determining the efficiency of the differentiation, and in obtaining a high yield of neurons. Furthermore, do not exchange hiPS cell medium with NIM until hiPS cells have reached 100% confluence.
7. During neural induction, cells are highly proliferative. If medium consumption increases during this time period, and a lot of cell death occurs, add double amounts of NIM, and exchange medium twice a day.
8. The neuroectodermal sheet has to be broken up into small aggregates. The concentration and incubation time of Dispase solution needed, to fully dissociate the neuroectodermal sheet, may vary. If cells do not detach within 10 min, incubation times can be extended up to 30 min. If the Dispase incubation period is too short, the cells will undergo a lot of mechanical damage when using the cell scraper.
9. If the neuroectodermal aggregates do not attach to the PLO- and laminin-coated 6-well plates within 24 h, repeat washing steps, and transfer aggregates into freshly prepared PLO- and laminin-coated 6-well plates.
10. Before changing media, do not prewarm NPM, as FGF2 will lose its activity if incubated for long periods of time at 37 °C. This will result in poor proliferation rates and increased spontaneous differentiation of NP cells into neurons.
11. If the differentiating neurons start to detach from the PLO- and laminin-coated cultureware, the culture can be overlaid with higher concentrations of laminin (up to 20 μg/mL) to reduce the likelihood of the cells detaching from the plate. This higher concentration of laminin should be applied during frequent media changes. Furthermore, for long-term maturation experiments, SureBond+ReadySet (cat# ax0052; Axol Bioscience; Cambridge, UK) can be used, following the manufacturer's instructions, instead of PDL and laminin coatings.

12. RT-PCR lysates can be stored at −20 °C for half a year. Thaw lysates on ice, and proceed to RT-PCR quickly, as the RNase inhibitor gets inactivated during the freezing process.
13. PFA-fixed cells can be kept in 1× DPBS at 4 °C for 2–3 weeks.
14. The optimal antibody concentrations have to be determined experimentally, but the ratios given here can be used as a starting point.

Acknowledgments

The study leading to these results has received support from the Innovative Medicines Initiative Joint Undertaking under (grant no. 115439), resources of which are composed of financial contribution from the European Union's Seventh Framework Programme (FP7/2007-2013) and EFPIA companies. This publication reflects only the author's views and neither the IMI JU nor EFPIA nor the European Commission is liable for any use that may be made of the information contained therein. We thank Marcel Leist, Dirk Stenkamp, Daniel Bischoff, Robert Ries, Carsten Hecker, Stefan Jäger, Natascha Piede, Michael Traub, and Teresa Bluhmki.

References

1. Heilker R, Traub S, Reinhardt P et al (2014) iPS cell derived neuronal cells for drug discovery. Trends Pharmacol Sci 35:510–519
2. Moors EHM, Cohen AF, Schellekens H (2014) Towards a sustainable system of drug development. Drug Discov Today 19:1711–1720
3. Takahashi K, Tanabe K, Ohnuki M et al (2007) Induction of pluripotent stem cells from adult human fibroblasts by defined factors. Cell 131:861–872
4. Koch P, Opitz T, Steinbeck JA et al (2009) A rosette-type, self-renewing human ES cell-derived neural stem cell with potential for in vitro instruction and synaptic integration. Proc Natl Acad Sci U S A 106:3225–3230
5. Traub S, Stahl H, Rosenbrock H et al (2017) Upscaling of hiPS cell-derived neurons for high-throughput screening. SLAS Discov 22:274–286
6. Falk A, Koch P, Kesavan J et al (2012) Capture of neuroepithelial-like stem cells from pluripotent stem cells provides a versatile system for in vitro production of human neurons. PLoS. One 7:e29597
7. Traub S, Stahl H, Rosenbrock H et al (2017) Pharmaceutical characterization of Tropomyosin receptor kinase B-agonistic antibodies on human induced pluripotent stem (hiPS) cell-derived neurons. J Pharmacol Exp Ther 361:355–365
8. Barde YA, Edgar D, Thoenen H (1982) Purification of a new neurotrophic factor from mammalian brain. EMBO J 1:549–553
9. Bothwell M (2014) NGF, BDNF, NT3, and NT4. Handb Exp Pharmacol 220:3–15
10. Adachi N, Numakawa T, Richards M et al (2014) New insight in expression, transport, and secretion of brain-derived neurotrophic factor: implications in brain-related diseases. World. J Biol Chem 5:409–428
11. Longo FM, Massa SM (2013) Small-molecule modulation of neurotrophin receptors: a strategy for the treatment of neurological disease. Nat Rev Drug Discov 12:507–525
12. Jia F, Wilson KD, Sun N et al (2010) A nonviral minicircle vector for deriving human iPS cells. Nat Methods 7:197–199
13. Chambers SM, Fasano CA, Papapetrou EP et al (2009) Highly efficient neural conversion of

human ES and iPS cells by dual inhibition of SMAD signaling. Nat Biotechnol 27:275–280

14. Borghese L, Dolezalova D, Opitz T et al (2010) Inhibition of notch signaling in human embryonic stem cell-derived neural stem cells delays G1/S phase transition and accelerates neuronal differentiation in vitro and in vivo. Stem Cells 28:955–964
15. Shi Y, Kirwan P, Livesey FJ (2012) Directed differentiation of human pluripotent stem cells to cerebral cortex neurons and neural networks. Nat Protoc 7:1836–1846

INDEX

A

Albumin 6, 9, 34, 50, 82, 105, 149–155, 235, 236
Alpha fetoprotein (AFP) 51, 149–155
Assessing neuronal excitability 12, 219–225
Automation 6
Autophagy .... v, 11, 165–173

B

Bile acids 9, 141–146
Bioreactors .... v, 7, 55–69, 79–90, 150, 153
Bulk culture .... v, 6, 41–52

C

Calcium imaging 38, 219, 220, 222, 225
Cardiac cell outgrowth assay 185–192
Cardiac differentiation 7, 71–76
Cardiac spheroids 185–192
Cardiomyocytes .... v, 7, 12, 55–69, 71, 72, 75, 76, 80, 197–206
Cell culture metabolomics 119–129
Cell expansion in stirred tank bioreactors 79–90
Cholesterol 142, 143
Compound testing 3, 8, 12, 101, 197–206

D

Definitive endoderm cells 42, 43, 46–50
Differentiation into definitive endoderm 6, 41–52, 101–114
Drug discovery .... v, 3–5, 7, 8, 13, 17, 55, 79, 159, 243, 244
Drug testing 8, 11, 32, 228, 235, 236
Drug toxicity testing of 3D cells 235–240

E

Electrostimulation 12, 219–225
Evaluating the effect of drug compounds 12, 185–192
Excitability 12, 219–225

F

Fabrication of microfluidics 12, 227–232
Fluorometric imaging plate reader (FLIPR) 12, 219–225

H

Hepatic differentiation 41–52
Hepatocytes .... v, 6, 9, 41–44, 50, 141–146, 149–155, 235
High content analysis 165–173, 175–183
High-content screening 175
High-throughput screening 3, 4, 6, 7, 13, 101–114, 185, 243, 249, 251, 254, 255, 258, 259
Human induced pluripotent stem cells (hiPSC) 5–8, 32, 41–52, 93–99
Human iPSC-derived astrocytes 31–39
Human iPSC-derived cardiomyocytes 12, 197–206
Human iPSC-derived cortical neurons 209–214
Human iPSC-derived hepatocytes 6, 141, 143
Human iPSC-derived podocytes 7, 93–99

I

Image analysis 11, 166–173, 176, 177, 179–181, 183, 187, 190
Induced pluripotent stem cells (iPSC) .... v, 1–9, 23, 44, 71, 94–97, 131–138, 141, 159–163, 165, 175, 212, 214, 222, 224
In vitro disease modelling 6, 8, 31 39, 79
In vitro models 5, 41, 102, 141

L

Liposomes 143, 144, 146
Liquid chromatography (LC) 150
Liquid chromatography-tandem mass spectrometry (LC-MS/MS) based metabolomics 9, 119–129

M

Microbioreactors 235
Microfluidic devices 12, 186, 228, 235–240
Microfluidics 9, 12, 186, 191, 227–232, 235–240

Carl-Fredrik Mandenius and James A. Ross (eds.), *Cell-Based Assays Using iPSCs for Drug Development and Testing*, Methods in Molecular Biology, vol. 1994, https://doi.org/10.1007/978-1-4939-9477-9,
© Springer Science+Business Media, LLC, part of Springer Nature 2019

Mitochondrial function .... v, 11, 175–183
Monolayer generation..17–28
Multielectrode array (MEA)........ 12, 197–206, 209–214
Multipotent progenitor cell..101

N

Neural progenitor cells (NPCs) 32, 36–38, 211, 212, 214, 221, 222, 224
Neurodegenerative diseases 8, 11, 13, 31, 159, 165, 175, 176, 210, 243, 259
Neurons .... v, 8, 11–13, 31, 36, 159–163, 165–173, 175–183, 198, 202, 209–214, 220, 223–225, 243, 249, 251, 254, 255, 258, 259
Nontargeted metabolomics 119–129

O

Organ-on-a-Chip 11

P

Perfusion feeding 80, 88
Photolithographic techniques 12, 227–232
Plasma proteins 9, 149, 150
Pluripotency .... v, 9, 23, 45, 73, 102, 132, 133, 136, 137
Pluripotent..131
Pluripotent stem cells..4, 32, 44, 45, 71–76, 103
Podocytes... 7, 93–99
Process development...79
Proteomics .... v, 9, 131–138

R

Renal podocytes v

S

Safety pharmacology 7, 79
Seeding 23, 27, 28, 34, 36, 39, 45, 50, 61, 62, 67, 85, 90, 95, 96, 103, 104, 107, 114, 159–163, 169, 180, 191, 201–203, 206, 213, 214, 225, 236, 239, 249, 261
Soft lithographic technique 227–232
Spontaneous network activity........................12, 209–214
Stem cell differentiation 41, 43
Stirred tank bioreactors (STBRs) 6, 55, 56, 79–90
Suspension culture 7, 32, 55, 56, 80

T

Targeted metabolomics 9, 119, 121–125, 127, 128
Targeted proteomic analysis 9, 131–138
Three-dimensional (3D) cell-based assays 186, 236
Three-dimensional (3D) cell culture 185, 186, 235–240
384-well plates 12, 159–163, 259

V

Vascular endothelial cells .... v, 6, 7, 17–28

MIX
Papier aus verantwortungsvollen Quellen
Paper from responsible sources
FSC® C105338

If you have any concerns about our products,
you can contact us on
ProductSafety@springernature.com

In case Publisher is established outside the EU,
the EU authorized representative is:
Springer Nature Customer Service Center GmbH
Europaplatz 3, 69115 Heidelberg, Germany

Printed by Libri Plureos GmbH
in Hamburg, Germany